TOOLING FOR WAR

Tooling for War

Military Transformation in the Industrial Age

Edited by

Stephen D. Chiabotti

Imprint Publications

Chicago

1996

Library of Congress Catalog Card Number 96-75943
ISBN 1-879176-24-6 (Paper)

Military History Symposium Series of the United States Air Force Academy, Vol. 4
Carl W. Reddel, Series Editor

Printed in the United States of America

*For the creators of the core curriculum at the
United States Air Force Academy
—in particular, Brigadier General Robert F. McDermott—
the most extensive and balanced
undergraduate offering of required courses
in engineering, science, the social sciences, and the humanities
in American higher education.
And for the special young Americans
who pass through this program to the end of
defending democratic values in a dangerous world.*

Contents

Contributors

Stephen D. Chiabotti was born in 1950 and grew up in northern Minnesota. He is a Distinguished Graduate of the United States Air Force Academy, where he majored in physics. After graduation, Lt. Col. Chiabotti spent several years training Air Force pilots and earned his M.A. and Ph.D. from Duke University with fields in the history of science and technology and military history. He then returned to the United States Air Force Academy as an instructor and assistant professor. He was instrumental in establishing the Academy's current curriculum in the history of science and engineering and the history of technology and war. Following his teaching assignment, Lt. Col. Chiabotti served as the Chief of Aircraft Program Management at Air Training Command Headquarters and was the focal point for overhauling the Air Force's undergraduate pilot training through the acquisition of nearly $5 billion in aircraft and training systems. He returned to the Air Force Academy to direct military history and manage the Airpower Project. In the latter capacity he coordinates and focuses resources from the National Air and Space Museum, the French Air Force Historical Service, the Air Force's Training System Program Office, the private sector's multimedia producers, and the United States Air Force and civilian academic community.

William H. McNeill was born in Vancouver, British Columbia, Canada in 1917. He studied at the University of Chicago, receiving his B.A. in 1938 and M.A. in 1939, and at Cornell University, where he received his Ph.D. in 1947. McNeill, a former president of the American Historical Association, is an internationally recognized scholar who has received seventeen honorary degrees. Before his retirement in 1987 he served as the Robert A. Millikan Distinguished Service Professor at the University of Chicago. His book, *The Pursuit of Power: Technology, Armed Force, and Society since A.D. 1000* (1982) explores the effects of increasing commercial activity on technological innovations in warfare. Other works include *Greek Dilemma, War and Aftermath* (1947), *America, Britain and Russia, Their Cooperation and Conflict, 1941–1946* (1953), *Past and Future* (1954), *Greece: American Aid in Action, 1947–1956* (1957), *Rise of the West: A History of the Human Community* (1963; winner of the 1964 National Book Award), *Europe's Steppe Frontier, 1500–1800* (1964), *The Ecumene: Story of Humanity* (1973), *Venice, the Hinge of Europe, 1081–1797* (1974), *The Shape of European History* (1974), *Plagues and Peoples* (1976), *Metamorphosis of Greece since World War II* (1978), *The Human Condition, An Ecological and Historical View* (1980), *The Great Frontier* (1983), *Mythistory and Other Essays* (1986), *A History of the Human Community* (1986), *Polyethnicity and National Unity in World History* (1987), *Arnold J. Toynbee: A Life* (1989), and *Population and Politics Since 1750* (1990). McNeill is the Robert A. Millikan Distinguished Service Professor in History and has received the Erasmus Prize, the most prestigious cultural award made by the Netherlands.

Dennis E. Showalter was born on 12 February 1942. He received his B.A. from St. John's University in 1963, his M.A. in 1965 and Ph.D. in 1969 from the University of Minnesota. He has served on the faculty of the Colorado College since 1969, and from 1991 to 1993 was Distinguished Visiting Professor of History at the United States Air Force Academy. He has published numerous books and articles on the subject of military history including *Railroads and Rifles: Soldiers, Technology, and the Unification of Germany* (1975), *German Military History Since 1648: A Critical Bibliography* (1983), Voices of the Third Reich: An Oral History (1989), "Prussia, Technology and War: Artillery from 1815 to 1918," in *Men, Machines, and War*, edited by R. Haycock and Keith Nelson (1989), Tannenberg: Clash of Empires (1990), and "Caste, Skill, and Training: The Evolution of Cohesion in European Armies from the Middle Ages to the 16th Century," in *Journal of Military History* (1993).

Jon Tetsuro Sumida is an associate professor of history at the University of Maryland, College Park. He has been a fellow-commoner of the Archives Center at Churchill College, Cambridge, and a fellow of the Wilson Center and the Guggenheim Foundation. He received his B.A. from the University of California, Santa Cruz (1971), M.A. (1974), and Ph.D. (1982) from the University of Chicago. His many articles have appeared in such journals as *International History Review, Journal of Modern History*, and *Journal of Military History*. His books include *The Pollen Papers: The Privately Circulated Printed Works of Arthur Hungerford Pollen, 1901–1916* (1984) and *In Defence of Naval Supremacy: Finance, Technology and British Naval Policy, 1889–1914* (1989).

Timothy H. E. Travers received his B.A. and M.A. from McGill University in 1963 and 1967, respectively, and his Ph.D. from Yale University in 1970. He is Professor of History at the University of Calgary. He has published widely on the effects of technology on warfare including *How the War Was Won, Command and Technology in the British Army on the Western Front, 1917–1918* (1992), *The Killing Ground: The British Army, the Western Front and the Emergence of Modern Warfare, 1900–1918* (1987), and *Men at War: Politics, Technology and Innovation in the 20th Century (1982)*. He is also author of "Could the Tanks of 1918 Have Been War Winners?," *Journal of Contemporary History* (1992) and "The Evolution of British Strategy and Tactics on the Western Front in 1918: GHQ, Manpower, and Technology," *Journal of Military History* (1990).

James S. Corum is Professor of Comparative Military Studies in the School of Advanced Airpower Studies at Air University at Maxwell Air Force Base, Alabama. He received his B.A. in history and German at Gonzaga University (1975), his M.A. in history at Brown University (1976), his M.Litt. in History at Oxford University (1984), and his Ph.D. from Queen's University (1990). Dr. Corum has made numerous presentations on air power. His publications include *The Roots of Blitzkrieg: Hans von Seeckt and German Military Reform* (1992), "The Old Eagle as

Phoenix: The Luftstreitkrafte Creates an Operational Air Doctrine" in *Air Power History* (1992), "The Development of the Luftwaffe's Army Support Doctrine, 1934–1941" in *Journal of Military History* (1994), and *German Air Doctrine in the Interwar Period* (1995).

John F. Guilmartin, Jr., Lieutenant Colonel USAF, Retired, is an Associate Professor of History at Ohio State University, Columbus, Ohio, where he teaches military history, maritime history, and early modern European history. He holds a B.S. from the USAF Academy and an M.S. and Ph.D. in history from Princeton University. He served two tours in Southeast Asia in long-range combat aircrew recovery helicopters during 1965–66 and 1975. He was a Navy Research Fellow at the U.S. Naval War College during 1986–87, and served as the leader of Task Force II, Weapons, Tactics and Training, for the Secretary of the Air Force/Gulf War Air Power Survey. He has published and spoken widely on premodern military, naval history, and on the theory of war.

Alfred Price received his Ph.D. from Loughborough University in England and is a Fellow of the Royal Historical Society. He served in the Royal Air Force as an aircrew officer in several aircraft types including the Vulcan bomber and specialized in electronic warfare and air fighting tactics. Since 1971 he has been a full-time writer on aviation subjects and has published 41 books and more than one hundred articles. His works include *Instruments of Darkness,* his two volume *The History of US Electronic Warfare,* a history of the Battle of Britain entitled *The Hardest Day,* and *Aircraft versus Submarine.* With Jeffrey Ethell he coauthored *One Day in a Long War* on the air war over North Vietnam, and *Air War South Atlantic* on the Falklands conflict.

Alex Roland, Professor of History at Duke University, received his B.S. from the United States Naval Academy in 1966, his M.A. at the University of Hawaii in 1970, and his Ph.D. from Duke University in 1974. From 1966 to 1970 Roland served in the United States Marine Corps, leaving the service at the rank of captain. Before assuming his position at Duke University, he served as an historian with the National Aeronautics and Space Administration. Dr. Roland has published widely in the area of military history including: *Men in Arms: A History of Warfare and Its Interrelationships with Western Society* (1991), *Model Research: The National Advisory Committee for Aeronautics, 1915–1958* (1985), and *Underwater Warfare in the Age of Sail.* His many articles include such works as "Technology and War: The Historiographical Revolution of the 1980s," in *Technology and Culture* (1993) and "Technology, Ground Warfare, and Strategy: The Paradox of American Experience," in *Journal of Military History* (1991). Dr. Roland is the Vice-President of the Society for the History of Technology and Vice-President of the Naval Undersea Museum Foundation.

Jacob W. Kipp received his Ph.D. in Russian history from Pennsylvania State University in 1970. He is Research Coordinator for Geo-Strategic Studies with the Foreign

Military Studies Office of the U.S. Army Combined Arms Center at Fort Leavenworth, Kansas, Adjunct Professor of Russian and East European Studies at the University of Kansas, and serves as United States editor for *European Security*. Dr. Kipp served on the faculty of Kansas State University from 1971 until 1985 and as associate editor of *Military Affairs*. His research has focused on Russian military, naval, and air power history. His many publications include *Soviet Aviation and Air Power*, "The Zhirinovsky Threat," *Foreign Affairs* (May–June 1994), and the Garland Series of International Bibliographies in military history, which he coedited with Robin Higham. He is one of the editors and a member of the editorial board for the University Press of Kansas series, *Modern War Studies*.

Sam C. Sarkesian served for over twenty years as an enlisted man and officer in the United States Army. He received his B.A. from The Citadel and his M.A. and Ph.D. from Columbia University. He is currently Professor of Political Science at Loyola University of Chicago. Dr. Sarkesian is Chairman of the Academic Advisory Council of the National Strategy Forum and is a member of the International Institute for Strategic Studies. He has published widely in the area of United States military policy, including: *America's Forgotten Wars: The Counterrevolutionary Past and Lessons for the Future* (1984), *The New Battlefield: The United States and Unconventional Conflict* (1986), *U.S. National Security: Policymakers, Processes, and Politics* (1989), and *Unconventional Conflicts in the New Era: Lessons from Malaya and Vietnam* (1993).

Foreword

The papers contained in this volume were originally presented at the Sixteenth Military History Symposium, held at the United States Air Force Academy, 21–23 September 1994. The symposium brought together leading scholars and military officers to examine the technological transformation of war in the century from 1849 to 1949. Aside from the sheer complexity of the phenomenon examined, no clear thesis or consensus emerged. There was, however, a general recognition of interaction between the military community and those of science, engineering, and social science in shaping equipment, operations, and outlook. Moreover, elevated thinking by committed individuals emerged as a common ingredient in the successful application of new military technologies.

The Military History Symposium Series at the United States Air Force Academy began in 1967 as an annual event, sponsored by the Academy and its Association of Graduates, with the Department of History as the planning and directing agency. Since 1970, the symposia have been held biennially. The purpose of the series is to provide a forum in which recognized scholars may present the results of their research in the field of military affairs. In this way the Academy encourages interest in a vital subject among civilian and military scholars, members of the armed forces, and the cadets who will be the future leaders of the United States Air Force.

The Sixteenth Military History Symposium benefitted greatly from the financial support received from the Association of Graduates, the George and Carol Olmstead Foundation, and the Major Donald R. Backlund Memorial Fund. A generous contribution from the Secretary of Defense's Office of Net Assessment also made the symposium possible.

This volume owes a great debt as well to members of the Department of History at the United States Air Force Academy. Under the leadership of the department's head, Colonel Carl Reddel, the symposia have become a world-renowned gathering of outstanding scholars, high-ranking military officers, and important policymakers interested in achieving a keener understanding of the military past. Although the symposium shares priority with the department's primary mission of educating cadets, Colonel Reddel ensures dedication of the human, material, and financial resources necessary for execution of the event and publication of the proceedings. In that regard, Major John Farquhar provided the intellectual leadership for the conference as its Executive Director. He coordinated topics with the presenters, arranged travel and lodging, and looked after the budget essential to execution. In these tasks he was most ably assisted by Captain John Terino, while Captain Mike Grumelli deserves special mention for adroit management of finances. Without the efforts of these individuals, there would have been no symposium and precious little to edit.

Similarly, Mrs Jennifer Glover was absolutely essential to the publication of this volume. Jennifer was the focal point for all communications related to publishing. She became an expert on translating various word processing programs to a standard code, meticulously transcribed footnotes and bibliographic entries, and cheerfully oversaw a myriad of administrative details related to the conference and this publication. Any errors that remain after her steadfast work are solely the responsibility of the editor.

Finally, the cover art symbolizes the emergence of new technologies like rockets and aircraft above the foundation of seapower and against the ominous backdrop of nuclear weapons. The collage is a computer composition by Matthew I. Chiabotti. While my son contributed this modicum of his vast talent to the volume, the other members of my family offered both patience and support. I am grateful to them, and for them.

SDC

Introduction

Stephen D. Chiabotti

No military historian has made sense of the twentieth century. The standard texts are brilliant in their exposition and conceptual formulation of classical, medieval, and neo-classical warfare; and the French Revolution appears well understood in terms of its social impact on war. Walter Millis went so far as to credit the Americans and French with a political revolution in the "democratization of war." Millis went on to ascribe to the Germans a "managerial revolution" in the planning and manipulation of war through the auspices of a general staff. The third of Millis's revolutions, the technological, is perhaps less well understood, and yet has been the primary agent of change for warfare in the past century.[1] Recently, analysts have proclaimed a "military technical revolution," which quickly became a "Revolution in Military Affairs." The etymology reflects a deeper understanding of threat, finance, and politics as companion ingredients to technology in the military stew. The last nonetheless changes most quickly and begs adjustment from the others to keep the dish palatable. Thus, technological change is the subject of this volume.

The title is a "triple entendre." Tooling refers at once to the implements of war as well as the process of creating them. Moreover, building weapons and fashioning doctrine for the their intended use tends to invite, if not determine, war. And entrepreneurs who "tool about" the landscape of weapons technology, looking to equal or forge ahead of their international competitors, are no more or less guilty of precipitating war than diplomats who have seen it as the aegis of national unity, or sociologists who have viewed warfare as an expression of intraspecific dominance. All were in one way or another tooling for war.

Transformation in the tools of war embraces a set of paradoxes wherein the inherent conservatism of military leadership confronts the search for relative advantage through change. Lewis Mumford attributed the conservatism of military practitioners to blatant stupidity and resorted to his own etymology for verification: "war stimulates invention, but the army resists it. . . . No wonder that in English *to soldier* means to withhold efficiency in work."[2] Robert O'Connell, on the other hand, maintains that new weapons have to fit the narrow range of behaviors deemed appropriate for the practitioners of war—Homeric confrontation in the case of western combatants. According to O'Connell, when new weapons, such as crossbows, submarines, poison gas, or nuclear weapons threaten to change this normative behavior, they are eschewed by the military establishment as part of a grand scheme to preserve the species.[3] This ostensible resistance to change appears to break down in the middle of the twentieth century when new technologies emerge as the panacea for an otherwise bloody and prolonged stalemate. Ironically, armies, navies, and air forces throughout the century do battle with essentially the same weapons. And technical symmetry amid deliberate efforts to gain advantage through development that is both covert and aberrant is indeed paradoxical. More so, as notes William McNeill in this volume, manifold efforts to manage and control technical change have led increasingly to a global situation that is unmanageable and out of control.

To make sense of modern weapons development, however paradoxical it may be, we must examine the roots of the phenomenon in the middle of the nineteenth century. And, since the resolution of history is blurred by proximity, the middle of the twentieth century appears to be a reasonable point of chronological termination. Hence, the century from 1849 to 1949 will be the focus of the discussion. Each of these dates is associated with technical developments that not only exerted dramatic effect on the conduct of war but also tend to illustrate the changes that occurred in how weapons were invented, procured, and employed. As a point of departure, in 1849 a French captain named Claude Minié invented the cylindro-conoidal bullet. Minié balls had almost immediate impact in the Italian Wars of Unification and the American Civil War by expanding the killing zone of infantry weapons nearly tenfold, eliminating the bayonet charge as the raison d'être of infantry tactics, and forever banishing equestrian cavalry as an instrument of shock. Minié's invention is illustrative of a craft tradition in the military arts. The better gadget is discovered by a member of the guild of practitioners, and tactics evolve in a period of trial and error that spans battles from Solferino to the Somme.

The year 1949 illustrates perhaps a different kind of benchmark in military-technical development. In that year, the Soviet Union detonated a nuclear weapon and altered the strategy of war to perhaps an even greater degree than Minié had altered tactics. Moreover the Soviet blast demonstrated that even the iron curtain was osmotic with regard to lethal technology, despite profound efforts to keep fission weapons exclusively within the province of the West. The atomic bomb symbolizes a century of military technical transformation. The bomb was the product of intensive scientific research in a climate of deliberate state investment and planning. It was issued to the military and mandated for use under a veil of secrecy, while some of the very scientists who developed the bomb conspired to lift the veil on a weapon too powerful to use. The trial-and-error method of craft in proofing weapon effectiveness on the battlefield had given way to the scientific method of hypothesis and experimentation in the laboratory.

Meanwhile, the cost of weapons and their impact on the political landscape may have facilitated the birth of a new kind of scion—the social scientist. Economists, political scientists, and business managers have put their own twist on the path of weapons development, and behavioral scientists have tried to measure and regulate the impact on human performance in the field. While disciplines that advertise themselves as science by title seldom are, they nonetheless provide a valuable bridge of understanding between the physical framework of the weapons and the psychological framework of the users. If Henri Antoine Jomini and Carl von Clausewitz represent the theoretical ying and yang of this dichotomy, then social scientists live on the curved boundary that bisects the circle.

Similarly, the engineer haunts the boundary between science and craft. Engineers are people interested in the application of scientific principle to achieve practical ends. Renaissance engineers proved valuable in the construction of fortifications to forestall the new gunpowder weapons, and their modern contemporaries exerted a large influence in the fabrication of artillery, ships, motors, and rockets during the century under examination.

Hence, the structure of this volume is defined not so much by chronology as by the cumulative impact of four traditions on weapons development and warfare: craft, engineering, science, and social science. Although elements of all four could be ascribed to any weapon developed in the century between 1849 and 1949, the relative influence of each appears to fall in chronological order with the listing. The craft tradition dominates early, but receives a substantial augmentation from science and engineering in the middle, while all three are enveloped by the porous membrane of social science as the twentieth century matures.

William McNeill provides majestic oversight for the whole process in his Harmon Memorial Lecture, "The Structure of Military-Technical Transformation." His global perspective is punctuated with an acute awareness of regional differences, as he blends technology, finance, and operational art with international politics. McNeill evokes irony as he portrays a world in which the quest for dominant weapons technologies in pursuit of power has left us powerless.

The transition from a craft tradition of military-technical development to one more heavily influenced by engineering is exposed in three essays that constitute the first section. All three deal with artillery. Big guns are perhaps the defining technology of the half-millennium that began in 1500. Cannon brought down castle walls, unhorsed the feudal array, and redefined polity in terms of the nation state. They stimulated mining and metallurgy, banking and finance, as well as chemistry and ballistics. Cannon presaged cylinder boring techniques for steam engines and presented a paradigmatic vision of an internal combustion engine, albeit one with a free-floating piston. They were also the principal killers in two world wars. It is perhaps fitting that Louis XIV had his cannon inscribed with the words, *Ultima Ratio Regis* (the last argument of the King).[4] Within fifty years a young artilleryman demonstrated the power of a "whiff of grape" in supplanting the Bourbon dynasty with one of Bonapartes. A century later the guns of August drummed the dirge for the Romanovs, Habsburgs, and Hohenzollerns. And if the Windsors remained, albeit astride a nation on the verge of bankruptcy, they were at somewhat a loss for dialogue with new common rulers of Europe. Perhaps Hitler's remark to the British ambassador during the Munich affair was most revealing of the new aristocracy, "If any more people in tired suits call on me, I'll send my ambassador in London to see your king in a pullover."[5]

World War I was indeed the last argument of kings, and the dialogue on land and sea was punctuated by artillery. The big guns gave new meaning to combat, as the urge to close gave way slowly to the calculus of trajectory and distance. The craft of killing eye-to-eye or yard-arm-to-yard-arm was replaced by engineering applications of force at distances beyond the vision of gunners, who seldom saw what they were shooting at, let alone whom they were killing.

On land, the evolution of distance killing by artillery occurred gradually, beginning with the wars of unification in Europe and America and progressing through the colonial wars in South Africa and Manchuria. The initial impetus was supplied by the range of rifled infantry weapons whose evolution proceeded rapidly from Minié ball to Maxim gun, but within two decades the sheer capability of rifled, breech-loading artillery began to push ranges over the horizon. By the time of the Great War, indirect fire provided a form of "parallel" warfare at the tactical level that stealth and precision-guided munitions currently provide at the theater and strategic levels. It was no longer necessary to peel the onion of the enemy's defense from the outside in. Command, control, and communications were targeted as often as the front-line trenches, and artillery became the queen of battle. Dennis Showalter pays careful attention to this evolution and documents the concurrent social transformation that occurred within the artillery branch in the half-century that concludes with World War I. He also chronicles the legitimation of artillery as a bona fide branch of land armies as the gunners learned to "march in step" with infantry and cavalry.

The maritime analog to the Minié ball was provided by the Whitehead torpedo, which parallels the Frenchman's conoidal bullet in shape, invention, pattern of employment, and impact, while the necessary tactical evolution toward distance and nonconfrontation is remarkably similar to what occurs on land. For the first time in naval history, a considerably smaller boat armed with automotive torpedoes could inflict lethal damage on a ship of the line. Naval officers responded to the danger by seeking ever increasing range from their main batteries. But here the transformation was more sporadic and accelerated than on land. At Tsushima in 1905, the Russians and Japanese fought at ranges of 4,000 yards or less, despite the threat of torpedoes.[6] The latter, however, developed rapidly over the next decade and forced Royal Navy officers to consider engaging at over 10,000 yards with the German High Seas Fleet, whose doctrine and construction were more closely geared for a short-range battle. Jon Sumida reveals the torpedo as the driving force behind the Royal Navy's "quest for reach" and provides a painstaking account of how engineering procedures and apparatus clashed with a craft tradition of fire adjusted by observers pushed increasingly higher into the superstructures.

For no lack of effort or expense, His Majesty's Navy shot poorly at Jutland. They might have gathered as much from their performance at Gallipoli, where neither the navy nor the army could hit the broad side of Turk. Timothy Travers rounds out this trio of essays oriented toward the admixture of craft and engineering in operational art with the supreme failure of technology and tactics on both land and sea in the Gallipoli campaign. He demonstrates convincingly that communication had by 1915 become the limiting factor for gunnery on land and at sea. Not only did gunners have to think over increasing distances, they had to think in three dimensions by involving the air service. Lacking high ground or a superstructure tall enough to provide line-of-sight inland, army and navy gunners had to appeal to aircraft for a man on the hill or in the crow's nest. Unfortunately, it took another world war and a couple smaller conflicts in East Asia for techniques of aerial spotting and interservice communication to mature. The result at Gallipoli was poor shooting and a general failure in technology and tactics.

Gallipoli, Jutland, and the Somme could be viewed as failures for artillery to live up to its promise. On the other hand, they demonstrate that technology requires a careful fit with the remainder of the operational infrastructure and the social ethos of the operators before it can be successful.

If artillery, machine guns, and shovels were the defining technologies of World War I, then it was the internal combustion engine that defined its sequel at mid century; and the quest for a steady supply oil to fuel engines dominated the origins of the Pacific War and the conduct of World War II in the European theater. Tanks, trucks, and aircraft became the modern manifestations of cavalry and restored shock and mobility to the often stagnated battlefields of World War I. The new machines demanded technical proclivity from a branch steeped in tradition and renowned for its technical conservatism. Napoleon's hussars wore breastplates and carried lances at Waterloo, while the uniforms of tank and aircraft crew were painfully slow to evolve away from the high boots and tucked britches of the cavalry, and horseback riding was still an important part of the curriculum at the U.S. Army Air Corps Tactical School in the 1930s. These equine traditions provided a poor platform for conceptualizing the three-dimensional envelopments that characterized most of the successful operations of World War II. It is not surprising that the enduring synthesis came from a German signals operator who combined the better observations of two British historians.

If the Guderian—Liddell-Hart—Fuller formula was not adopted by France, there were good reasons that grew out of historical experience and the peculiar socioeconomic setting of the Third Republic in the 1930s. James Corum exposes the interrelationship of the French doctrine for Methodical Battle and the German penchant for what British correspondents called *Blitzkrieg* in the cultural settings of the two countries during the interwar period. He also echoes McNeill's sentiments by demonstrating that personalities did indeed make a difference in the formulation of doctrine for armored and aerial warfare at mid-century. And, while engineers played a major role in the development of light, powerful engines for aircraft and tanks, they had little to do with their employment. Here success or failure with symmetric weapons appears to rest on doctrine. Corum highlights the differences in doctrine between the Germans and French at the onset of World War II and provides considerable insight into the reasons behind the outcome of the Battle of France.

As World War II progressed, so did rapid developments in rocketry of all scales. The bazooka and *Panzerfaust* emerged as modern equivalents of the crossbow and longbow in providing infantry a foil for the power of armored cavalry, while *Katyusha* bombardment rockets added their frightening howl to many a Russian attack. Most significant, however, was the work of Walther Dornberger's team on the A-4 exoatmospheric rocket. If the V-2 vengeance weapon was a colossal failure in producing a little over 5,000 casualties for the billions of marks expended, its progeny that carried men to the moon twenty-five years later were not. And no technology evokes more of the ethos of engineering and the vision of science than does rocketry.

John Guilmartin provides a broad and penetrating analysis of the evolution of this alternative to conventional artillery. He moves from black powder to ballistic missiles with breathtaking pace in chronicling the attempt to engineer better artillery. Guilmartin provides a technical cornucopia that demonstrates physical calculations of mass, impulse, and range as dominating factors in rocketry, whether solid- or liquid-fueled. The paper accompanies Corum's because rockets represent an alternative form of combustion engines that reached watersheds in technical and doctrinal evolution during World War II. In many ways, the combination of exoatmospheric rockets and thermonuclear weapons, that were developed at least in concept near the end of the war, circumscribed international politics for the next fifty years. And the reluctance or willingness of soldiers, sailors, and airmen to adopt these new artilleries provided the internal dynamic for a reorientation of roles and missions among the three services. Francis Scott Key was perhaps accidentally prescient when he included the rockets' red glare in the American anthem; because it was rockets that sent airmen into deep bunkers, set soldiers and sailors about the business of flying ballistic artillery, and sequestered diplomats in chambers for the purpose of avoiding what had for millennia been the fruition of their work—war.

Science involved in warfare reaches full exposition in the next section. No two communities appear more antithetical than those of science and the military. The former is international, effusive, and typically undisciplined in personal habit and physical appearance. The latter is fiercely nationalistic, characteristically taciturn, and wantonly disciplinarian. How did they come to such collusion in the art of war? The obvious answer lies in the increasing technical sophistication of the implements and practice of war. Without science, soldiers would be at a loss for smokeless powder, canned food, and antibiotic drugs. They would be blind and deaf for want of radar and the codebreaking techniques that typify modern signals intelligence. They would be dumb without the communicative devices extracted by science from the electromagnetic spectrum and for the most part lost without the precise positioning information afforded by satellite reference systems. Their weapons would pale in comparative power and accuracy and seem little advanced from those of a hundred years ago. In other words, science over the last hundred years has redefined the framework of soldiering. Logistics, medicine, communications, intelligence, and weapons all bear the unmistakable marks of scientific development.

The transition was painful at first. In a profession where failure generally results in death, soldiers had a natural mistrust for new gadgets that could succumb to the extreme friction generated by combat or fail to penetrate the thick fog of war. Napoleon was reluctant to try Fulton's steam ships to tow troops across the English Channel and disbanded the balloon corps. But four decades later his nephew was encouraging French scientists to invent everything from better artillery to time-keeping devices. Science responded to cash incentives, but the record of successful employment is spotty until the natural watershed of World War II. Here, deliberate application of scientific research to warmaking technologies paid dividends that mattered. Wireless, radar, sonar, and the proximity fuse were unparalleled successes and fully defined a fourth electromagnetic dimension for soldiers to consider and master. Mathematical techniques broke the German Enigma and Japanese Magic codes, providing Allied planners an unprecedented advantage in strategic, operational, and sometimes even tactical intelligence. And the atomic bomb punctuated what for many came to be viewed as a war of the wizards.

Alfred Price illuminates a short chapter of the wizard war with an episode on German attempts to bomb English cities by navigating along radio beams. The interface between scientists and soldiers appears wanting in this particular episode and may have doomed employment of a technique with great promise. The natural mistrust of German aviators and their awareness of the inevitable countermeasures testify to the difficulties inherent in the operational employment of science. In any event, Price sheds new light on a reasonably well-known story. And some fortuitous research demonstrates that a bit of moonlight can turn the backroom work of the boffins into little more than sophistry.

Alex Roland, on the other hand, demonstrates by implication that the boundary between science and the military has grown porous indeed—with goods and services flowing in both directions at a copious rate. He dispels the notion of military technical conservatism by portraying soldiers as principal agents in the revolution of nanosecond technology that embraces radar, nuclear weapons, and digital computation. He posits these nanotechnologies, particularly the last, as the defining elements of the post–World War II era. He conceptualizes the Fermi, a unit of measurement considerably smaller than the imaginations of nineteenth-century military practitioners, as the new measure of war because it is common to both computers and nuclear weapons. According to Roland, the former have prevailed over the latter as the defining technology of the post-war era; and this section might just have easily been titled "Science: Scion of Soldiery."

The final section of this volume deals with social science and its rivalry with the humanities as a proper vehicle for understanding and perhaps forecasting war. Civil wars are the most bitter, and the one between social science and the humanities is no exception. In fact, the climate of mutual disrespect rivals that between science and engineering. Despite dissimilar subject matter, it would appear that humanities bond more closely to the hard sciences than the social sciences, and no one wants to party with the engineers—socially or otherwise. The pretense of practical application appears to divide the intellectual communities along hard lines. All have nonetheless contributed to the art of war, and the social scientists in equal measure with rest, if only in the last century.

Henri Antoine Jomini gave us the first interpretation of war that could be called social science in that his instructions were largely prescriptive in nature. His approach was rational and a natural legacy of the intellectual climate of the previous century. Clausewitz, cleaving to the intellectual tradition of the German romantics, was probably more representative of the humanities. That both published (the latter by his wife posthumously) at the inception of railroads in Europe is purely coincidental, since neither took much account of technology as a significant variable in the art of war. Railroads and technology were nonetheless significant in the formation of their intellectual progeny. The Prussian and later German General Staffs were to a large degree necessitated by the complex timetables generated by the mobilization and rail movement of troops and the need to remain abreast of the changes in technology proposed by the Dreyses and Krupps of the world. Hence, management came to the military. Similarly, the sheer cost of rail construction or refitting an entire army with rifles, as the French did between 1866 and 1869, brought men of finance to the table of Mars.

Although Lev Tolstoy wrote at mid-century in the middle of the railway revolution, his epic on war and peace had little to do with finance and management of railroads. Tolstoy focused instead on the more romantic issues surrounding the great events of 1812. On the other hand, Polish banker and railway entrepreneur Jan Bloch was a worthy representative of both management and finance. He showed a true disposition toward social science in his attempts to forecast the future of war in technical, political, and economic features. Although it is hard to visualize Bloch as a subject of Nicholas II, he was indeed a Russian and a key figure in the industrialization of Eastern Europe. Jacob Kipp accurately places these two Russians, Tolstoy and Bloch, on the teeter-totter ridden by social science and the humanities and demonstrates that the fulcrum perhaps moves toward the advantage of the former with the introduction of railroads. In Eastern Europe railroads became the natural equivalent of A. T. Mahan's sea lines of communication, and trade in goods and services could serve military as well as commercial ends. No wonder the two were intermixed so thoroughly in planning and execution on both sides of the Oder.

Bloch pioneered the application of social science to war, and he warrants scrutiny if for no other reason than he was right. The title of Bloch's four-volume work on *Future War: In Its Technical, Political and Economic Relations* suggests a groundbreaking framework for analysis to complement prescient thinking on the likely outcome of a first world war. Yet Bloch's work is fraught with irony. The independence of Poland that he sought to achieve through commercial means came from war. But the kind of war he predicted—the consumptive sausage grinder of attritional stalemate—occurred in the western and southern European theaters of World War I. The war in the East was more fluid and devolved along more traditional lines: the loss of a battle meant the loss of a province, albeit a province the size of Poland or the Ukraine. Both Bloch and Kipp find the precursors of the Great War in the Russo-Turkish War of 1877–78, when the preparadigmatic social sciences of management and economics began to exert perhaps as much impact as intuition and leadership.

But some remain unconvinced of the relevance of social science to war. "There is, dammit, no such thing as military science," railed Theodore Ropp once on the cover sheet of one of my obviously misguided seminar papers. After handing it back, he explained why—to the entire seminar. "The essence of science is predictability and the governance afforded by laws. This piece of chalk will fall at the same velocity and impact the table with the same momentum tomorrow as it does today. There are no equivalent laws for war because it involves the free will intrinsic to human behavior. Warfare is an action-reaction phenomenon that defies counterfactual analysis as well as laboratory experiments. Since neither the canons nor the methods of science apply to war, there can be no military science."

What then, replies Kipp, do you call the study of war? The Soviets called it military science, and its roots in the work of Jan Bloch are well developed in Kipp's paper. Ropp called it history, which pretends to convey only an interpretation of the past as one tool for understanding the present. The remainder of the present and certainly the future are left to the social scientists, whom Sam Sarkesian chronicles in his paper.

Sarkesian is skeptical of the efficacy of social science as a military tool and presents a comprehensive bibliographic essay on efforts to date. Here the military has become apprentice to the sorcerer of social science. Sarkesian advocates a healthy dose of humanities, with all of its grey areas and manifold uncertainties, for the pretenders to military science and warns of the incalculable nature of military friction and the impenetrable thickness of the fog of war. What better solo from the choir for a gallery crowded with historians?

But the influence of social science on the modern American military is undeniable. The dominant advanced degree in the officer corps is management. And social science principles infuse a succession of staff rubrics that run from Zero Defects, through Management by Objectives, to Total Quality. Military training is defined in terms of Instructional Systems Development, while the lesson objectives and desired learning outcomes of the educationists often define professional military education and are making substantial inroads at undergraduate institutions like the Air Force Academy. "I have a lesson learned here," is legitimate discourse in drafting requirements for weapons systems, forming personnel policy, or targeting for a bombing campaign. The lesson learned legitimates history for the social scientist without accounting for the inevitable differences in context. And when the lessons become laws for the moment, as so often they do, we might wonder as to what apprentice is serving which sorcerer.

Tooling for War then comprises ten essays excerpted from the proceedings of the Sixteenth Military History Symposium that took place at the United States Air Force Academy on 21–23 September 1994. While the essays themselves highlight product in the form of military technologies, the organizational pattern of the volume, in conjunction with William McNeill's Harmon Memorial Lecture, seeks to illuminate process. Product and process are intercausal in the transformation of organized violence and world polity. Cannons and railroads had considerable influence on the formation of nation states, while national policy directed improvements to artillery and rail transportation networks throughout the nineteenth century. There can be little doubt that missiles and computers are given to similar influences in the century that began in 1950. The new artilleries and networks of communication that circumscribe the present are perhaps best understood in terms of their historical antecedents. And this volume is constructed with the hope of all history: informing an understanding of the present from the study of the past.

Notes

1. Walter Millis, *Arms and Men: A Study in American Military History* (New Brunswick, N.J.: Rutgers University Press, 1956).

2. Lewis Mumford, *Technics and Civilization* (New York: Harcourt, Brace & Jovanovich, 1963), 95–96.

3. Robert O'Connell, *Of Arms and Men* (New York: Oxford University Press, 1989).

4. Ibid., 159.

5. David Irving, *Hitler's War* (New York: Avon Books, 1990), 133.

6. Eric Grove, *Fleet to Fleet Encounters: Tsushima, Jutland, Philippine Sea* (London: Arms & Armor, 1991), 21–25.

The Structure of Military-Technical Transformation

William H. McNeill

We live under an extraordinary cloud of uncertainty. Technological changes alter human experience in far-reaching ways many times over in a single lifetime. I, for instance, can remember when radios were squawking toys boys built at home in hope of hearing broadcasts from Schenectady, and then, a few years later during the Battle of Britain, came the wonder of Edward Murrow's trans-Atlantic voice, clear as a bell, with bomb explosions muffled in the background. More recently, computers began to sprout around me everywhere; and an array of other novelties that did not exist when I was born, or were unavailable to ordinary people, have altered our daily routines— cars, airplanes, frozen food, plastics, TV, e-mail, fax machines, antibiotics, and many more.

Because technological change is so pervasive and powerful among us—and not least among the military—we are tempted to assume that restless technological transformation is natural and normal. But the historical record shows that this is not so. In times past, most people lived out their lives in accustomed fashion, using the same things their forefathers had used, and making no deliberate effort to alter or improve them. Human inventive capabilities, however real, came into play only occasionally and exceptionally; whereas we must adjust to an avalanche of innovation, some of it planned and deliberate, some of it unforeseen and unwelcome.

This essay asks how and why we find ourselves in such an unusual circumstance, and, in particular, explores what it was that provoked the extraordinary military-technical transformation of the industrial age that started in the 1840s and, despite some subsequent slowdowns, has spread and accelerated, rather jerkily, ever since.

Let me begin by pointing out that on the face of things, any significant military-technical change is undesirable simply because it makes trouble and increases risk. To use a new weapon effectively, fighting men have to change their habits and learn new skills. This is bothersome in itself; and, in practice, success is never sure ahead of time. In war, sensible persons therefore shy away from compounding the risk and uncertainty created by the enemy, by the weather, and by all the other friction of war, and rigorously refrain from trying anything new. Instead, prudent fighting men rely on experience, adhere carefully to time-tested ways, and, in short, behave exactly like Colonel Blimp. He became an object of cartoon ridicule in the 1930s, yet Colonel Blimp's frame of mind constituted the norm of sane military management in past ages. How did he get so out of step with our times?

The changeability of military technology since the 1840s is all the more surprising because this was a time when military men encased themselves in ever thicker layers of bureaucracy. And bureaucracies are not usually inclined to innovation. After all, agreeing with one's bureaucratic superiors is the way to get ahead, while conforming to precedent can keep the unambitious out of trouble. And when an awkward problem arises, it can always be referred to a committee, thereby postponing action indefinitely. Routine is therefore at the heart and core of bureaucratic behavior; yet from the 1880s important segments of the military bureaucracies of the most powerful nations of the world systematically began to encourage radical technical innovation. This occurred in spite of obvious risks and ever-mounting costs, as one improved weapons system after another displaced its predecessors in rapid and apparently endless succession. Odd behavior indeed, and all the more so since many expensive innovations were soon scrapped as obsolete and never used in action. Moreover, when new weapons were employed in World Wars I and II, they did not bring easy victory, but instead magnified destruction enormously, hurting winners as well as losers. Why did it happen? How did long-standing national rivalries boil over into such a risky and unsettling arms race?

An historian is always tempted to look for similar experiences in the deeper past, and plausible historical parallels to the arms races of the industrial era can be found. Two eras in particular occur to me as faint foreshadowings of the modern experience. One came in the Hellenistic Age, when rival rulers employed a handful of military engineers to build increasingly powerful siege engines and larger and larger warships. A second period of rapid and deliberate technological change occurred in China under the Sung Dynasty (960–1279) when gunpowder weapons and a galaxy of other military inventions (especially naval) burst upon the scene.

But both these outbreaks of technical instability turned out to be relatively short-lived. Hellenistic engineers quickly reached technical limits of size, strength, and resilience set by the wood and fiber available for their catapults and ships; and the political rivalries that had stimulated that arms race disappeared once the Romans established their military supremacy throughout the Mediterranean coastlands. In the Far East the pattern of events was different. Sung officials' efforts to encourage military invention in hope of warding off barbarian assault failed. Instead, after borrowing some high-tech, up-to-date weaponry from their enemies, the Mongols were able to complete their conquest of China in 1279; and after the Mongols were driven out of China in 1368, subsequent Chinese regimes regularly preferred diplomacy to war, and were cautious—though never completely inflexible—in investing in new military technology.

But, of course, the Mongol storm was not confined to China; and their conquests spread knowledge of the explosive force of gunpowder throughout Eurasia. Among the peoples whose traditional ways of war were thus affected, the European response was by far the most radical and persistent. As a result, China's initial flirtation with systematic pursuit of technological improvement under the Sung Dynasty was soon overshadowed by Europeans' enduring enthusiasm for more and better guns, large and small.

In a sense, the modern arms race dates back to the reckless way rival European rulers set out to build wall-destroying artillery in the fourteenth and fifteen centuries. To begin with, critical skills for casting large metal objects were narrowly circumscribed around the city of Liège, but just as guns were becoming really powerful and comparatively mobile, the breakup of the Burgundian lands after Charles the Bold's death in 1477 divided Europe's gun-casting capabilities between French Valois and German Hapsburg rulers. As a result of this happenstance of dynastic politics, no single monarch or state was ever able to monopolize big guns in Western Europe, whereas in all the other civilized lands of Eurasia, when effective artillery arrived on the scene it was swiftly monopolized by a single ruler. Comparatively vast empires resulted—Ottoman, Safavid, Mughal, Muscovite, and, of course, the Chinese, where, however, big guns were less important than elsewhere simply because the Chinese rulers had no wish to destroy walls their soldiers defended against the continuing nomad danger.

Thereafter, Western Europe remained technologically innovative, largely because state rivalries persisted in nourishing deliberate efforts to improve weaponry and military organization. Any new practice or superior weapons design spread very rapidly from one army to another. This had the effect of maintaining an ever-shifting and ever-precarious balance of power within Europe, whereas in all the rest of Eurasia, once rulers succeeded in monopolizing heavy guns, they saw no reason to tinker with a weapons system that enabled them to break into the stronghold of any defiant local potentate or potential rival who lived within range of their artillery and field army.

The modern history of Japan offers a particularly vivid and convincing example of how dispersed access to guns accelerated military-technical change for about half a century, until a single victor emerged whose policy of restricting access to guns stabilized Japan's new political-military order for the ensuing two hundred years.

Samurai swordsmen and archers found nothing to admire in clumsy guns when this Chinese invention first came to their attention, and the success with which the Japanese repelled massive Mongol invasions in 1274 and 1281 confirmed Japanese warriors in their disdain for newfangled weaponry from abroad. But these attitudes changed abruptly after 1542, when local military leaders realized that the (by then much improved) guns, large and small, that European sailors carried on shipboard as a matter of course offered enormous advantages in the local feuds that had long simmered among them. Japanese craftsmen quickly learned to produce muskets and larger guns like those Europeans employed, and when military rivals hurried to acquire these new weapons in ever-larger numbers, the scale and decisiveness of warfare escalated very quickly. Commoners armed with muskets proved able to overwhelm the most expert swordsmen, and within half a century, a low-born warlord, Toyotomi Hideyoshi (d. 1598), was able to crush all rivals and establish his authority throughout the country.

His successors, the Tokugawa Shoguns, sought to maintain their sovereign power and stabilize Japanese society by weaving a complex network of alliances and agreements with local clan leaders throughout the country. All concerned were eager to reaffirm the prestige and privileges of samurai swordsmen whose traditional role in society had been seriously compromised by the sudden importance of musketeers. Accordingly, after repressing a serious revolt (1637–38), the Tokugawa Shoguns proceeded to disarm commoners by confiscating guns and prohibiting their manufacture. In addition, by cutting off contact with the outside world, except for a single Dutch ship permitted to anchor off an island in Nagasaki harbor once a year, the Shoguns made sure that unauthorized weapons and other subversive novelties (like Christianity) could not be smuggled into the country.

These measures allowed samurai swordsmen to retain their traditional primacy in Japanese society for the next two hundred years, even though a lasting peace deprived them of their function as fighting men. This paradoxical situation was eventually upset when in 1854 Commodore Oliver Perry, largely on the strength of his naval guns, compelled the Japanese government to change its policy of excluding foreigners, thus inaugurating a new era of tumultuous military-political upheaval that climaxed in World War II.

Japan's fluctuation between extremely rapid, violent accommodation to new weapons and a no less remarkable, deliberately contrived stability exaggerated a parallel fluctuation in European accommodation to gunpowder weapons. For the radical political-military upheaval that prevailed in Europe, when guns were new, slowed down very perceptibly after 1648, when comparatively well-consolidated states and bureaucratically organized standing armies emerged from the Thirty Years' War. Political rivalries did not disappear, and military-technical change did not come to a complete halt. But the uniformity of equipment and training that made large standing armies more efficient also increased the cost of introducing new weapons very sharply, since many thousands of any new model were required if the benefits of uniformity were not to be lost. This became a very persuasive consideration for all European military administrators. As a result, the small, successive changes of design (cumulatively important, though often trifling in themselves) that had come very quickly in earlier centuries slowed almost to a halt. The fact that the British army used the same musket from 1690 to 1840 aptly illustrates the resulting stabilization of Old Regime armies, for during all those years unchanging muzzle-loading muskets were the principal infantry weapon, and infantry remained the undisputed queen of battles.

Naval design also attained remarkable stability during these decades, and international rivalries simmered down as well. When the ideological fires fed by Protestant-Catholic controversy subsided, war became little more than the sport of kings, reinforced by the rivalries of merchants along Europe's Atlantic face. By 1750 or so it certainly looked as though Europe too, like Japan after 1636, had adjusted to the shock of gunpowder weaponry and was settling down toward comparative stability in matters of military technology and political structure.

But, as we all know, that was not the way things went. Instead, international rivalries intensified, beginning with the Seven Years' War, 1756–63, followed by the wars of the American Revolution, 1776–83, and rose to a notable crescendo with the wars of the French Revolution and Napoleon Bonaparte, 1791–1815. This succession of wars, in turn, provoked unprecedented efforts to mobilize human and material resources, transforming the economy and society of Europe and inaugurating the industrial age in which we live.

Nonetheless, although all the years of war between 1756 and 1815 stretched Old Regime military-political management to the limit, they did not alter weapons in any notable fashion. To be sure, the French had responded to their defeats in the Seven Years' War by improving the design of their field artillery, and enjoyed perceptible advantages at Valmy against the Prussians (1792) and at Toulon (1793) against the British as a result. But other armies soon caught up, while other innovations of the war years, such as rockets, observation balloons, and field telegraph, had only marginal importance. The same was true of navies, although the British resort to larger caliber guns, the so-called carronades, prefigured what was later to happen to naval armament, without, however, transgressing the limits set by sails, muzzle-loading cannon, solid shot, and wooden ships.

Despite the Revolution, an almost unbending technological conservatism prevailed in the French armed forces as well as among their politically conservative enemies. This is nicely symbolized by the fact that Napoleon disbanded the balloon observation corps that civilian initiatives had introduced to the French army in the revolutionary year of 1793; and Wellington, after witnessing a trial rocket-firing during which the missiles' erratic course endangered him and other observers, refused to have anything more to do with weapons that, when all went well, doubled the range of existing artillery. (Nevertheless, as the Star Spangled Banner may remind us, the British navy and several continental armies continued to employ rockets, abandoning them only in the 1840s when radically improved guns had begun to match the range and improve upon the accuracy of rocket fire.)

What eventually upset military-technological conservatism was nevertheless an indirect effect of the mounting intensity of warfare that distracted Europe between 1756 and 1815. Demand for iron, uniform cloth, and other commodities assumed unprecedented scale when millions of men were mobilized into armies and navies and had to be equipped. When war ended, this demand suddenly ceased, facing the mills, factories, and artisan shops that had supplied Europe's armed forces with a crisis of survival. Many closed down; especially on the continent, where state arsenals had played the principal part in war production. In Great Britain, however, a host of civilian enterprises had supplied both British and continental forces with iron, cloth, and other materials on an unprecedented scale; and many of the forges and factories that had sprouted luxuriantly during the war years succeeded in finding new civilian markets for their products after 1815, though not without undergoing a difficult postwar depression.

The fate of the iron industry was especially important, for the cheapening of iron, thanks to efficient new furnaces built to supply the British navy's voracious appetite for cannon and other hardware (anchors, chains, and the like), permitted the rapid development of new civilian markets. In particular, steam engines, steamships, and railroads soon were constructed largely of iron; while bridges and new forms of heavy machinery also expanded the civilian demand. What we have learned to call the first industrial revolution, based principally on coal and iron, thus moved into high gear; and with it dawned the industrial age with which this volume of essays is concerned.

At first, the military market was noticeably absent. After 1815, demobilization and military cut-backs everywhere prevailed. Military men were not inclined to experiment with novelty of any kind, and civil administrators were interested mainly in reducing the cost of the armies and navies that each government chose to maintain. For a while, efforts at making the Concert of Europe into a Holy Alliance against revolution affected diplomacy and perhaps helped dampen the rivalries that had emerged from the peace settlement. At any rate, peace and stability were widely wished for after the storms and strains of revolutionary war, and no responsible authority entertained the notion of trying to upset the balance of power by trying to improve upon existing weapons systems.

This postwar era ended abruptly in 1841 when key figures in the French navy came to feel that their nation and service had been humiliated by failure to support the French protégé, Mehmed Ali of Egypt, in his collision with the Ottoman Sultan and the British navy. Mehmed Ali (1769–1848) was an Albanian soldier of fortune, who seized control of the Ottoman province of Egypt in 1805, and then relied mainly on French advisers to help him modernize the country. His army, trained and equipped along European lines, soon proved far superior to any rivals in the eastern Mediterranean; and when the Ottoman Sultan imprudently attacked his over-mighty subject in 1838, Egyptian victories quickly threatened to topple the Ottoman regime. But the British were unwilling to see a French protégé installed in Constantinople, and by using their Mediterranean fleet to blockade Egypt made it impossible for Mehmed Ali to supply his army by sea. Land communications were inadequate, so the Egyptian army had to withdraw and submit to a settlement dictated by the European powers. French assent to this upshot was very grudging, and came only after King Louis Philippe refused to risk war in support of Mehmed Ali, thereby provoking the angry resignation of his fiery, patriotic prime minister, Adolph Theirs.

Memory of this humiliation rankled, and one of Louis Philippe's sons backed French naval officers when they proposed a simple way to counter Great Britain's galling naval preponderance. Their plan was to install steam engines in French warships, thus allowing them to move against the wind without having to tack. The British immediately felt exposed to cross-Channel invasion, since by choosing a time when the direction of the wind would prevent sailing vessels from matching the mobility of steam-powered ships, even a few of the remodeled French ships could neutralize the Royal Navy's numerical superiority. With this, the fat was in the fire. Not surprisingly, the British Admiralty swiftly matched the French by installing steam engines of their own; and the superior industrial base and political tradition that Britain had inherited from the Napoleonic Wars made it comparatively easy for them to maintain superiority at sea despite a succession of other French efforts to renew the challenge by launching further technical innovations one after another.

Until the 1880s, British responses to French initiatives remained reluctant. Any significant change in naval technology meant that the Royal Navy's existing stock of battleships, and the skills of sailing and fighting them, lost part of their value. Change was troublesome and costly. It was also distasteful. Spic-and-span sailing vessels had to take dirty coal on board so that nasty steam engines could spew the sails with even dirtier smoke. Equally distressing was that aristocratic naval officers had to accept uncouth mechanics as colleagues in managing their ships.

But despite heart-felt regret, by the 1840s Britain's traditional reliance on wooden walls was no longer possible. Another French technical invention made that evident to all concerned. As early as 1822 a French army officer, Henri J. Paixhans, succeeded in designing a gun that could fire explosive shells safely, and published a book explaining how his shell guns could easily destroy any wooden warship. In a trial firing two years afterwards Paixhans' guns did indeed destroy an old hulk, just as he had predicted. Thereupon, after appropriate deliberation lasting some thirteen years, the French navy decided in 1837 (just before the humiliation of 1841) to install the new shell-firing guns on shipboard. The Royal Navy and other European navies, including the Russian, swiftly followed suit.

From that time onwards, naval officers realized that sea battles, as they had known them, were a thing of the past. Lying yardarm to yardarm in the approved Nelsonian fashion, and firing broadsides of solid iron shot until the less efficient (or merely unlucky) ship was pounded into submission had become impossible. One or two hits from exploding shells sufficed to cripple any ship, (and set it on fire) as the Russians demonstrated at Sinope in 1853 by shelling the Turkish navy into oblivion in a few hours. That meant, all of a sudden, that the Russian navy could sail to Constantinople unopposed; and it was this prospect that persuaded the French and British governments to cooperate in sending ships and soldiers to help the Turks, thus launching what became known as the Crimean War (1854–56).

From many points of view, this short and half-forgotten war marks the point in European history when the systematic technological conservatism that had dominated military management since 1648 broke down. The French and British navies both accepted the premise that wooden ships had become obsolete, and competed in building steam-powered, armored vessels of wildly diverse designs, intended to carry enormous mortars and other heavy artillery for attacking the Russians' fortified naval base of Sevastopol. But existing steam engines were comparatively weak and consumed a great deal of coal so that naval vessels still had to rely on sails for long-distance cruising. Awkward hybrids therefore prevailed in naval design for the next thirty years before sails could be abandoned.

On land the technological impact of the Crimean War was rather more significant. In the first place, acute problems in meeting sudden increases in the demand for hand guns provoked Europeans to imitate what came to be known as the American system of manufacture for small arms, thus introducing mass production methods to the armaments industry on a really large scale. This was, I suppose, the most fundamental step yet taken in the industrial arms race, liberating an important segment of military supply and design from the shackles of artisan modes of production. Thereafter, it became comparatively cheap and easy to modify small arms while still retaining the benefits of uniformity even when supplying hundreds of thousands or even millions of men. In addition, the Crimean War brought civilian entrepreneurs to the forefront of artillery design and manufacture. Thereafter, private pursuit of profit began to reinforce national rivalries in encouraging technological change in armaments, thus assimilating military technology into the competitive model that already prevailed in the civilian marketplace.

Details of these transformations demonstrate vividly the indirect and unpredictable paths of change in human affairs. Mass production, for instance, began with the War of 1812 when the American government found itself desperately short of muskets, largely because the French had supplied American forces in the Revolutionary War from across the Atlantic. A corps of artisans able to turn out standard muskets therefore did not exist in the United States when war with Britain in 1812 created a sudden need for such weapons. Ingenious Yankees in the Connecticut river valley eventually responded by inventing automatic and semi-automatic machines—the so-called American System of Manufacture—that could make gun parts accurately enough to allow a workable weapon to be assembled from interchangeable parts. Invention and installation of such machinery took a while, and assembly lines were not fully operational until about 1850. The new machines were costly and often wasted material, but turned out gun parts far faster than had previously been possible. Moreover, workers tending the machines needed no special skill. European gunsmiths, by contrast, used simple hand tools—hammer and file for the most part—and were economical of raw materials; but since hand-made parts were never exactly the same, each gun had to be carefully fitted together with delicate filing and other time-consuming adjustments.

American gun-makers, of course, had only a relatively modest market for their standardized products at home. Hoping to expand his sales, one of them, Samuel Colt, brought his wares to the London Exhibition of 1851, where he astonished the public by disassembling revolvers, scrambling the parts, and then reassembling and firing his pistols. The possibility of mass production of standardized gun parts was therefore familiar in Western Europe when the Crimean War provoked a sudden surge in demand for small arms. But established artisan methods set sharp limits on how quickly production could be increased, since training skilled gunsmiths took time. On top of that, both English and French armies were experimenting with muzzle-loading rifles, using a new bullet invented by another Frenchman, Capt. Claude Minié, in the mid-1840s. Rifles were more accurate and carried considerably further than smooth bore muskets, but to attain these advantages existing smooth-bore muskets had to be rifled—another exacting task for the limited number of gunsmiths who could do the job. Under these circumstances, British gunsmiths tried to take advantage of their situation by raising prices, with the resulting public controversy delaying instead of accelerating output.

This experience persuaded the British government to mechanize small arms manufacture by importing milling machines from the United States. Accordingly, a new arsenal to manufacture military rifles was set up at Enfield in 1855 and became fully operational four years later, after the war was over. Other forms of mass production were easier to organize, so that, for instance, a machine set up for the purpose in the Woolwich Arsenal began to turn out 250,000 Minié bullets a day and a second machine combined bullet and cartridge into a single package at a comparable rate. The advantages of mass production were just as obvious to other governments so that within a decade of the time mechanized rifle production at the new Enfield Arsenal came on line, similar establishments arose in all the other leading countries of Europe, and spread to Turkey and Egypt as well.

Increased rates of manufacture made the introduction of new designs for small arms feasible again. The difference was enormous, for when first France (1866), and then Prussia (1869), decided to reequip their armies with modern up-to-date rifles, it took only four years to provide every soldier with the improved weapons. By contrast, when in 1840 the Prussians had decided to reequip their army with an older design of breech-loading rifle—the so-called needle gun, invented by Johann Nicholas von Dreyse—it took twenty-six years to complete the changeover. The artisans Dreyse employed to manufacture the new weapons could not produce more than 10,000 a year; and even when the resources of the state arsenals were brought to bear, production only increased to about 22,000 per annum. By comparison, in 1863, when the Prussians were still straining to complete their 1840 program, the new Enfield Arsenal turned out 100,370 rifles in a single year, routinely and without any exceptional emergency to spur extra effort.

Long-standing obstacles to technological change of small arms were thus swept away, and inventors rapidly developed increasingly effective rifle (soon, also, machine gun) designs so that, from time to time, European armies continued to reequip their infantrymen—by the millions. Each such change required new drill, new tactics, and new logistics to match the guns' increasing appetite for ammunition. Under the circumstances, familiar routines began to blur and established rules for the conduct of battle became obsolete as the experience of World War I eventually showed. But until 1914 most army officers refused to admit that anything had happened to upset their battle plans. Instead they left technological change to a handful of specialists and assumed that radically improved infantry weapons would have no important effect on how soldiers would have to behave in battle.

But military men were not left to their own devices when it came to changes in artillery design. Instead, the yawning gap between what they expected and what turned out the be the case in 1914–18 widened even further because the manufacture of artillery and other heavy weapons became inextricably entangled with the pursuit of private profit.

This began when individual industrialists decided to apply civilian skills to the manufacture of guns, believing that it would be easy to produce a better weapon than the muzzle-loading cannon that government arsenals turned out. On the continent, it was Alfred Krupp of Essen who pioneered the manufacture of technically

superior breech-loading steel artillery. Like Samuel Colt, he announced his technological prowess by exhibiting a few samples at the London Exhibition of 1851. But at
first he had difficulty persuading governments to buy his product, partly because
steel guns were expensive, partly because they sometimes suffered from casting
flaws and fractured unexpectedly, and mostly because military procurement officers
were accustomed to acquiring artillery from state arsenals and distrusted the crass
and selfish motives of an upstart manufacturer like Alfred Krupp who, after all, expected to make money from selling his guns. Egypt was his first customer (1855);
Prussia rather reluctantly followed with a trial batch of 300 guns; but only when the
Russians placed far larger orders after 1863, did breech-loading steel artillery really
begin to displace bronze muzzle loaders. Range and rate of fire for field artillery began
to increase accordingly; and a long series of improvements came very rapidly thereafter as private firms and state arsenals competed with one another in introducing
new designs.

But field artillery was limited by the fact that guns had to be light enough for
horses to pull them cross country. No comparable limit affected naval artillery; and
the race between ship's armor and big guns therefore became more technologically
significant than anything happening to field artillery. The pace of naval change was
enormously enhanced by the fact that during the Crimean War two venturesome
private manufacturers in England, William Armstrong and Joseph Whitworth, decided it was time to bring military engineering up to the level of civil engineering by
showing the government's arsenal how to make bigger and better guns. They both
had the resources at their command to design and build prototypes that were in fact
superior to existing arsenal products. But persuading military procurement officers
to buy newfangled weapons was another matter. Armstrong and Whitworth trum-
peted rival claims for the superiority of their guns; and public tests of armor-piercing
capability showed both strengths and weaknesses in their competing designs. Intense controversy arose between Whitworth and Armstrong; as well as between
those who preferred to entrust the manufacture of big guns to state employees and
those who argued that private manufacturers, tainted though they might be by greed,
ought to be preferred if their products were indeed superior. Official policy waffled,
and, after a brief flirtation with William Armstrong, (persuading Whitworth to give up
gun-making) the British armed services reverted in 1864 to arsenal production. Thereupon, thanks initially to vigorous sales in the United States, where the Civil War
created sudden demand for his big guns, Armstrong developed an international
market for his wares, rivaling (and ere long also collaborating with) Alfred Krupp's
parallel enterprise.

After his dismissal from official appointment as engineer of rifled ordnance,
Armstrong's relation with the Admiralty became intensely ambivalent. He always
nourished the hope of selling guns, turrets, and other heavy equipment to the Lords
of the Admiralty once again, yet more than once he whipsawed the Royal Navy into
unwelcome new expenditures in the arsenal by equipping foreign navies with guns
(or complete warships) that out-performed existing British models. From the point of

view of the managers of the Royal Navy he thus matched at home the continued challenge from the French, who from time to time invested in new weapons that promised to make the existing British fleet obsolete.

By the 1880s, resulting uncertainty had become acute. British arsenal designs and production persistently lagged behind innovation originating privately or in French arsenals. In particular, two technical changes—one French, one German— made the latest 80-ton muzzle-loading monster guns that British battleships were carrying hopelessly obsolete. First, between 1881 and 1887 the French chose to challenge British naval preponderance by concentrating their naval construction entirely on fast, long-range cruisers, designed for commerce raiding, supplemented by even faster torpedo boats for short-range operations. British ships were too slow to catch the new French cruisers on the high seas; and British naval guns fired too slowly to be able to hit an approaching high-speed boat before it came into torpedo range. Thus, in spite of all its expensive efforts at technological modernization, the Royal Navy once again faced the prospect of being unable to safeguard the Channel or to protect British commerce from the French.

The second problem was equally intractable. In 1878–79 Krupp introduced a new line of big steel artillery pieces, suitable for naval use, and designed to take advantage of slow-burning smokeless propellants that had recently been perfected. Demonstration firings showed military observers from all the leading countries of the world that Krupp's new guns completely outclassed muzzle loaders like those Woolwich Arsenal was producing. Obviously, from a British point of view, something drastic had to be done and quickly to preserve the Royal Navy's power and effectiveness.

This was the setting in which the naval arms race assumed a new character, becoming far more expensive, far more radical, and far more important for the national economy of Great Britain and for other countries that chose to challenge British naval preponderance. In a word, what happened was that the modern military-industrial complex came to birth when a small group of technically minded British naval officers, of whom Capt. John Arbuthnot Fisher was the ringleader, began to foment and hasten technological changes, believing that if official funds and policy actively promoted improvement in weapons systems, British skill and industrial capacity would suffice to keep the Royal Navy ahead of all rivals indefinitely.

The effect was drastic. Instead of responding sluggishly and regretfully to innovations arising privately or in French arsenals, the Admiralty began to challenge inventors and manufacturers to come up with appropriate new devices, and, before long, helped them to meet development costs for particularly promising innovations. In a sense this was no great departure from established routines. The Admiralty had long been accustomed to specify the size, shape, and other characteristics of warships constructed in official dockyards, and the Woolwich Arsenal built naval guns to specification as well. Ever since iron and steam had begun to supplant wood and sails, specification for new ships involved departure from former patterns—sometimes very drastic. But by and large, before the 1880s specified innovation merely transferred (with adjustments) existing civilian technologies to naval construction. Naval technology had consistently lagged behind, largely because those in charge were so loathe to abandon old ways and accustomed routines.

But a reckless new spirit, welcoming and accelerating innovation, took root in the 1880s, when Fisher and others like him inaugurated what may be called "command technology," and soon applied it across the entire spectrum of naval purchases. In effect they reversed older relationships between inventors and military procurement officers. Instead of waiting until someone came along with a new device, as army and navy officers had been accustomed to do, and challenging the innovator to prove that the cost and trouble of change over was worth it, the British Admiralty began to define what it wanted in the way of new performance characteristics, and then required arsenal personnel to compete with private manufacturers to see who could most nearly match their desires. Invention thus became deliberate and organized, with the result that innovation in naval technologies soon outstripped civilian engineering in important fields like the development of hydraulic machinery, steam turbines, diesel engines, optical glass, radio communications, and electrical control systems—not to mention more obvious matters like steel metallurgy and the chemistry of explosives.

In 1886, when the Admiralty was first authorized to buy materiel from private manufacturers whenever the arsenal could not provide an equivalent item, no one foresaw that the Royal Navy would become as intimately intertwined with heavy industry as it did. But in fact, the arsenal was critically handicapped after 1886 because the massive investment needed to go over to using steel as raw material for guns and ships was never made. Krupp had shown in 1887 that long-barreled, breech-loading steel guns were indisputably superior. Armstrong and the French—both a new private firm, Schneider-Creusot, and arsenal gun-makers—responded by investing in steel-making and gun-manufacturing plant; but the Woolwich Arsenal was never granted the necessary funds for this radical changeover, so that naval procurement increasingly went to private sources.

Naval officers were prepared neither to buy abroad nor to depend solely on Armstrong for supplying steel guns and other heavy equipment for their ships. They solved that problem by inducing England's leading steel-maker, Vickers, to enter the armaments market in 1888, and sought to play one firm off against the other thenceforth. As the scale of successive naval-building programs increased—and increase they did thanks to foreign competition, first from the French, then from Germany, the United States, and Japan—price ceased to be decisive in more and more instances. Often only one supplier had the capability of making a particular item. Oftener still, decision of which contracts to award to which firm became an overtly political act. Naval-building became a recognized way to counteract the business cycle by keeping men at work in periods of depression. Even more telling, naval contracts exempted English steel-makers and other heavy engineering firms from having to compete on world markets with cheaper American and German producers. Navy expenditures (supplemented, but on a comparatively modest scale, by army purchases) became a critical balance wheel for the entire national economy. Indeed, on the eve of World War I as much as a sixth of the male workforce of Great Britain was employed by the Navy or by prime contractors for the Navy.

Similarly powerful military-industrial complexes swiftly formed in France and Germany, and emerged in the United States as well, without, however, attaining comparable weight in the economic-political life of our nation until during and after World War II. In Japan, on the contrary, the military-industrial complex had been of prime importance for the national economy and for politics ever since the Meiji Restoration in 1868. But before World War I, the Japanese were still catching up with European technology, and their version of command technology therefore involved less outright invention and more borrowing (with minor adaptation) than was the case in England, Germany, and France.

I will not attempt to deal with more recent perturbations and turning points in the history of the wars, arms races, and recurrent military-technical transformations that have followed. Other essays in the volume will shed light on diverse aspects of that vast subject. Instead I wish to conclude with some brief reflections on the process as a whole, aimed at addressing the theme assigned to me: the structure, or perhaps better structures, of society and politics that provoked and sustained the radical changes of the past hundred and fifty years.

First and most obvious: rivalries among sovereign states were an essential ingredient, and when such rivalries provoked actual warfare, the pace of technical change regularly intensified. This needs no argument. Without the French-British rivalry, the naval history of the nineteenth century, with which I have been mainly concerned in this essay, would be inconceivable; and this rivalry was what created the world's first autocatalytic military-industrial complexes—on both sides of the Channel. The French always relied more on technologically proficient engineers in state service, and never gave their successive naval building programs the consistent political support the Navy enjoyed in Britain, largely because the French army, with far less technologically varied demands on industrial production, always came first. But despite these differences, military purchasing also played a critical role in the development of the French economy in the nineteenth century, and by the 1890s, the French, like the Germans, had brought private industry into an increasingly close partnership with the state.

At the same time, the actual expression of the state rivalries of Europe depended on what a few key personalities decided at particular times and places. Thus, the almost whimsical way William Armstrong decided to use the resources of his engineering firm to build better guns after reading a newspaper report about how a single field artillery piece had affected the outcome of the Battle of Inkerman had consequences far beyond anything he conceived of when he first sketched how he proposed to build bigger and better gun barrels by sweating layers of wrought iron around one another. Similarly, if Fisher had been more scrupulous in obeying his naval superiors, the public outcry that arose when he secretly primed a well-known journalist with facts about the sorry state of the Royal Navy's armaments in 1884 would not have resulted in the passage of an expanded naval budget—the first of a series of escalating budgets, each supported by a carefully contrived publicity campaign in which newspapermen, naval officers, industrialists, politicians, and other interested parties soon learned to cooperate.

Were all these interested parties preordained to coalesce into Great Britain's military-industrial complex? And was that complex preordained to provoke parallel structures in France, Europe, and other countries, including the United States? Or did individual decisions and the happenstance of particular response to specific situations have the unintended and unforeseen effect of bringing them together? No one can answer that question with certainty. What happened, happened; but it seems to me that personal decisions, with a heavy freight of unforeseen consequences, were what drove the process as a whole. If key personalities had been different, the course of events would surely have been different too—perhaps diverging only slightly, perhaps fundamentally. For instance, would Germany have set out to rival England's navy without Adm. Alfred von Tirpitz, Kaiser Wilhelm II, and the writings of Capt. Alfred Thayer Mahan? Would World War I have been fought had the policies of the Kaiser's government not persuaded France, England, and Russia to bury their differences and form the Triple Entente? Or would we have atomic warheads today if refugees from Germany had not persuaded Albert Einstein to sign a letter that alarmed President Franklin D. Roosevelt in 1939? These and many other accidents of human encounter have sustained the arms race ever since it took on its modern form in the 1880s.

Would things have turned out approximately the same anyway? I find it impossible to believe that personal decisions in critical situations did not alter outcomes in detail, and, through the cumulation and conjunction of details, shape and reshape the arms race fundamentally. I also find it obvious that what key individuals hoped and expected to achieve by their decisions seldom or never matched up with what happened. Instead, unexpected and unforeseen responses to particular decisions prevailed—universally and perpetually. The reactive process was enormously complex, limited only by the diffusion of information (and misinformation) among participants. Purposes were essential inasmuch as they governed everyone's actual decisions. But results were always surprising—sometimes radically, embarrassingly different from what had been intended. After all the Kaiser lost his throne, the Royal Navy bungled the Battle of Jutland; and Americans find themselves burdened by nuclear warheads and afraid of what others may do if and when they acquire access to these almost unimaginably powerful explosives.

The effect so far has been to make international relations more dangerous and unpredictable than they were when wars were fought with weapons long familiar to all concerned. In addition, costs have escalated sporadically but ineluctably—matched in our time only by the escalation of medical costs. Eventually, limits to both forms of extravagance will surely assert themselves. Conscious policy is likely to remain ineffective in shaping long-term results, as hitherto. Changes in the process itself will have to occur, perhaps through the involvement of competing interests and groups that are now largely inert, or, alternatively, by some sort of (presumably atomic) catastrophe that might end human life entirely, or merely end the industrialized arms race by establishing a world monopoly of capital weapons.

As I said at the beginning, we live in an exceptionally uneasy age and need to reflect on how the process of weapons development became so unmanageable exactly when deliberate invention of specific improvements of particular weapons became routine. We need to confront the irony whereby the rational triumphs of deliberate, organized invention became increasingly irrational in their aggregate effect. We even need to wonder about unending technical change and our capacity to endure it—individually, collectively, and ecologically.

I therefore leave you with much to think about and no ready answers. Time will tell, as always. That is mildly comforting to an elderly historian like me. I fear it will merely irritate technically proficient, can-do officers, and politicians, trained, as they are, to take command of the situation and to solve most problems by ordering up new, more powerful machines.

Marching in Step: Technology and *Mentalité* for Artillery, 1848–1914

Dennis E. Showalter

Between 1848 and 1914 the Atlantic world witnessed an artillery revolution. At mid-nineteenth century, gunners were stepchildren, manning weapons essentially unchanged since the days of Napoleon and officered by social undesirables. Less than three-quarters of a century later, artillery had become a principal focal point of intellectual and scientific thought everywhere in military Europe. The guns and howitzers that went to war in 1914 epitomized high-tech weaponry, their refined complexities far overshadowing the box-kite aircraft and the primitive motor trucks that were their principal rivals at the technological summit. The gunners themselves had lost their marginal status and become part of the military establishment. Acceptance, however, came at a price. This essay analyzes the parallel developments of technology and *mentalité* that brought artillery into the mainstream of the West's armed forces.

From Minié to Krupp

The first stage in the process lasted from 1848 to 1871. It involved a complex four-way interaction among weapons, organization, tactics, and institutions. Artillery technology at midcentury was undergoing parallel revolutions in response to changes in infantry tactics and armament. The general introduction of rifled muskets and the corresponding adoption of open-order tactics increased artillery's vulnerability while decreasing its opportunities for confronting mass formations. An obvious method of restoring the artillery's preeminence at long ranges was to rifle *its* barrels, giving the gunners an extended killing zone. But the rifles tested in Europe's arsenals and issued to Europe's armies had significant drawbacks. Whether muzzle-loaded like the majority of the designs or loaded at the breech like the 6-pounders Alfred Krupp was attempting to market in Prussia, rifles had an uncomfortably slow rate of fire. Their muzzle velocity was so high that it reduced the blast effect of individual shells, which often buried themselves so deep in the ground that they failed to explode at all. The rifles' relatively small caliber also diminished the effect of their canister rounds—the ultimate weapon against a close assault.

What was needed as well or instead, critics argued, were cannon capable of firing shell, shrapnel, and canister directly at medium and short ranges. Since the greatest possible blast effect for single rounds was desirable, the gun should be of relatively large caliber: preferably a 12-pounder. Since accuracy at long range was

not particularly important, the muzzle velocity of these proposed new designs could
be reduced. This meant their barrels could be shorter and lighter, offering a corresponding improvement in mobility over existing smoothbore systems.[1]

For ten years gunners debated the question of which innovation represented
artillery's future. France, the continent's leading military power, and Prussia, its leading military pretender, compromised on 50 percent solutions: half rifles and half shell
guns. Russia began rifling its existing smoothbores after the Crimean War, then
produced an interim generation of bronze smoothbores before adopting a system of
bronze rifled breechloaders in 1867. On the other side of the Atlantic the Union's
Army of the Potomac, which had first call on the North's supply of modern weapons,
began with a two-to-one balance in favor of rifles, but changed to a rough parity
between rifles and shell guns in the Civil War's later years.

These proportions were consistently modified in the light of tactical experience.
Much of the hardest fighting in the American Civil War took place in broken terrain
at relatively short ranges. Both Union and Confederate armies were heavily committed to shock tactics: repeated assaults by masses of infantry. In such contexts the
shell-firing smoothbores emerged as the dominant field artillery weapon on both
sides. In Bohemia, on the other hand, during the campaign of 1866 Prussian shell
guns were so consistently outranged and outshot by Austrian rifles that after the
war they disappeared entirely from the order of battle. The French, believing the
combination of the Chassepot and the *mitrailleuse* could repel most attacks, converted their shell guns to rifles as an interim measure prior to the Franco-Prussian
War.

Gun ratios were almost as important as gun models. The number of pieces required relative to infantry and cavalry under modern conditions remained a consistent subject of debate. The Army of the Potomac averaged between three and four
guns per thousand men. A Prussian army corps in 1870 had at full strength about
three-and-a-half guns to a thousand rifles, depending on how many of the corps's
three organic horse artillery batteries were detached to the cavalry.

The problem lay in maintaining those proportions in the field. By the end of the
Civil War, William T. Sherman, hardly a proponent of head-on frontal attacks, declared one or two guns to a thousand veterans was more than enough support for
offensive operations.[2] In contrast Robert E. Lee's Army of Northern Virginia, fighting
largely from fixed positions, had as many as nine or ten guns per thousand infantry.
This, however, reflected manpower losses rather than tactical doctrine. The Prussians
in their campaigns against the Republic in the south and east of France confronted a
similar problem. Infantry battalions with an official strength of a thousand shrank to
eight, sometimes even six, hundred under the harsh conditions of winter campaign-
ing. As a result, the proportion of artillery grew at times to six or seven guns per
thousand rifles—a ratio generally regarded as both hampering the army's mobility
and the infantry's offensive spirit. The twentieth-century argument that fire support
can become addictive was heard on the Loire and in the Vosges a century earlier as

Prussian colonels and generals complained their men had become unwilling to attack unless the artillery did most of the work.[3]

Organizationally, artillery during this period suffered from its nature. The basic tactical unit, the battery of from four to eight guns, was self-sufficient in a way an infantry battalion or cavalry squadron was not. Higher formations tended to be correspondingly regarded as a nuisance. One gunner in the Confederate Army of Tennessee declared "a captain under . . . battalion management is a perfect automaton."[4] This mind-set was congenial to senior officers on both sides of the Atlantic, who tended at the beginning of a war to decentralize part of their artillery, attaching batteries by ones and twos to divisions and even brigades while holding the rest in a general reserve. As late as 1870, the French artillery virtually abandoned its peacetime regimental organization, with some army corps receiving batteries from as many as five different regiments.

Experience nevertheless overcame entropy in the Atlantic world's midcentury wars. Individual batteries contributed too little to either attack or defense, while masses of guns held in reserve too often came into action late or not at all. The Union Army of the Potomac made the best use of what might be called a traditional system that in its developed form gave each corps an artillery "brigade" of five or six batteries, with another twenty or so held in army reserve. The system's success, however, owed more to the extraordinary abilities of Gen. Henry J. Hunt than to any inherent merit. By the time Grant moved into the Wilderness in 1864, the artillery reserve was little more than a replacement pool for exhausted batteries. It was the Army of Northern Virginia that established the precedent for modern artillery organization by simultaneously concentrating and decentralizing its guns. As part of the general reorganization after Stonewall Jackson's death at Chancellorsville, the general reserve was abolished. Each of Lee's three corps received no fewer than five battalions of artillery. Three were usually attached to the corps's infantry divisions with two in a general-support role; all five could be concentrated under the corps's artillery commander.[5]

Prussia took this structure a step farther. Each of its corps included a field artillery regiment of three "foot" and one horse battalions. Prior to the Seven Weeks' War, one battalion was assigned as direct support to each of the corps's two divisions, with the balance (less attachments to the cavalry divisions formed on mobilization) contributing to artillery reserves that frequently saw no action at all. The guns that did reach the battlefield tended, moreover, to deploy by batteries, when not by sections. The result was an embarrassing ineffectiveness, particularly during the war's decisive battle of Königgrätz, where Prussian infantry found itself relying on breechloading needle guns not only to repel Austrian attacks, but to support their own as well.

After 1866 nomenclature and doctrine changed significantly. The name "reserve artillery" disappeared from Prussia's military vocabulary. The same guns were now called "corps artillery." Nor was this a mere semantic change. Artillery officers were now encouraged to think in masses, concentrating and using guns in large numbers from the very beginning of a battle. During the Franco-Prussian War, battalions, sometimes even regiments, were deployed as units and handled as entities by colonels and majors who were no longer regarded as glorified staff officers. From the war's first days, gun lines of up to and over a hundred barrels anchored Prussian positions and destroyed French ones.[6]

This development reflected a fundamental shift in artillery tactics. From its initial appearance on the battlefield through the first half of the nineteenth century, artillery was more effective in defense than attack. Napoleon's "artillery charges," bringing masses of guns to point-blank range against enemy formations, were more the exception than the rule. The heavy, cumbersome pieces that represented state-of-the-art technology required both open ground and an obliging enemy to perform that particular *pas de dance*. Sénarmont's achievement at Friedland amounted to one of a kind. Napoleon's artillery did most of its best work from set positions.[7]

The artillery on both sides during the American Civil War was similarly more successful holding lines than breaking them. From Malvern Hill to Gettysburg, Union guns set at naught the boldest Confederate assaults, while at Cold Harbor massed Confederate artillery did far more than Minié balls to smash Grant's head-on attack. In the western theater the artillery of the Army of Tennessee, arguably the least efficient and certainly the worst-equipped of any major field army during the war, was the army's backbone during the Atlanta Campaign. The smoothbore 12-pounders in particular had fearsome reputations as man-killers, with a shock power and a deterrent effect extending far beyond the actual casualties inflicted. The inscription on the monument to the 1st New York Battery at Gettysburg might well stand for all the Civil War's cannoneers: "Double canister, at ten paces."

During the Civil War, in short, the long arm was really the short arm. This reflected the offensive limits of midcentury artillery technology against even a moderately alert enemy—particularly one who strengthened his positions by field fortifications.[8] The range of smoothbores was too short, the fire-power of rifled guns too limited, to make the risks acceptable except in desperate circumstances. On the other side of the Atlantic as well, neither Prussian, Austrian, nor Piedmontese artillery showed particular capacities for maneuvering and deploying under fire in 1866.

At the same time, neither in the old world nor the new were artilleries particularly effective in their traditional roles. It had long been an article of faith that a major prerequisite of victory involved winning the artillery duel: silencing the enemy's guns so that one's own infantry could advance relatively unmolested.[9] The development of long-ranged infantry rifles, whether breech- or muzzle-loaders, tended to modify the concept significantly. Infantry fire, by itself or with minimal help from the big guns, was usually quite sufficient to see off even the most determined bayonet charges. Artillery was, moreover, disconcertingly difficult to silence in practice. Batteries could shift positions. Gun crews could be replaced. Artillery pieces at midcentury prefigured machine guns during World War I. Enough of them survived enemy action to defeat enemy attacks.

Nor was artillery particularly effective in directly preparing attacks. Whatever claims might be made for rifled cannon on the testing ground, in practice artillery was still far from a precision weapon. Overhead fire was as dangerous to one's own men as to an enemy seldom obliging enough to deploy in masses on open ground, as the Prussians had done at Ligny in 1815. Midcentury infantry tactics encouraged moving faster through the killing zones. Both Prussia's company columns and Austria's battalion-sized close formations were configured for maneuverability under rifle and artillery fire. Casualties, even heavy casualties, were expected, but accepted as the price of victory in modern war.

The limitations of this concept were demonstrated time and again in the American Civil War, where lines of infantry deployed one behind the other repeatedly failed even to reach enemy positions except at abortively high cost. Europeans who believed Union and Confederate armies were too ill-trained and poorly disciplined to implement their tactics properly were disabused of that notion in 1866, when Austrian columns were pulped by Prussian rifle fire alone.[10] From the first days of the Franco-Prussian War, Prussia's infantry faltered against an enemy whose breechloading Chassepot rifles outranged and outshot the now-obsolescent needle gun, and whose doctrines emphasized the tactical defensive. Prussian officers relearned the fact that once an attack stopped for any reason it was difficult to restore momentum. A man who halted to shoot tended to remain where he was. To compound the infantry's problems, the wooded, broken terrain of eastern France disrupted formations even during approach marches, to say nothing of advances.[11]

It was then that the guns came to the fore. Their success was based on deploying as many pieces as possible as soon as the decision to attack was made and the area of advance determined, then advancing to effective ranges even when the positions were exposed to enemy fire. As part of the new order, batteries were expected to hold their positions to the last man even if the last rounds had been fired. The new tactics worked in good part because French artillery was so consistently overpowered and outshot that it posed only a marginal threat. The French had fewer guns than the Prussians—a function of the ill-fated decision to make the short-ranged, rifle-caliber *mitrailleuse* an artillery responsibility. Their 4-pounder rifles were obsolescent. Their 12-pounders were significantly less accurate than their Prussian counterparts. French time fuses were notoriously unreliable compared to the percussion-fused rounds in Prussian caissons.

The French continued as well to hold guns in reserve, bringing them into action by batteries as needed. The usual result was that these formations found themselves in positions already ranged, and came under fire before even having a chance to unlimber. French artillery doctrine further contributed to the Prussian sense of superiority by encouraging batteries to withdraw or change positions under those circumstances. To the Prussian gunners, it seemed that their French opponents were all too willing to concede the day after losing a man or two.

By the time of Vionville and Gravelotte, Prussian artillery was concentrating on killing French infantry wherever it appeared—targeting encouraged by the fact that their long-range rifle fire was inflicting heavier casualties on German gun lines than shell or shrapnel rounds. After the Second Empire's collapse, the new emphasis on preparing and supporting infantry attacks was further encouraged by the relative inefficiency of Republican France's field artillery. Infantry might be improvised. Replacements for the batteries trapped in Metz or surrendered at Sedan could not spring so readily out of the ground. Guns were not lacking. Efficient crews and competent battery officers were other matters altogether.[12]

Institutionally, artilleries at midcentury were joining the military mainstream. Austria's gunners had considered themselves an elite since the era of Prince Johan von Lichtenstein and the Seven Years' War. In Northern Italy and Bohemia, above all at Königgrätz, they consistently outperformed their comrades of the infantry and cavalry. Benedek's *Nordarmée* owed its escape in large part to the cannoneers who kept the victorious Prussians at a respectful distance. In France the artillery benefitted consistently from Napoleon III's interest in its technical and tactical aspects. Its traditional status as a "learned arm" gave it solid representation on the various commissions that met in the Second Empire's final years to evaluate reforms in defense policy. Its officer corps was mostly professionally trained, in contrast to the infantry and cavalry with their high proportions of former enlisted men. The Artillery Committee was responsible for passing on proposed technical innovation for the other arms as well—a position facilitating institutional power plays but affirming for the other arms the artillery's integration into France's military system.[13]

Prussian artillery improved its social image during the same period. Under its new and hard-driving inspector-general, Gustav von Hindersin, the gunners' reputation for energy and efficiency attracted increasing numbers of aristocrats and *Abiturenten*. Regimental selection boards had a wider choice of candidates than at any time in their history. A future chief of staff, Alfred von Waldersee, began his service in the artillery of the Prussian Guard. Prince Kraft zu Hohenloe-Ingelfingen was the scion of one of Prussia's greatest noble families, and in a long career demonstrated that noble birth and high status were no obstacles to technical expertise, professional skill, and a cool courage that earned the admiration of everyone who saw the Prince under fire in Bohemia and France.[14]

Across the Atlantic, gunners were prominent at high command levels on both sides of the Civil War. In part this reflected the makeup of the prewar United States army, with four regiments of artillery as opposed to only ten of infantry. George Meade and George Thomas, William Sherman and Ambrose Burnside—all began as subalterns in one artillery regiment, the 3rd.[15] Given the massive demand for trained officers combined with the limited opportunities for promotion within the artillery, especially in the Union army, it was scarcely surprising that officers like John Gibbon, John Reynolds, and Charles Griffin ended their careers commanding divisions or corps. The Confederacy drew fewer senior officers from the artillery, in good part because its organization offered better chances for achieving field rank, and gunner majors and colonels were reluctant to transfer. Nevertheless Braxton Bragg, Jubal

Early, and A. P. Hill were only three of the gunners who led Confederate corps and divisions with varied degrees of success and distinction.

Into the Breech

The second phase of the modernization of Europe's artillery, from 1871 to 1897, was dominated by a single factor: mission. The defensive role so important for artillery in the American Civil War and in 1866 was being increasingly assumed by the infantry rifle fired from entrenchments. The Franco-Prussian War had shown events in North America were no anomaly. Even improvised field works defended by men armed with modern rifles were something more than a challenge to be overcome by training and discipline. Writing in 1874, Helmuth von Moltke observed that improvements in firearms had given the tactical defense such a great advantage that offensive operations were best undertaken only after defeating several enemy attacks.[16] That recommendation, however worthy, required an extremely obliging adversary. As for the great German's encouragement of flanking maneuvers as an alternative, these operations as well ran the proximate risk of encountering field fortifications. With Europe's armies growing increasingly symmetrical, systematically reading each others' professional literature and taking detailed account of each others' doctrines, the possibility of facing even subordinate commanders feckless enough to waste their own troops in unprepared assaults seemed remote.

Any doubts as to the dominant tactical problem of the nineteenth century's third quarter faded in the Russo-Turkish War of 1877–78. The Ottoman army proved anything but a formidable opponent in the field. But forced back on the city of Plevna, Osman Pasha established a hastily dug line of fortifications north of the town. Armed with breechloading single-shot and magazine rifles the Turks smashed attack after attack, strengthening their positions between rounds. The Russian artillery was the last word in bronze breechloading technology, but its 4- and 9-pounder rifles did little more than shift Turkish dirt. On a single day, 22 August 1877, the gunners fired almost 5,000 shells in support of their attacking infantry. Yet it was the bayonets and the raw courage of the men in the front lines that broke the Turkish positions. Neither an extensive preliminary bombardment nor an immediate two-and-a-half-hour preparation proved of significant assistance on 30 August, when casualties were so heavy that the Russian/Rumanian high command finally decided to starve the city into submission.[17]

Plevna was a warning. If entrenched positions were not overrun in a first assault, opportunities for later success at reasonable cost were almost certain to diminish exponentially. That fact in turn reinforced the concept that had proved so useful for the Prussians in 1870: using guns in masses—and using them primarily to prepare infantry attacks. Experience suggested that this process was likely to take time. It was also likely to require large numbers. An individual round from a rifled field gun had at best limited blast and fragmentation effect. Heavier weapons had been long since relegated to the siege trains for the sake of mobility. Therefore, it seemed not merely logical but necessary to bring as many guns as possible into action against the same target.

The events of the Franco-Prussian War suggested another possibility in the same context. Rifled cannon had become just accurate enough, and their ammunition just reliable enough, to make effective close support of infantry attacks just possible. Cannon might no longer be able to gallop to pointblank range and blast open a hostile position, but they could now fire over the heads of their own infantry, or from flanking positions, until the final assault went in. Infantry officers were initially unenthusiastic. Whatever its present claims might be, artillery had never been known as a precision weapon. In particular, overhead fire seemed as risky for one's own men as dangerous to an enemy. But modifying existing designs in the direction of flatter trajectories and higher muzzle velocities, thereby enhancing the ability of individual guns to range precisely, would make them more effective as a support weapon, better able to keep enemy positions under fire until the last minute.

In this context artillery's traditional role of silencing enemy guns took second place. More and more of Europe's gunners dismissed it as making no more sense under modern conditions than chess players opening a game by exchanging queens. Artillery did not best fulfill its functions by trading shots at 5,000 yards and leaving the rival infantries to fight it out by themselves. Guns were disconcertingly difficult to disable permanently. Personnel casualties could be replaced from the wagon lines, and in emergencies from nearby infantry units. The occasional chance hit that blew up a caisson was spectacular enough but seldom likely to disable even one or two guns, to say nothing of a whole battery. Finally, the sheer numbers of guns involved in a modern battle meant that most of a day could be wasted in silencing them even under the most favorable circumstances. Hohenloe, who after 1871 was for Europe's artillerymen the intellectual inspiration Moltke was for Europe's armies in general, instead described a scenario in which artillery initially sought only to draw enough fire away from its own infantry and sought to silence opposing guns as part of a coordinated final assault.[18]

Implementing the new concept required new material. In the immediate aftermath of the Franco-German War, the Kaiser's gunners talked of a 96-mm barrel firing an 8-kg shell. Such a gun would require an eight-horse team, but veterans of the fighting in eastern France considered the probable results in combat well worth the increased weight. Critics' concern for cross-country mobility combined with the costs of extra horses to produce a compromise. The *System 73* was based on an 8.8-cm gun less powerful than the original proposal, but light enough to be drawn by a six-horse team. The new guns had the highest muzzle velocity, the longest range, and the best accuracy in Europe—at least temporarily.[19] The French, the Russians, and the Austrians introduced similar systems within the next decade. Great Britain's imperial commitments, on the other hand, led to an emphasis on light pieces suitable for mountain warfare—the "screw guns" immortalized by Rudyard Kipling—and to the early abandonment of breech loaders in favor of muzzle-loading rifles considered more reliable in the circumstances of warfare outside Europe. As a result during the Egyptian expedition of 1882, British batteries were embarrassingly outclassed by Egyptian batteries armed with Krupp breechloaders and trained in continental techniques of fire control. Nevertheless, not until the late 1880s did the Royal Artillery take delivery of breechloading 12-pounders similar to those long in use by European armies.[20]

The problem of dealing with field entrenchments was addressed from another angle as well. Modern artillery seemed destined to play much the same role in battle it traditionally performed in sieges. What it therefore needed was the ability to blast open redoubts and blow in trenches, to search dead ground and make covered positions unsafe. In short, modern artillery needed a high-angle capacity.[21] Howitzers had been in use since the eighteenth century, only to be replaced by smoothbore shell guns in the 1850s. In the 1870s and 1880s, continental armies reintroduced heavy howitzers to their orders of battle. The typical design was a 12-cm or 15-cm piece, usually manned by foot or siege artillery formations rather than field gunners, and intended as much for work against permanent fortifications as for service *en rase campagne*. The new guns nevertheless laid the foundations for integrating heavy artillery units into field armies. The "gambardiers" brought with them a bias in favor of scientific gunnery and heavy metal that never quite disappeared in the coming era of heroic vitalism.[22] Nor did the high-angle pieces remain siege weapons *manqué*. By the century's final decade lighter designs, true field howitzers, were entering into service as well: the British 5-inch, the Austrian 10-cm and, by far the best, Germany's 10.5-cm *Feldhaubitze 98*, ancestor of a long line of similar weapons around the world.

European gunners during the 1870s and 1880s benefitted from a broad spectrum of lesser technical innovations as well. Gun carriages were lightened, their durability enhanced by the introduction of metal parts on the traditionally all-wooden vehicles. Fixed ammunition and improved breech mechanisms increased rates of fire to three or four times what they had been at midcentury. Fuses became steadily more reliable— a response to the demands of long-range bombardment.[23]

As a result, gunners began reevaluating the mix of ammunition in their caissons. Shrapnel had been introduced in the Napoleonic War. A thin container filled with metal balls and a small explosive charge, it was designed to burst above enemy troops and positions. Its theoretical value against targets caught in the open had never been questioned. Its unreliable fuses continued to make it unpopular in practice—until artillerymen began coping with the ramifications of using common shell to prepare attacks. These were still black powder rounds with limited blast and splinter effects. Achieving any results with them was a slow process. Shrapnel, on the other hand, was a potentially effective large-scale man-killer, particularly against exposed targets like gun crews. Its bursts would not destroy trenches or redoubts. But if rounds of proper design, with effective fuses, fired from a hundred or more guns could simultaneously spray tens of thousands of lethal balls into defense systems, the occupants were at least likely to keep their heads down while the infantry closed in. It seemed a reasonable alternative to the excruciatingly slow process of dismantling the defenses themselves. And by the mid-1880s shrapnel fuses had in fact improved enough to justify increasing the proportion of shrapnel rounds in a German battery to almost half of the total. Other armies, even the British, followed suit.[24]

The precise nature of the ammunition carried by field guns, however, seemed less important as artillerymen continued to evaluate the probable result of going up against first-line troops occupying prepared defenses. The image popularized by Hohenloe, of lines of guns standing in the open and blasting away with their low-quality black powder rounds, seemed to promise the approximate results of flinging handfuls of dried peas against a wall. But what were the alternatives? German theorists emphasized methodical, precise shooting. Battalion and battery commanders were expected to reconnoiter firing positions carefully and make relatively few changes of position once in action.[25] Fire from concealed or partly sheltered positions was considered significantly more effective than direct fire. This theory was reinforced in the early 1890s, as improved propellants and explosives enhanced the blast effect of cannon shell while reducing the clouds of smoke that protected batteries from precise observation by obscuring the target. And even in the nineteenth century, a battery seen was a battery at risk.[26]

Indirect laying remained, however, a German shibboleth. Elsewhere, on the continent and in Britain, gunners continued to accept the premise that future battles would be decided over open sights. The captain immortalized by Kipling, who celebrates his acceptance to the elite Royal Horse Artillery by replacing the shells in his gun limbers with beer, was more anachronism than archetype even by British standards. Nevertheless the complexity imposed on gunnery by the new technology of range finding, clinometers, telescopes, and dial sights, was widely considered excessive under battlefield conditions.[27] Particularly as armies moved towards short service, French, Russian and Austrian artillery officers questioned the ability of peasant conscripts to handle instruments whose fragility made their utility questionable.[28] Similar objections would be widely made a century later to a new generation of battlefield technology intended to be used by ordinary soldiers.

Artillerymen's reaction to indirect-fire technologies also reflected a significant change in the internal dynamic of European armies. Modern war was increasingly perceived as demanding the systematic, integrated cooperation of the three major combat arms, infantry, cavalry and artillery. This in turn meant the artillery had to sacrifice much of its traditional status as a "learned" arm, somewhat separate from the vulgar close-combat types who went to war afoot or on horseback. In Germany, Hohenloe was a consistent proponent of the idea that gunners knew how to die as well as troopers or riflemen.

Horses provided another point of integration. The nineteenth-century gentleman was expected to be at home on horseback as a mark of his social status, as opposed to vulgar utilitarian considerations. Changing modes of transportation increasingly made the riding horse a vestigial means of getting from one place to another. "Keeping a stable" involved steadily growing opportunity costs. The horse, in short, was becoming a symbol of conspicuous consumption. For impecunious aristocrats and ambitious bourgeoisie, the artillery might not have the cavalry's social cachet, but it did provide access to a primary status symbol of polite society. Individually and collectively, artillery officers began challenging cavalrymen on their own ground in race meets and games. In the British army, an artillery battery defeating a cavalry regiment in a polo match might live for years on the triumph. In Germany the number of noblemen in a field artillery regiment depended increasingly on its garrison assignment. Even in the caste-conscious Prussian Guard, the 2nd Uhlans and the 1st Field Artillery maintained a "couleur" relationship: honorary membership in each others' messes, mutual hosting of parties, and similar social interaction.[29]

Expansion was a third facilitator of integration. Prior to the mid-nineteenth century the artilleries of Europe faced glacially slow promotion structures, particularly at field-rank levels. As late as 1870, for example, 94 percent of the artillery majors in Austria-Hungary were over 43 years old—relatively long in the tooth for field operations.[30] In the quarter-century after the Franco Prussian War, however, Europe's armies enlarged their artillery branches significantly. In 1870 a Prussian corps had one regiment of field artillery, consisting of four battalions with a total of fifteen batteries. By 1890 a standard corps included a field artillery brigade of two regiments each of three three-battery battalions, plus a two-battery battalion of horse artillery. Similar patterns obtained elsewhere, creating fresh career opportunities, particularly at those levels from major to colonel that continue to represent reasonable, achievable career goals in present-day armed forces.

Integration took place at lower levels as well. For the first half of the nineteenth century the artillery of continental Europe tended to recruit disproportionately from urban sources, artisans and craftsmen, and to be concentrated disproportionately in certain areas. Its political reliability was correspondingly—and often legitimately—questioned.[31] After 1875, the expansion of artilleries led to a much more representative cross-section of recruits being assigned to the batteries and regiments. By 1900 the enlisted personnel of the field and heavy artillery in most armies was not significantly different from the infantry. Russia and to a degree Austria-Hungary were partial exceptions, yet even in these semi-industrialized states the artillery no longer absorbed most of the literate, technically skilled conscripts.[32]

The process was by no means complete. Not until 1912, for example, did an artillery officer rise to corps command in Germany. Even in France, infantry and cavalry officers were perceived as having some advantage in obtaining command assignments in combined-arms formations. Nevertheless as the turn of the century approached, artillery officers were well on their way to full-fledged membership in the brotherhood of arms. Their new status, however, contributed to a paradox. European armies were increasingly committed to the principle that modern war demanded offensive action for the purpose of seeking a decisive battle. This approach was in part the consequence of a quarter-century of peace—and of peacetime maneuvers. When only blanks were fired the results were purely theoretical—particularly when compared to highly visible, correspondingly spectacular, infantry assaults and cavalry charges. An army's annual maneuvers were both its principal opportunity to impress foreign observers and domestic politicians and the principal chance for officers to show they possessed "the right stuff." Dash, energy, *coup d'oeil*—these were the qualities parliamentary deputies and promotion boards liked to see.[33]

The cult of the offensive was favored by institutional dynamics stressing decisiveness, initiative, and aggressive action over reflection and calculation. It was reinforced by post-Darwinian pessimism: the widespread belief in military circles that modern societies were too complex, modern men and women too fragile, to stand the strain of modern war. It was not least a consequence of the sense of weakness pervading the armies of France, Austria-Hungary, Germany, and Russia—their generals and their general staffs had in common a corrosive sense of their own forces' actual shortcomings compared with their potential enemies' perceived strengths. In such a context offensive strategies and tactics offered the advantage of the initiative, of getting inside the opponent's decision-making loop and imposing one's own game plan on a desperate situation. The alternative was to stand in place and either be hammered into the ground like a tent stake, or see one's army and home front erode under the unbearable pressure of attritional war.[34]

Flexibility and Firepower

This mind set left Europe's artilleries stepping off on the wrong foot by the 1890s. For twenty years and more the gunners had concentrated on tactics and technologies favoring attrition. To sustain their new place in their respective armies now required contributing to a quick decision. The hint of a solution came from the gun designers. For light field pieces the practical limits of chemistry, metallurgy, ammunition design, and crew efficiency had been reached in 1890.[35] The major possibility of improvement involved recoil mechanisms. Gun carriages still were impelled backwards every time a round was fired. Moving the piece back into battery and readjusting the aim now took more time than reloading.

The recoil problem was initially addressed by various combinations of trail spades and buffer springs, the best example being Germany's *Feldkanone 96*. These, however, were temporary solutions. Everywhere in Europe, designers and engineers addressed the problem of adapting the hydraulic mechanisms already in use by naval and fortress guns for field service. The German firm of Heinrich Ehrhardt and Russia's Putilov factory were well on their own ways to designing a true quick-firer when France unveiled in 1897 *Mademoselle Soixante-Quinze*. Its hydraulic recoil system returned the barrel to firing position in seconds, with the carriage remaining stable during the entire process of firing and reloading. The gun captain needed to make only minor adjustments to the sights to keep his piece on target. A well-trained crew could deliver twenty rounds a minute as a matter of course—thirty under emergency conditions, as long as strength and shells remained. And since the crew members now remained in place instead of moving around, it became practical and worthwhile to protect them by adding a steel shield to the carriage.[36]

The "seventy-five" made Europe's artillery parks obsolescent overnight. Germany found itself compelled to redesign its *FK 96,* producing over a decade virtually a new weapon. Russia, despite bureaucratic and military anguish at the cost, adopted a 76-mm quick-firer in 1900. Austria followed suit, more or less, in 1905. The new guns offered several possible lines of tactical development. Their accuracy enhanced the possibilities of indirect fire, directed by a forward observer using a telephone or visual signals. Accuracy and stability meant quick-firers could be used to deliver surprise concentrations from concealed positions, without the ranging rounds that traditionally warned the target what was coming. This "predicted firing" required careful calculations based on accurate maps and survey procedures, but offered particular chances for using artillery in the night operations currently fashionable among theorists. A third option was the direct approach: taking advantage of the quick-firers' shields to bring them into advanced positions in large numbers and overwhelm an enemy by massed fire at close range before he could react.[37]

Combat experience provided no clear indication of the best alternative. Artillery had played no role on either side during the Spanish-American War. In South Africa, British artillerymen had been taught the importance of covered positions by Boer marksmen, and Boer gunners who used their few cannon in ones and twos as virtual snipers' weapons. Far from diminishing the gunners' fighting spirit, concealment enhanced morale. The real psychic damage occurred when crews deployed in the open came under fire from enemies they could not see.

On the other side of the ledger, field howitzers proved of limited use on the veldt—partly from defective ammunition and poor training, but also because Boer entrenchments were difficult to observe and range in under South Africa's normal conditions of terrain and weather. Even before the conflict dissolved into guerrilla war, British artillerymen tended to agree that the material effect of artillery fire on both sides had been far less than its moral consequences. Boer guns, while employed in small numbers, tended to shake the confidence of infantry, particularly at long ranges. Being fired on without either the opportunity to retaliate or the numbing effect of the adrenalin rush accompanying a real attack left even experienced regulars more cautious and more jittery than their commanders approved.[38]

The potential impact of artillery fire on fighting spirit suggested by the Boer War particularly engaged the attention of continental general staffs increasingly concerned with the reliability under fire of their conscript infantries. The Germans worried about socialism and the Austrians about nationalism. The most republican of French generals questioned the training and discipline of the reservists on whom their army would have to rely in order to come near matching German numerical superiority. Russian theorists questioned whether the traditional stolidity of their *muzhik* riflemen would survive on a quick-firing battlefield.

The Russo-Japanese War raised rather than settled this and related lines of questioning. Japanese field artillery was significantly inferior to the Russian in both numbers and quality. Japanese officers compensated by concentrating fire on specific targets. They kept their guns out of sight, deploying them on reverse slopes whenever possible. Japanese batteries made effective use of forward observers who maintained contact by telephone and visual signals—a process facilitated by the inability of both Russian cannon and Russian snipers to put the observation posts out of action. Finally, the Japanese employed heavy artillery under field conditions for the first time in modern war. Massed at army level and controlled by telephone, these pieces proved particularly useful in the set-piece battles that characterized the war in Manchuria.

Considered in hindsight, the Japanese experience seems a clear signpost to the future. However, closer inspection of events in a conflict assiduously studied by every military power in Europe suggested mixed results. Japanese tactical superiority did not mean consistent overwhelming of the Russian artillery. Once the Russians learned, like the British in South Africa, the perils of leaving guns in exposed positions, Japanese counterbattery fire lost much of its effectiveness. The Russian guns always seemed able to fire back. The usual Japanese response was to push their infantry forward at all costs, trusting élan to bring them across potential killing zones. The tactic worked about as well as anything works in modern war. More Russian batteries were put out of action at bayonet point than were destroyed by shell and shrapnel fire. Not infrequently Russian gunners were taken by surprise in their concealed positions, overrun by Japanese riflemen while trying to silence Japanese artillery on the other side of another hill. Perhaps in theory the Russian artillery should have concentrated against the Japanese infantry. But was this tactically sensible given the kind of covered positions seemingly required in the face of even the relatively mediocre Japanese material?[39]

Europe's armies drew significantly different conclusions from the same events. Russia and Austria-Hungary were sufficiently handicapped by underdeveloped industrial bases and truncated military budgets that their approaches were essentially footnotes to decisions taken by the three Western powers. The French approached the quick-firing revolution with Cartesian logic. After 1871 French artillery had tended to emphasize indirect fire from concealed positions, keeping guns and their crews protected as far as possible. Even before 1897 this concept had been questioned, and the seventy-five made possible a paradigm shift in defining the artillery's mission. The doctrines developed for the new guns stressed bringing them into action by batteries and battalions that would range a target quickly and, instead of destroying it with the systematic methods favored by the Germans, paralyze the enemy's will and his ability to move with intense bursts of fire—*rafales* that would shock and demoralize until the bayonet-tipped *furia francese* should make itself felt.[40]

The French artillery significantly adjusted its organization to fit its new weapon. Since midcentury, a standard field battery everywhere in Europe had consisted of six guns and six or nine caissons. These strengths were determined partly by the number of guns one officer could control in action, but also by the effective length of a march column under combat conditions. With more than twelve or fifteen six-horse teams and their correspondingly bulky vehicles, a battery became too unwieldy to maneuver. The French conceded the point. They also accepted the reasoning that with quick-firing artillery the decisive factor was not the number of barrels in a gun line but the number of rounds available to each barrel. Making a virtue of necessity, the French reduced the number of guns in a battery to four and increased the caissons to twelve. They further institutionalized their belief in field artillery as a shock weapon by holding large numbers of guns in reserve—a major challenge to conventional wisdom. By 1914, while each of the two divisions in a French corps had only thirty-six guns, the corps commander had forty-eight under his personal command.[41]

The existence of a strong corps artillery only heightened French concern that artillery and infantry be closely integrated in practice. Doctrine and regulations alike stressed the subordination even of senior artillery officers to the infantry commanders that their guns supported. This approach reflected a widespread conviction that French infantry was of such relatively poor quality that it needed all the help it could get. The French army, however, was also concerned with maximizing its direct impact against a numerically superior German enemy. *Rafale* and bayonet charge were symbiotes, combining to make each one geometrically more effective.[42]

Concern for impact also affected French evaluations of high-angle fire and heavy artillery. The French army was by no means as committed to undergunned mobility as some postwar mythologies indicated. Its basic problems were doctrinal. No one in France was quite able to determine exactly what heavy artillery pieces were supposed to do. The French had enough access to Russian reports on 1904–5 to be well aware that Japanese heavy artillery had at best limited effect against even improvised entrenchments. Fire control posed a related problem. French doctrine fully recognized that "fire kills." It also concluded that the best counter to modern weapons technology was flexibility. In future combat, French infantry and their supporting seventy-fives were likely to pop up almost anywhere. What use in that context could be made of pieces apparently able to do little more than blast away at fixed targets? Give the *Boche* time to entrench, and the battle was in any case likely to be lost. The latter point also shaped French reluctance to adopt light field howitzers, as other European armies were doing. The kinds of targets such howitzers were expected to engage were exactly the kind of targets French artillery doctrine expected to keep to a minimum. If the seventy-fives could do their job, field howitzers were redundant. If they could not—but that was a subject better left unaddressed, except perhaps in the late hours of the night, when sleep would not come.[43]

More than any continental power, Germany regarded its approaches to war as vindicated by both Russian and Japanese experiences in Manchuria. In general terms the moral and physical impact of the offensive at all levels seemed confirmed beyond doubt. In the specific contexts of artillery doctrine and practice, theorists continued to insist that artillery was best used in masses, under unified control, to deliver well-aimed, well-directed fire from concealed positions. They paid increasing attention to predicted shooting from maps, to techniques of observation, and to problems of maintaining communications in battle among the guns, their observers, and the infantry they were supporting.[44]

Concentration did not mean, as in France, grouping the largest possible number of guns at the highest possible level of command. In 1899 the German army abolished corps artillery altogether. Instead, each of the two divisions of an army corps received its own artillery brigade—two regiments, each of two three-battery battalions. This new structure had several advantages. With its relatively large number of staffs it provided for the control German doctrine emphasized. It offered opportunities for expansion; regimental headquarters could at need handle three or even four battalions temporarily attached in battle or permanently assigned. And it enabled officers to be made available on mobilization for the artillery formations of the reserve divisions that were increasingly important to the German army's order of battle.[45]

This compromise between firepower and flexibility was reflected in equipment as well. In contrast to their French counterparts, an increasing number of German artillerymen argued that the light field howitzer could still fulfill a new version of its historic role by blowing in trenches and rifle pits and by knocking out guns with shell fire protected by their shields from the effects of shrapnel. Give the howitzer a shield and a modern recoil mechanism, and it might even evolve into the principal weapon of the field artillery—particularly if entrenched machine guns became, as they seemed increasingly likely to do, the backbone of defensive positions.

The field howitzer's status in the German army continued to rise. Originally introduced on a scale of one battalion to each army corps, in 1912 its numbers were doubled. One battalion of the four in a division was now armed with howitzers. This expansion also reflected design evolution. In 1909 the howitzer was mounted on a shielded recoil carriage, making it the best weapon of its type on the continent. Since 1905 the howitzers had also been equipped with a Krupp-designed general-purpose round that could function with time or contact fuses, and could be used either as shrapnel or shell by virtue of its combination of shrapnel bullets and TNT explosive. Originally criticized as a complex compromise, the new round proved itself sufficiently effective by 1911 to inspire the construction of an equivalent for the light guns.[46]

That three-fourths of the German army's light artillery still consisted of shrapnel-firing, flat-trajectory, high-velocity cannon by no means showed technical indifference or institutional commitment to existing and expensive equipment. Both French and Russian field guns could outshoot Germany's light howitzers by thousands of yards. The latter's superior shells would be of no use if the guns, teams, and crews were put out of action before they were in position to reply. The German field artillery's weapons mix also reflected growing emphasis on the need for coping with modern defense systems by more precise coordination of infantry and artillery operations. A battlefield expected to feature entrenchments and concealed battery positions offered a correspondingly high risk that artillery preparation would degenerate into pointless area bombardment, no matter what kind or weight of ammunition was used. German tactical doctrines therefore increasingly stressed using the infantry attack to force the defender's hand, bringing his riflemen onto the parapets and forcing his guns to open fire and expose themselves to counterbattery. Only then would the field gun and its shrapnel round come into their own, doing what they could do best by destroying exposed human targets.[47]

Accepting this specialized role for the flat-trajectory cannon was facilitated by the foot artillery's growing commitment to mobile operations. In 1902 its regiments received organic draft horses instead of drawing them from a general pool. At the same time the army introduced a new 15-cm heavy field howitzer. It lacked a shield, but unlike its lighter counterpart it had a modern recoil carriage with springs and buffers. And it could throw a 90-pound explosive shell 7,500 meters. In its improved and shielded version it would prove the backbone of Germany's artillery during World War I, the famous "crump" or "Jack Johnson." By 1914 each active army corps had a battalion of them: four four-gun batteries whose cannoneers prided themselves on being able to take their three-ton weapons anywhere their comrades of the light artillery could go and demolish anything that blocked the path of the infantry.[48]

The German artillery went to war with an integrated system of weapons and doctrines based on a defensible evaluation of past experience. Developments in Britain offer a useful counterpoint. In the fifteen years after the Boer War, the British army underwent a paradigm shift. Haldane's reforms might in principle have been intended to produce a general-purpose expeditionary force able to operate anywhere in the world, but wherever the new model British army went, its opponents were expected to be from the first division. In practice, no serving officers with any influence denied that the era of colonial wars in the traditional style was over. The army's new focus was preparing for war on the continent of Europe against Germany.[49]

In specific terms the obsolescence of British artillery material demonstrated during the Boer War offered corresponding opportunities for a *tabula rasa*: new weapons and new doctrines. Adoption of a quick-firing gun was delayed for several years by an intense debate between advocates of a design emphasizing mobility and one favoring firepower. The Russo-Japanese War finally threw the weight of professional opinion inside and outside the Royal Regiment of Artillery on the side of the famous 18-pounder. Introduced in 1903, this gun was outranged by its French counterpart but fired a significantly heavier round. In 1908 Britain also adopted a 4.5-inch field howitzer that at the time was the world's best in its class. This denial of the Boer War experience owed much to battlefield reports from Manchuria emphasizing the potential of high-angle fire against both entrenched infantry and camouflaged artillery. It owed something as well to the influence of German doctrines on the use of artillery in the offensive. The British army's divisional artillery in its final form had the same strength, seventy-two pieces, and the same three-to-one ratio of guns to howitzers as did the Germans. On the other hand the British approach to heavy artillery might well have been copied from France. The British Expeditionary Force went to war with no modern medium howitzers, while the battery of 60-pounder guns organic to each division were usually viewed as long-range snipers' weapons, to be used against targets of opportunity in the manner of the Boer War.[50]

Each of these decisions in armament and organization were the product of prolonged debate. As a result, the British found themselves committed to new material and new structures without having a clear idea of how best to use either one. Douglas Graham and Shelford Bidwell exaggerate both the British artillery's anti-intellectualism and its commitment to pragmatism at the expense of theory. What happened in the decade prior to 1914 was just the opposite. British gunners were so concerned with getting it right that they found virtually equal validity in the French notion of direct, rapid fire as the best means of ensuring cooperation and close support, and in the German arguments for indirect fire from concealed positions, massed and coordinated by senior artillery officers. One school of thought, strongly advocated by the infantry as well, favored fire at short range, with guns kept in sight of the target and conforming to the movements of the infantry rather than the configuration of the ground. Their opponents favored accuracy at long ranges, saving ammunition and maximizing effect by comprehensive systems of observation and larger staff and command elements. By 1914 neither the gunners nor the army had decided which path to follow—not from inanition, but because the arguments seemed so closely balanced.[51] In that context "British Empiricism" offered a line of conduct that was by no means the same as muddling through. The Royal Artillery went to war ready to use whatever techniques seemed most likely to work at a given time—a mind set that contributed not a little to the final emergence of British artillery as, by a whisker, the dominant arm on the Western Front by 1918.

Conclusion

The differences in artillery doctrine and practice among Europe's major military powers in 1914 were less the product of differing military cultures than of alternate responses to a common problem. These responses were the products of a common technology. In contrast to the rifles of 1866 or the tanks of 1940, there was not the proverbial dime's worth of difference among the artillery equipment that took the field in 1914. And that equipment could support a wide variety of techniques. The same pieces, and their accompanying fire control and supply systems, could be used to overwhelm an enemy quickly or destroy him methodically. The crucial question was not which technique was preferable in some abstract sense, but rather which one would contribute most effectively to the successful sustained offensives necessary to the short wars that Europe's generals firmly believed were all their armies and societies could sustain.

Less obvious, but no less significant, was the artillery's enhanced place in the military pecking order. The gunners by 1914 had become key players in their respective armed forces—bidding fair, indeed, to replace a cavalry whose operational roles were becoming increasingly vestigial as the focus for an emerging cult of the horse. Concern for integration, for remaining part of the team, encouraged artillerymen in all countries to focus on the combined-arms aspects of doctrine and training at the relative expense of branch-specific, technical concerns. Their ideal relationship to the infantry was not supportive but symbiotic. The postwar observation of one

French officer that the rhetoric of the infantry's and artillery's hearts beating as one could have benefitted from a few more miles of telephone wire makes one point, but misses another. The artillery was still too uncertain of its status to risk stepping outside paradigms established by military systems as a whole. The gunners reinforced; they did not challenge. Perhaps, ironically, the institutional integration of artillery into Europe's armies had as much to do with the stalemated slaughters of 1914 as did the technical innovation that is the stated theme of this volume.

Notes

1. See the general discussion in Dennis E. Showalter, *Railroads and Rifles: Soldiers, Technology and the Unification of Germany* (Hamden, Conn.: Archon, 1975), 166 passim.

2. Paddy Griffith, *Battle Tactics of the Civil War* (New Haven and London: Yale University Press, 1989), 137 passim; and T. V. Moseley, "The Evolution of Civil War Infantry Tactics" (Ph.D. diss., University of North Carolina at Chapel Hill, 1967).

3. William T. Sherman, *Memoirs*, 2 vols. in 1, reprint ed. (New York: Reprint Service, 1990), 887.

4. Quoted in Larry J. Daniel, *Cannoneers in Gray: The Field Artillery of the Army of Tennessee, 1861–1865* (University, Ala.: University of Alabama Press, 1984), 130.

5. Cf. L. Van Loan Naisawald, *Grape and Canister: The Story of the Field Artillery of the Army of the Potomac* (New York: Oxford University Press, 1960); and Jennings Wise, *The Long Arm of Lee*, reprint ed. (New York: Oxford University Press, 1960).

6. Cf. Kraft Karl zu Hohenloe-Ingelfingen, *Aufzeichnungen aus meinem Leben*, 4 vols. (Berlin: Mittler, 1897–1907), 3:357ff.; and C. von Hoffbauer, *Die deutsche Artillerie in den Schlachten und Treffen des deutsch-französischen Krieges, 1870–71*, 3 vols. (Berlin: Mittler, 1873–78), vol. 1.

7. Griffith, *Battle Tactics*, 175–76; Gunther E. Rothenberg, *The Art of Warfare in the Age of Napoleon* (Bloomington: University of Indiana Press, 1978), 117–18; and B. P. Hughes, *Open Fire: Artillery Tactics from Marlborough to Wellington* (Strettington, Sussex: A. Bird, 1983).

8. See Edward G. Hagerman, *The American Civil War and the Origins of Modern Warfare. Ideas, Organization, and Field Command* (Bloomington: University of Indiana Press, 1988), 175 passim.

9. Kraft Karl zu Hohenloe-Ingelfingen, *Letters on Artillery*, tr. N. L. Walford, reprint ed. (Quantico, Va.: U.S. Marine Corps, 1989), 275ff.

10. Perry Jameson, "The Development of Civil War Tactics" (Ph.D. diss., Wayne State University, 1970); Showalter, *Railroads and Rifles*, 125ff.

11. Jacob W. Meckel, *Ein Sommernachtstraum: Erzählt von einem älteren Infanteristen* (Berlin: Mittler, 1888), is a vivid and accurate reconstruction of the realities of an infantry attack during the Franco-Prussian War. Cf. Albert von Boguslawski, *Tactical Deductions from the War of 1870–71*, tr. L. Graham (London: H. S. King, 1872).

12. Michael Howard, *The Franco-Prussian War* (London: Hart-Davis, 1961), incorporates sound general discussion of artillery's roles in the principal battles. Hoffbauer, *Die deutsche Artillerie*, provides the details; Hohenloe, *Aufzeichnungen*, vol. 4, furnishes the personal touch. A study of French artillery during the war would be a welcome addition to the literature; in its absence Charles A. Romain, *Les Responsabilités de l'artillerie française en 1870* (Paris: Berger-Levrault, 1913), remains useful. Jean G. M. Roquerol, *L'Artillerie dans le bataille du 18 août* (Paris, 1906), more or less complements Hoffbauer.

13. Richard Holmes, *The Road to Sedan: The French Army, 1806–1871* (London: Royal Historical Society, 1984), 33 passim.

14. See the general discussion in Karl Demeter, *Das Deutsche Offzierkorps in Gesellschaft und Staat, 1640–1945*, 2d ed., rev. (Frankfurt: Bernard & Graefe, 1962), 15; and Hohenloe, *Aufzeichnungen*, 3:182ff.

15. In practice the pre–Civil War artillery regiments usually functioned as infantry; the U.S. Army had only a few true field artillery batteries. Nevertheless the gunners never regarded themselves as just "red-legged infantrymen."

16. Quoted in *Moltke on the Art of War: Selected Writings*, ed. D. J. Hughes (Novato, Calif.: Presidio, 1993), 52.

17. Bruce W. Menning, *Bayonets before Bullets: The Imperial Russian Army, 1861–1914* (Bloomington: University of Indiana Press, 1992), 60ff., is the definitive English-language account of Plevna. E. S. May, *Field Artillery with the Other Arms* (London: Low Marsten, 1898), 152ff., stresses the relative inefficiency of Russia's artillery, as opposed to the strength of Turkish defenses. Thilo von Trotha, *Tactical Studies on the Battles Around Plevna*, tr. C. Reidmann (Kansas City, Mo.: Hudson-Kimberly, 1896), is a good staff analysis.

18. Hohenloe, *Letters on Artillery*, 318ff. Cf. also his *On the Employment of Field Artillery in Combination with Other Arms*, tr. F. C. H. Clarke (Woolwich: n.p., 1872).

19. W. H. G. von Mueller, *Die Entwickelung der Feldartillerie in Bezug auf Material, Organization und Tahtik von 1815 bis 1892*, 2d ed., 3 vols. (Berlin: Mittler, 1893–94), 2:9ff.

20. Robert H. Scales, "Artillery in Small Wars: The Evolution of British Artillery Doctrine, 1860–1914" (Ph.D. diss., Duke University, 1976), 74ff.

21. Cf. "Wurffeuer im Feldkriege," *Jahrbücher für die deutsche Armee und Marine* (hereafter cited as *JAM*) 54 (1886); R. Wille, *Das Feldgeschütz der Zukunft* (Berlin: C. R. Wille, 1891); and "Kanone und Haubitze in der Feldartillerie," *Militär-Wochenblatt* (hereafter cited as *MW*), no. 5 (1892).

22. The German army took this process further than any of its counterparts. Cf. C. von Sauer, "Über Gefechtsmässige Schiessübungen der Artillerie," *JAM* 24 (1877); "Fussartillerie mit Bespannung," *MW*, no. 4 (1895); "Was tut der Fussartillerie not?" *JAM* 125 (1903); and H. H. Friedrich, *Die Taktischen Verwendung der Schweren Artillerie* (Berlin: Gisenschmidt, 1912).

23. W. Heydenreich, *Das moderne Feldgeschütz*, 2 vols. (Leipzig: Goschen, 1906), is a contemporary overview of these developments.

24. Cf. *inter alia* Carl von Sauer, "Was wir vom Shrapnel hoffen," *JAM* 23 (1877); B. Ernestus, "Kritische Betrachtungen über die Zukunft der Feld-Artillerie" (Leipzig, 1875); and H. Rohne, "Zur Munitionsausrüstung der Feldartillerie," *JAM* 130 (1906). Scales, "Artillery in Small Wars," 120ff.; discusses British experiences in the same period.

25. See "Zur gegenwärtigen Artillerie-Taktik," *JAM* 14 (1875); "Die Taktischen Grundsätze der deutschen Feld-Artillerie verglichen mit denen der benachbarten Grossmächte," ibid. 64 (1887); "Der 'Entwurf' eines neuen Exerzier-Reglements für die preussische Feld-Artillerie," ibid. 69 (1888) and 70 (1899).

26. "Anderes Pulver, andere Taktik," *JAM* 71 (1889); "Das rauchschwache Pulver und die Feldartillerie," *MW*, no. 93 (1889); "Der Einfluss des rauchschwachen Pulvers auf die Tätigkeit, Verwendung und Führung der Feld-Artillerie . . . ," *JAM* 76 (1890).

27. Scales, "Artillery in Small Wars," 105 passim; and C. D. Bellamy, "The Russian Artillery and the Origins of Indirect Fire," *Army Quarterly and Defence Journal* 112 (April and July 1982).

28. The military effectiveness of short-service conscripts was a particular issue in France. See Douglas Porch, *The March to the Marne: The French Army 1871–1914* (Cambridge: Cambridge University Press, 1981).

29. Cf. Otto Pfeil und Klein-Ellguth, *Fünfundvierzig Jahre im Dienst für König und Vaterland: General der Artillerie Franz Graf von Pfeil und Klein-Ellguth; Ein Lebensbild 1855–1937* (Karlsruhe: n.p., 1973), 25ff.; Franz von Lenski, *Lern- und Lehrjahre in Front und Generalstab* (Berlin: Bernard & Graefe, 1939), 318ff., 401ff.

30. Gunther Rothenberg, *The Army of Francis Joseph* (Lafayette, Ind.: Purdue University Press, 1976), 81.

31. Douglas Porch, *Army and Revolution: France 1815–1848* (London: Routledge & Regan Paul, 1974), 79 passim.

32. A rough rule of thumb in continental armies was to assign batteries a mix of country boys who presumably knew how to handle horses and townsmen believed to be more familiar with machinery. The success of the blend depended heavily on the man-management skills of the officers and NCO's.

33. "Die Artillerie im Mänover," *MW*, no. 1 (1884); "Artilleristische Manöverbetrachtungen," *MW*, no. 97 (1893); "Manövergedanken eines Feldartilleristen," *MW*, no. 87 (1892); C. von Hoffbauer, *Massengebrauchs der Feldartillerie* (Berlin: Mittler, 1900), 185ff.; "Die Taktischen Formen unserer Feld-Artillerie," *JAM* 56 (1885), discusses this subject from a German perspective.

34. Cf. Jack Snyder, *The Ideology of the Offensive: Military Decision-Making and the Disasters of 1914* (Ithaca, N.Y.: Cornell, 1984); and Stephen van Evera, "The Cult of the Offensive and the Origins of the First World War," *International Security* 9 (Summer 1984).

35. Bruce Gudmundsson, *On Artillery* (Westport, Conn.: Greenwood, 1993), 6; Boyd L. Dastrup, *The Field Artillery: History and Sourcebook* (Westport, Conn.: Greenwood, 1994), 40–41.

36. Michel de Lombares, "Le '75,'" *Revue Historique des Armées* (no. 2:1975), is an excellent, well-documented survey of the gun's development, adoption, and career. Paul Lintier's memoir, *Avec une batterie de 75. Ma pièce: souvenirs d'un cannonier 1914* (Paris: Plon-Nouritt, 1916), is an example of the almost sexual emotions the gun could inspire.

37. Douglas Graham and Shelford Bidwell, *Fire-Power: British Army Weapons and Theories of War, 1904–1945* (Boston: Allen & Unwin, 1985), 9.

38. Scales, "Artillery in Small Wars," 211ff., is the best modern survey of artillery in South Africa. The development of field artillery in the U.S. after 1865 is an historical footnote, but excellently presented in Vardell E. Nesmith, "The Quiet Paradigm Change: The Evolution of the Field Artillery Doctrine of the United States Army, 1861–1905" (Ph.D. diss., Duke University, 1977).

39. Gudmundsson, *On Artillery,* 17ff. is an excellent brief summary. O. von Estorff, *Taktische Lehren aus dem Russisch-Japanischen Feldkrieg im Lichte unserer neuesten Vorschriften* (Berlin, 1909), is also useful on artillery issues. For a recent general discussion of the Russo-Japanese War from an operational perspective see R. M. Connaughton, *The War of the Rising Sun and the Tumbling Bear* (London: Routledge, 1989). Cf. as well Gary P. Cox, "Of Aphorisms, Lessons, and Paradigms: Comparing the British and German Official Histories of the Russo-Japanese War," *Journal of Military History* 56 (July 1992).

40. Cf. Gen. L. A. H. Langlois, *L'Artillerie de campagne en liaison avec les autres armes* (Paris, 1892); and J. Challéat, *L'Artillerie de terre in France pendant un siècle*, vol. 2, *1880–1910* (Paris, 1938). Robert Ballagh, "The Development of French Field Artillery Doctrine, 1870–1914" (M.A. thesis, Duke University, 1972), is a solid survey.

41. H. Rohne, "Zur Organisation der deutschen und der französischen Feldartillerie," *JAM* 136 (1909), is an excellent comparative overview. Cf. "Vier oder sechs Geschutze?" *MW*, no. 20 (1896); "Sollen die Feldbatterien in Zukunft zu sechs oder zu vier Geschützen formirt weiden?" ibid., no. 13 (1896); "Aenderung der Batteriestärke oder der Feuerordnung?" ibid., nos. 59, 80, 81 (1897); Marx, "Batterien zu 6 oder zu 4?" *JAM* 126 (1904).

42. Gen. Alexandre Pércin, inspector-general of the French artillery from 1906 to 1911, discusses these concepts in *Artilleurs et fantassins, facheuses polémiques* (Angoulème, 1912), and *Le Combat* (Paris, 1914). Cf. as well H. A. Niessel, *Combination des efforts de l'infanterie et de l'artillerie dans le combat* (Paris: Alcan, 1908); F. J. Aylmer, *French Views on the Tactical Employment of Field Artillery* (London, 1911); and more generally, Joseph C. Arnold, "French Tactical Doctrine, 1870–1914," *Military Affairs* 42 (April 1978); and Gudmundsson, *On Artillery,* 22ff.

43. Cf. the discussion in Porch, *March to the Marne*, 232ff.; and Cadet Robert M. Rippergen, "The Development of the French Artillery for the Offensive 1890–1914" (unpublished paper, Dept. of History, U.S. Military Academy, 1994), based heavily on French archival resources.

44. Von Hoppenstedt, "Neue Kanonen—neue Taktik," *Kriegstechnische Zeitschrift* 4 (1903); Keller, "Vergleich der Schiessverfahren der deutschen, fränzösischen und russischen Feldartillerie," *Vierteljanrehshefte für Truppenführung und Heereskunde* 1 (1904). This approach was incorporated into the new artillery regulations of 1907, essentially the version the army took to war. See Richter, "Die Schiessvorschrift für die Feldartillerie vom 15. Mai 1907," *JAM* 133 (1907).

45. For contemporary evaluations of the reorganization from an operational perspective, see "Die Neuordnung der deutschen Feldartillerie," *JAM* 106 (1808); and "Der Einfluss der Neuorganisation der Feldartillerie auf die Truppenführung," *MW*, Nos. 58, 59, 61 (1899).

46. See *inter alia* Rüppel, "Zur Feldhaubitzefrage," *JAM* 126 (1904); Auwers, "Gedanken über die Weiterentwickelung der Feldartillerie," ibid. 137 (1909); and "Noch einmal die bedeutung der leichten Feldhaubitze," ibid. The new round is briefly discussed in Edgar Graf von Matuschka, "Organisationsgeschichte des Heeres 1890 bis 1918," in *Handbuch zur deutschen Militärgeschichte, 1648–1939,* vol. 5; W. Schmidt-Richberg and E. Graf von Matuschka, *Von der Entlassung Bismarcks bis zum Ende des Ersten Weltkrieges* (Frankfurt: Bernard & Graefe, 1968), 173.

47. H. Balck, *Tactics*, vol. 2, *Cavalry, Field and Heavy Artillery in Field Warfare*, 4th ed., rev., tr. W. Krueger (Ft. Leavenworth, Kans.: U.S. Cavalry Association, 1914), 401ff., and the more specific arguments by Rohne, expressed in "Über die Wirkung des Schrapnellschusses," *MW* 74 (1902); and "Warum Kann die Haubitze nicht das Hauptgeschütz der Feldartillerie werden," *JAM* 137 (1909).

48. H. Schirmer, *Gerät der Schweren Artillerie vor, in und nach der Weltkrieg* (Berlin, 1937), 52ff., describes the 15-cm howitzer's evolution. Von Bleyhoffer, *Die schwere Artillerie des Feldheeres* (Berlin, 1905), is a contemporary discussion.

49. Cf. John Gooch, "Mr. Haldane's Army: Military Organization and Foreign Policy in England," in *The Prospect of War: Studies in British Defence Policy, 1847–1942* (London: F. Cass, 1981); E. M. Spiers, *Haldane: An Army Reformer* (Edinburgh: Edinburgh University Press, 1980); and the comprehensive analysis by N. W. Summerton, "The Development of British Military Planning for a War Against Germany, 1904–14," 2 vols. (Ph.D. diss., University of London, 1970).

50. Scales, "Artillery in Small Wars," 267ff., is the most detailed survey. Both German and British armies averaged six field pieces per thousand rifles; the French average was five per thousand.

51. Graham and Bidwell, *Fire-Power*, 18ff. Cf. T. E. H. Travers, "The Offensive and the Problem of Innovation in British Military Thought, 1870–1915," *Journal of Contemporary History* 13 (July 1978).

The Quest for Reach: the Development of Long-Range Gunnery in the Royal Navy, 1901–1912

Jon Tetsuro Sumida

"[The] great rise in fighting ranges between 1900 and 1914 is a historical *problem* which needs to be solved."—Professor Charles H. Fairbanks, Jr. (1991)[1]

In 1901, the Royal Navy began to practice shooting at ranges of 6,000 yards, which was quadruple the distance of the annual prize firings that had been standard since 1884. At first, results were poor. But by 1912, annual battle practice was conducted at 7,000 to 8,000 yards with greater accuracy than had been achieved a decade previously at one-fifth of the distance. This improvement has been characterized as a "gunnery revolution," and its cause generally attributed to the development of better technology and technique.[2] Recent scholarship, however, has shown that Royal Navy gunnery of this period was less capable than had previously been supposed.[3] The same work has indicated, moreover, that the pace of invention was affected critically by the attitude of the Admiralty toward technological innovation. But although the revisionist literature examined the long-range firing story in greater detail than ever before, its main focus was on larger issues of policy and politics. As a consequence, the particular dynamics of the transformation of naval gunnery in the early twentieth century have yet to be explained.

This essay will attempt to repair this deficiency by asking the following questions: Why exactly did the Royal Navy seek to increase the effective range of naval ordnance when it did, what factors determined the course of its search for a solution to the long-range gunnery problem, and why did the Admiralty end its quest for greater capability before the full potential of available technology had been exploited, which was to have serious consequences during World War I? To address these inquiries, the following subjects were examined: gun-laying and sight-setting, the progress of torpedo technology, fleet tactics, naval finance, and the influence of intelligence assessments. The main argument of this study is that the leadership of the Royal Navy was concerned primarily with the immediate practical aspects of sea fighting and the task of investing limited resources in a way that maximized the value of weapons procurement, and not just advances in gunnery for their own sake. This meant that problems were complex, and choices difficult.[4]

The Long-Range Gunnery Problem

At the end of the nineteenth century, the effective range of naval artillery was generally believed to be 2,000 yards or less.[5] For the gunner, the target image at such distances was still big enough so that a relatively large degree of imprecise gun-laying did not preclude hits being made. This was important because ships in a seaway rolled, pitched, heaved, and yawed,[6] roll in particular causing guns that were trained across the broadside to elevate and depress with respect to the horizontal by large amounts. To overcome this effect, guns were fired at the end of a roll at the moment when motion in one direction had slowed to a stop before reversing. Judging the roll, and in addition allowing for the delays between the mental decision to fire, pulling the trigger, and actually discharging the gun, were difficult tasks that could not be performed exactly, which meant that gun elevation was practically always going to be at least somewhat wrong with respect to that indicated by the sight.[7]

Sight-setting error was, under these circumstances, of little importance. This was because when trajectories were nearly flat, which was the case at 2,000 yards or less, and the gun was aimed at a spot midway between the waterline and the top of the superstructure, the large size of the target image allowed a significant amount of leeway. In other words, even if the gun was given considerably more or less elevation than that which was required to hit the point of aim, the result would still be a hit somewhere on the target. Or to put it differently again, at 2,000 yards or less, even large misestimates of the target range would result in projectiles striking the enemy. And this being the case, the effect of sight-setting error was bound to be small in comparison with that introduced by the approximate nature of gun-laying.

As the distance between the firing ship and target rose above 2,000 yards, however, the combination of the diminution in the size of the target image and the effect of increasingly curved projectile trajectories reduced the margin of allowable error for gun elevation to extremely small numbers.[8] At 2,000 yards and when guns were aimed at a point midway between the waterline and top of the superstructure, hits could be obtained with the 12-inch gun (the standard main battery weapon of British battleships in the early twentieth century) even when sights were set with ranges that had been misestimated by as much as plus or minus 15 percent of the total distance (that is, plus or minus 311 yards), or the gun mislaid in such a way as to result in comparable error in elevation. At 3,000 yards, however, the margin for error fell to plus or minus 6 percent (185 yards), and at 4,000 yards, plus or minus 3 percent (130 yards). At greater ranges, tolerances became exceedingly fine. At 6,000 yards, they were plus or minus one-and-a-fifth percent (72 yards); at 10,000 yards, three-tenths of 1 percent (28.5 yards); and at 15,000 yards, less than one-tenth of 1 percent (13 yards).[9]

Given the imprecision of gunnery, it is easy to see why the late nineteenth century Royal Navy practiced shooting at 1,400 to 1,600 yards. At the turn of the century, however, improvements in the torpedo made an increase in gun range essential. The underwater weapon was potentially more effective than surface ordnance because the hulls of even the most powerful warships were not well protected below the waterline. Torpedoes and their launching apparatus, moreover, were light and inexpensive in comparison to heavy-caliber cannon. Their inaccuracy, however, meant that they had not posed a serious threat except at distances of less than 800 yards, which was well below the maximum effective range of warship artillery.[10] But the adoption of the gyroscope for improved guidance increased the effective range of torpedoes to 1,500 yards at top speed, and 3,500 yards at reduced speed.[11] If fired from stern tubes at a pursuing foe, the combination of the travel of the torpedo and the advance of their targets would give a range at full speed of close to 3,000 yards.[12]

For this reason Adm. Sir John Fisher, the commander-in-chief of the Mediterranean Fleet, Britain's premier naval command, argued in July 1902 that "on no account" would his fleet close "to within 4,000 yards" in order "to allow an ample margin clear of the torpedoes!"[13] During fleet maneuvers in 1903, ships that engaged at 1,800 yards or less were held to have suffered heavy losses from torpedoes.[14] Fisher's fear of them may have been the main reason why experiments with firing artillery at 6,000 yards had been carried out in the Mediterranean Fleet in 1899 and 1900.[15] This work prompted the Admiralty in February 1901 to order one long-range firing practice per year at 6,000 yards.[16] At this time, however, the maximum range to which the largest guns could shoot with some prospect of making hits was 3,000 to 4,000 yards,[17] where the margin of gun-laying error was from 3 to 6 percent. At 6,000 yards, the allowable margin for gun-laying error was just over 1 percent, which was simply too small, given the existing state of gun-laying and sight-setting, for hits to occur except by chance. Results, as a consequence, were poor.[18]

In 1904, the annual prize firing was replaced by the gun-layers test and battle practice. The first, which evaluated the ability of individual gunners to point their guns correctly, involved firing at a stationary target from a stationary ship at a range of 2,700 yards. The second, which measured the ability of a ship to work its guns under conditions resembling that of a battle, required shooting at a stationary target from a moving ship at ranges that varied from 3,500 to 8,000 yards. But although some progress had been made in the development of new gun-laying and sight-setting equipment and techniques, much remained to be done. This was reflected in the battle practice proportion of hits to shots fired, the best of which (10 percent) were one-third that of the Mediterranean Fleet average in prize firing in 1898 (30 percent). Performances in both exercises were indeed so disappointing that the range of the gun-layers' test was reduced back to the 1,600 yards of the annual prize firing, while future battle practices were to be carried out at 5,000 to 6,000 yards under much more strictly controlled circumstances.[19]

First Steps Toward a Solution, 1898–1904

Before turning to the response of the Admiralty to the failure of the 1904 gun-layer's test and battle practice, it is necessary to review the prior history of British long-range naval gunnery. In 1898, Capt. Percy Scott's invention of continuous-aim gun-laying—a system of moving guns up and down, and from side to side, to compensate for the roll, pitch, and yaw of the firing platform—provided a means of eliminating much of the gun-laying error introduced by ship motion. Scott also replaced the traditional open sights with telescopic sights, which by removing the optical difficulties inherent in open sights and enlarging the target image, greatly facilitated accurate gun-laying. In 1903, Scott was appointed captain of the Royal Navy's gunnery school with a mandate to teach his techniques to the service.[20]

In 1899 and 1900, long-range firing experiments were conducted in the Mediterranean Fleet in which fire was corrected by observing the fall of shot, which made large splashes that were visible from a distance.[21] The employment of this method was necessary because at ranges that were over 1,900 yards, holes made by even the largest projectiles were too small to be seen even with the assistance of a telescopic sight,[22] and explosions of shells on impact could be easily confused with the flash of the enemy's guns discharging.[23] The effect of these advances, however, was limited for several reasons. Control over the action of the hydraulic machinery required to move heavy-caliber ordnance was inadequate, and thus continuous-aim methods could only be applied to guns that were light enough to be worked by manual force. It took several years to procure enough telescopic sights to equip the fleet,[24] and then their value was considerably diminished by the fact that standards of sight calibration were extremely poor.[25] And because gun positions were located not far above the waterline, gunners lacked the elevation that was required to make accurate estimates of the distance between shell splashes and the target when ranges were greater than 2,000 yards.[26]

Meanwhile, the accuracy of sight-setting was improved by the procurement of new instruments. In 1899, the Royal Navy received its first practicable range-finder, which was manufactured by the Barr and Stroud Company. Its relatively short base length of 4½ feet, however, limited accurate range-taking to distances of 4,000 yards or less. Range measurement with the Barr and Stroud device, moreover, was a slow process and incapable of keeping up with even moderate—to say nothing of high—changes of range that could occur when opposing ships converged or diverged.[27] One method of dealing with change of range was to calculate a change-of-range rate. This was done by taking two range observations and keeping track of the time interval between them with a stop watch. Producing an estimate of the change-of-range rate in yards per minute was then a matter of simple arithmetic.

In theory, a change-of-range rate so obtained could be used to correct the sight-setting at 1-minute intervals. The application of a single change-of-range rate, however, could be an unsatisfactory guide because the relative motion of the firing ship and target was most likely to be such as to cause the range to change at a rate that was itself changing. This circumstance could occur even when the firing ship and target were steaming on unchanging courses and speeds. If courses were opposite, and speeds high, the change-of-range rate would not only be considerable, but the variation in the change-of-range rate would become increasingly rapid as the range decreased. Rapid variation in the change-of-range rate would also occur if the course or speed of either the firing ship and target were altered substantially.[28]

A partial solution to the change-of-range problem was found in 1901 by Lt. John Dumaresq, who invented a trigonometric slide calculator that, when set with the courses and speeds of the firing ship and target and the target bearing, indicated the change-of-range rate and deflection. The dumaresq, as it was called after its inventor, was capable of showing the correct change-of-range rate even when that rate was altering through adjustments of the bearing setting, because variation in the change-of-range rate were in proportion to changes in bearing.[29] Indications of the change-of-range rate by the dumaresq, when combined with an estimate of how far the last shot had missed the target derived from observation of the shell splash, could provide the basis for a sight correction.[30] The trustworthiness of the rates shown on the dumaresq, however, was always questionable because there were no means of measuring target course and speed, and these settings as a consequence had to be guessed. In addition, poor visibility conditions could prevent continuous bearing observation, and thus compromise accurate rate indication when the rate was changing.

By 1902, the advent of spotting and instrument-assisted range-finding had resulted in the invention of a new method of gunnery. In order to enable accurate spotting to take place at ranges that were greater than 2,000 yards, the spotting function was transferred from individual gunners to a single observer who was placed in a position located on the mast. By thus doubling or trebling the height of the spotter, the range at which accurate estimates could be made of the distance between the shell splash and the target was increased to over 4,000 yards. The spotter was joined by crewmen operating a range-finder and a dumaresq, and the trio were linked by voice-pipe to a transmitting station below,[31] from which spotting corrections, and estimates of the target range and deflection were sent by electrical signals or telephone to the several gun positions. This kind of centralized supervision, which replaced the autonomous action of individual gunners whenever the target was too distant to be hit consistently with traditional methods, became known as "fire control."[32]

Even observers at the top of the tallest masts in existence on British warships lacked sufficient elevation to make an accurate judgment of the distance from the splash to the target when ranges were 5,000 yards or more, which restricted the range of effective shooting based on spotting to one well below the distance at which naval guns could hit with reasonable consistency if well laid.[33] And when many splashes were generated by several guns firing rapidly and independently, individual gunners were incapable of knowing which had been created by their weapon and correcting their shooting accordingly.

These problems were mitigated by the invention of the salvo system. By 1902, it had been discovered that if several guns were fired simultaneously in what was known as a salvo, the fall of the shots could be assessed together. By spacing salvos so that the splashes caused by one had time to subside before the next fell, the main cause of confused observation of fire could be eliminated.[34] Because the ballistic paths of projectiles fired from the same gun differed slightly from shot to shot for a variety of reasons,[35] even projectiles fired from guns of the same caliber whose sights were set identically traveled distances that varied slightly. This meant that projectiles would fall over an area rather than a point, and the average of a scatter effect would serve as the basis for aiming. If all the splashes of a salvo were short of the target, or over or to the left or the right, correction was necessary. But if the target was straddled—that is, if the projectiles of a salvo fell both short of and beyond the target and more or less on line—and, with the spread of the salvo not too large—gunners could assume that some projectiles were falling in between and that hits were either occurring or would soon occur.

The practicability of the salvo system also depended upon the several guns having exactly the same elevation and train at the moment of firing. This was possible under average conditions of rolling with medium-caliber quick-firers, which were light enough to be laid by continuous-aim. In their case, individual gunners were notified of the moment of firing by a bell signal, and although firing would not be exactly simultaneous, this would not matter because the raising and lowering, and side-to-side motion, of the guns in time to roll and yaw largely prevented the intro-duction of unintended variations in elevation and train.[36] In the case of big guns that had to be laid at fixed angles of elevation and train, and fired on the roll, the likelihood of nonsimultaneity in the discharge of several guns after the bell signal—because of differences in the judgment and reflexes of individual gunners—was very high, and unless the sea was absolutely calm—which would minimize roll and yaw—gun el-evation and train were likely to differ substantially and shots of the salvo thus spread over too wide an area.

The invention of salvo firing coincided with the introduction of a new method of finding the target range through the fall of shot. In order to overcome the difficulty of estimating the distance between the shell splash and the target when ranges were long relative to the height of the observer above the waterline, gunners resorted to the practice of firing two shots, one of which was intended to fall well short of the target and the other to fall well over. If a short and over occurred, the size of the "bracket"—that is, the distance between the two ranging shots—was reduced and the process repeated until the bracket was small enough to justify salvos. Later on, bracketing was carried out by salvos in order to avoid basing aim on single shots, which occasionally could be eccentric. References to bracketing can be found as early as in 1903, and it was in use in conjunction with conventional spotting in battle practice perhaps by 1904 and certainly by 1905.[37]

The combination of fire control, the salvo system, and bracketing enabled quick-firing guns laid by continuous-aim to be aimed much more effectively at distances that were too great for direct evaluation of fire. In addition, the application of fire control alone—assisted perhaps by the bracket system—for the first time enabled big guns, whose hydraulic controls were too clumsy to allow the use of continuous-aim, to achieve a significant number of hits at ranges that were greater than 2,000 yards.[38] The capacity to hit at greater ranges in both cases, however, depended upon unarmored and thus highly vulnerable communications between gun positions and spotters remaining intact. For this reason, Capt. (R.M.A.) Edward Harding, writing under the pseudonym Rapidan, argued in a pamphlet of 1903, that the initial purpose of firing had to be the disruption of the target's internal communications before the enemy had wrecked one's own. In effect, this placed a high premium on having the ability to hit not simply at longer ranges, but at ranges that were greater than those from which the enemy could fire effectively.[39]

Harding also believed that range-taking through the observation of shell splashes was too slow and inaccurate to produce consistent hitting when ranges were long and changing. He furthermore was convinced that spotting would be compromised by the plethora of splashes that would be caused by two or more ships firing on one, a tactic that he encouraged. He thus argued that changes in the settings of the sight would have to be determined by a dumaresq.[40] Percy Scott, on the other hand, distrusted instruments and was apparently less concerned with the concentration of fire issue. In December 1903, he maintained for his part that "although great opposition is raised to spotting, no suitable alternative has, up to the present, been put forward to replace it."[41]

In order to settle these and other questions, fire-control trials were carried out in the spring of 1904 by special fire-control committees in the battleships *Victorious* of the Channel Fleet and *Venerable* of the Mediterranean Fleet. The joint report of the two committees stated that under "favourable conditions of wind and weather"—which meant a flat calm, and thus little roll, and clear visibility—and given a slowly varying change-of-range rate, centrally controlled fire could be opened carefully—that is, with salvos—at 8,000 yards. If the range was near constant, as would occur in a chase, the committees believed that effective firing could take place at 10,000 yards or even more. On the other hand, if the change-of-range rate was high, as in the case of two fleets approaching end-on, the report observed that effective hitting would not occur until the range was considerably less than 8,000 yards. Big guns, for reasons that will be explained, were found to be remarkably accurate.[42] There was, however, a measure of dissent expressed in the Channel Fleet by those who favored decentralized control and gun ranges of no more than 4,000 yards.[43]

The joint committee recommended that gun-layers practice at longer range,[44] and that sight-setting be assisted by instruments as well as spotting. These positions were probably the main cause of Scott's strong objections to the entire proceedings, although he may also have had serious reservations about firing taking place above 8,000 yards under any circumstances.[45] In May 1905, Scott denounced the trials in no uncertain terms in an interview with Lord Walter Kerr, the First Sea Lord.[46] In July 1904, Scott maintained in writing that "although it is necessary to have a range finder in order to obtain an initial range, and for a change of object, . . . spotting should be the sole guide for alteration of sights."[47] In October and November 1904, Scott criticized those who had called for gun-layers' practice at longer ranges, reiterating his conviction that practices at distances that were short enough for gun-layers to evaluate the effect of their fire—that is, 2,000 yards or less—was essential if they were to be trained to lay their guns accurately, which, he argued with good reason, was an essential prerequisite to effective shooting at longer ranges.[48]

Harding had joined the Naval Ordnance Department in December 1903, and it was he who summarized the findings of the trials in an Admiralty confidential print.

Harding resolved differences over whether fire control should be carried out by a completely centralized system, or one that was divided, through compromise. The primary system was to be one in which control was to be as highly centralized as possible, with direction from the mast-top. The secondary system was intended as a backup to the primary, and involved the subdivided control of groups of guns by turret or broadside battery from deck level. Harding also observed that a high percentage of hits to shots fired could be achieved at 5,000 yards or less, that some hits could be achieved at ranges that were as great as 8,000 yards if the range were not changing rapidly, and that big guns were much more effective than medium-caliber quick-firers. And finally, Harding made recommendations with regard to equipment that indicated a preference for sight-setting based on a combination of spotting, instrumental range-finding, and rate-correction via dumaresq.[49]

Scott could not have been happy about the provision for decentralized control and the important role assigned to instruments. On the other hand, Harding said nothing about longer range gun-layer's practice, which may have been in accordance with the stated principle of leaving "the settlement of controversial matter to further experience in the Fleet."[50] And by his strong support of centralized control, Harding implicitly was in agreement with Scott's position that the function of gun-layers was to point and not to aim—the latter being a task reserved for the spotters—which was the basis of Scott's insistence upon the appropriateness of short-range practice for gun-layers. The balance of Harding's report heavily favored Scott's views on gunnery. It indeed amounted to a manifesto for the replacement of traditional methods of individual aim by centralized fire control, and undoubtedly was a major contributor to the Admiralty's decision of late 1904 to increase greatly the amount spent on gunnery equipment.

Major Gunnery Improvement, 1904–1906

In his memoirs, Percy Scott maintained that Admiralty parsimony prior to 1905 precluded the development and supply of gunnery equipment that was essential if firing at longer ranges was to become practicable.[51] The internal draft accounts of Royal Navy spending on gunnery instruments support Scott's recollection. From fiscal year 1899–1903, annual expenditure on aiming accouterments was about £15,000 or less per fiscal year, and limited to the purchase of rang-finders and telescopic sights. In fiscal year 1903–4 spending rose to £27,000 and in 1904–5 to £35,000, of which less than a third—that is, £6,000 and £10,000, respectively—went to gear other than range-finders and telescopes.[52]

In October 1904, Adm. Sir John Fisher succeeded Kerr as First Sea Lord. Fisher had been a strong proponent of long-range gunnery, and his taking office as service chief undoubtedly contributed to a greater general willingness to allocate funds to the development of fire control, especially in light of the disappointing performance of the fleet in the 1904 gun-layer's test and battle practice, and the findings of the 1904 fire-control trials. That being said, it was also the case that Fisher opposed some of the recommendations of the Harding report on the fire-control trials on the grounds of excessive cost, and it was Lord Selborne who insisted in November 1904 upon the implementation of the full program.[53] The increase in expenditure on gunnery material was very substantial. In fiscal year 1905–6, the amount allocated to instruments was tripled to over £107,000, of which £99,000 was devoted to fire-control equipment other than range-finders and telescopic sights. In fiscal year 1906–7, spending on instruments rose to £122,500, of which £99,000 went to fire-control equipment other than range-finders or telescopic sights.[54]

The failures of the 1904 battle practice had been almost entirely attributable to poorly calibrated gunsights, haphazard gunnery training and testing conditions, and inadequate communications between spotters in the masts, the central control station, and the various gun positions. These defects were largely corrected over the next two years. Scott, as the first inspector of target practice, systematized the training and testing of gun-layers, publicized the results of the gun-layers' test and battle practice in order to encourage competition, and took the steps necessary to calibrate the sights of the fleet in accordance with the findings of a special committee that had subjected the question to intensive study. Capt. John Jellicoe, the director of naval ordnance, saw to the outfitting of all major warships with improved communications gear for gunnery purposes.[55] As a result of these changes, gun-layers' test and battle practice scores—albeit under unrealistically easy conditions—rose dramatically in 1906.[56]

In 1904, the reports of the Mediterranean and Channel Fleet committees had indicated that a few slower firing big guns were easier to control than many faster firing smaller pieces. Moreover, the flatter trajectory of heavy-caliber, in comparison with smaller caliber, projectiles produced by no means negligible differences in allowable margins of sight-setting or gun-laying error at the anticipated battle ranges of 4,000 to 5,000 yards. For example, at 4,000 yards, a 6-inch gun's margin of error was plus or minus 101 yards, while that for a 12-inch gun was 130 yards, a disparity of 29 percent.[57] And finally, the much greater destructive effect of heavier caliber projectiles, which had recently been improved by the adoption of a more effective armor-piercing shell, further compensated for the big gun's lower rate of fire and the necessity to lay big guns at fixed angles of elevation and train.[58] The greater effectiveness of big-gun hits was demonstrated in the engagement between the Japanese and Russian Fleets of 10 August 1904, which was witnessed by a British naval attaché and reported to the Admiralty.[59]

In March 1905, the Admiralty thus approved the design of the battleship *Dreadnought* and the three armored cruisers of the Invincible class—later to become known as battle cruisers—which were given all-big-gun armaments in place of the mixed batteries of large-, intermediate-, and medium-caliber guns that previously had been standard. By the fall of 1905, the new capital ship policy was further justified by information from the Battle of Tsushima, which, like earlier reports about the fighting in the Far East, attested to the superiority of big guns over medium-caliber quick-firers.[60]

There were, in addition, prospects of significant improvement in the laying of large-caliber guns. In early 1905, electrically powered heavy-gun mountings had been ordered for one of the new armored cruisers on the strength of a 1902 report that maintained that such equipment would significantly better the control of elevation and training.[61] By August, experiments with improved elevation and training controls for hydraulically powered big-gun mountings were underway.[62] These developments gave the Admiralty good reason to believe that Royal Navy capital ships would soon be able to lay their big guns by continuous-aim, which would mean that they could fire effective salvos under less than perfect conditions of wind and water.

There was also Percy Scott's advocacy of director firing,[63] a method of overcoming the problems of laying guns to fire at fixed angles of elevation and optionally train as well through centralized control of firing. This was to be accomplished through a mechanical master sight called a director, which by electrical signal indicated the correct elevation and train at each gun station, and by an additional electrical circuit connected directly to each gun caused their discharge with the pull of a single trigger. The operation of the director was intended to ensure that firing was exactly simultaneous, eliminating a major source of excessive and inconsistent dispersion, which would enable big guns laid at fixed angles of elevation and train to fire effective salvos. The director could also be located at the top of a mast, where view of the target was less likely to be obscured by smoke or sea spray. Firing guns at fixed angles of elevation and train through the use of the director was also to be preferred when rough seas produced heavy roll and yaw that could prevent even the most skilled layers from keeping guns steady through continuous-aim.

In August 1905, Rear Adm. John Jellicoe justified trials of director firing on the grounds that independent firing was ineffective at longer ranges and the practicability of salvo firing with continuous-aim had yet to be proved.[64] At about this same time, Scott produced a pamphlet that called for director-controlled firing of the *Dreadnought's* big guns in salvos. In this article, Scott also pointed out that the *Dreadnought's* eight-gun broadside could be divided into two groups of four, and fired alternately. This would enable the new model battleship to fire big-gun salvos twice as quickly as existing conventional vessels, which were armed with only four big guns, and thus give it the capability to find the range more quickly through spotting. And finally, Scott made it clear that he distrusted sight-setting that was heavily dependent upon corrections derived from instruments or formal consideration of various theoretical ballistical factors, and reiterated his preference for fire control based primarily upon the observation of the fall of shot.[65]

Meanwhile, other developments had improved the position of those who favored instrument-based sight-setting. In 1906, the Royal Navy began to receive production models of the Vickers clock. This device consisted of a dial marked with ranges and a pointer driven by a constant-speed clockwork motor and disc-roller drive[66] that when set with a change-of-range rate and a range from the range-finder, generated a succession of target ranges. The Vickers clock could thus provide gunners with range information independently of observation of the target, thereby compensating for the slow speed of optical range-taking and the probability that visibility of the target would be obscured in action from time to time by smoke or haze. Prototypes had been tested by the Royal Navy as early as in 1903.[67] Mechanical shortcomings may have delayed adoption, but the more likely reasons were probably a combination of the financial stringency that has already been described, and the fact that until 1905, no means were available to provide other than guessed settings of either the range, when it was greater than 4,000 yards, or the range rate.[68]

By 1905, however, there were prospects of major advances in both technical areas. In 1903 and 1904, promising trials had been carried out with an experimental Barr and Stroud 9-foot-base single-observer range-finder. Tests indicated that accurate range-taking would soon be possible at distances that were more than twice as great as those with the existing 4½-foot base instrument.[69] In May 1905, the Admiralty had ordered trials of a two-observer range-finding system and associated method of measuring target course and speed from simultaneously observed and plotted ranges and bearings that had been proposed by Arthur Hungerford Pollen, a civilian inventor. The perfection of this equipment would provide both an alternative and potentially more accurate solution to the range-finding problem and a means of determining the target course and speed with precision.[70] It was these developments in combination, perhaps, that finally convinced the Admiralty to order the Vickers clock in quantity.[71]

By the beginning of 1906, in short, both the spotting and instrument factions could argue that their respective approaches would soon be capable of solving the problem of setting sights with reasonable accuracy when firing at ranges of from 5,000 to 6,000 yards. These were moot points, however, because by this date, the prospect of very large extensions in the effective range of torpedoes had made it necessary to contemplate much longer effective gun ranges.

The Gunnery Problem Exacerbated, 1903–1906

As early as in March 1903, the Admiralty had considered proposals to increase the range of torpedoes by the application of heat to the compressed air used to drive the engine.[72] In August 1904, the Admiralty's Torpedo Design Committee reported that this line of development held considerable promise, and furthermore that the adoption of stronger compressed air storage chambers and a more powerful engine for the latest torpedoes would soon result in speeds at 2,000 yards that were nearly as great as the maximum speeds then attainable at half that distance.[73] In May 1904, Lord Selborne, the First Lord, maintained that in the near future, the range of torpedoes would reach 5,000 yards.[74] An additional 2,000 to 3,000 yards was required to provide time to turn away from an enemy that had decided to close the range rapidly in order to deploy his torpedo armament, and when in pursuit of a fleeing enemy, the safety margin had to be even greater.[75] It was probably for these reasons that by mid-1906, the Admiralty had increased its estimate of likely battle ranges from 6,000 to 8,000 yards,[76] ordered that battle practice be carried out at ranges of 6,000 to 7,000 yards,[77] and extended the notion of long battle range from 8,000 to 9,000 yards.[78]

At ranges in excess of 6,000 yards, sight-setting based on spotting alone when ranges were changing was highly problematical. This was because at the greater ranges observation of the fall of shot was more difficult, tolerances for error were much smaller, and time for bracketing salvos in any case were unlikely to be extended. In order for consistent hitting to take place when ranges were long and varying, the change-of-range problem had to be eliminated by the action of instruments, and the function of spotting restricted to the correction of constant residual error that was the product of various ballistical factors. Instrumentally assisted sight-setting, in other words, had to be regarded as an essential partner of spotting, not an alternative.[79] By 1906, the development of new-model range-finders and the invention of plotting offered at least the prospect of accurate measurement of the target course and speed data needed to set the dumaresq and the initial range required by the Vickers clock. But the generation of ranges from change-of-range rates remained problematical because of the inherent shortcomings of the dumaresq and Vickers clock.

In the case of the dumaresq, the indication of the correct change-of-range rate when that rate was altering continuously required that the target always be visible. Changes in the indication of the rate by the dumaresq had to be transferred manually to the Vickers clock, and because such transfers could not be carried out continuously, continuous change in the range rate was imperfectly represented by the clock to a greater or lesser degree depending upon the rapidity of change and the time intervals between resettings.[80] And because of the weakness of the clock drive,[81] which meant that it could not be mechanically connected to a transmitter, the generated ranges had to be read and then manually set on a separate mechanism that sent the data to the guns.[82] There were, in short, three potential sources of inaccuracy—inaccurate calculation of the rate because of poor visibility, inexact representation of the changing rate, and delays and outright mistakes in the manual transfer of data between the several components of the system[83]—which acting in combination could produce significant and unpredictable rate error.[84]

Matters were further complicated by the fact that at 7,000 to 8,000 yards, the time of flight of the projectile became long enough (about ten seconds)[85] for the relative positions of the target and firing ship to shift significantly between the moment of firing and the moment of shell impact on the target. In addition, wind, or temperature and density of the atmosphere, affected projectile direction and trajectory. Thus, generated ranges that represented the geometric range—that is, the distance between the target and firing ship at the moment of firing—had to be corrected for relative motion during the time of flight and the influence of ballistical factors to produce the gun range—that is, the range by which sights had to be set in order to bring the projectile exactly onto the target. Not to apply corrections, or to apply them manually was to add yet another source of error to those inherent to the dumaresq–Vickers clock combination.

Attempts to correct the clock periodically with range-finder observations to reduce the effect of rate-induced cumulative range error, or even to apply a rate correction in the hope of eliminating error altogether, were problematical because of the relatively slow speed of operation and inaccuracy of even the latest single-observer instruments. Correcting rate-induced range error or rate through the observation of the fall of shot was risky because of the difficulties of spotting, which will be discussed in detail later on. Spotting for rate in addition greatly complicated the task of making adjustments for constant time of flight and ballistical error. And heavy reliance on observation of the fall of shot for sight-setting was undesirable on tactical grounds because of the concentration of fire issue.

Harding, it will be remembered, had pointed out the significance of the subject in his pamphlet of 1903. In that year, tactical exercises had demonstrated that when two lines of battleships were separated by 5,000 yards, each ship would be capable of firing on any one of four to five vessels in the opposite line. Concentrating the fire of several ships on one was thus theoretically possible, and for this reason a 1904 report argued that if "several ships have concentrated on the same ship of the enemy it will no doubt count in their favor."[86] In July 1906, an Admiralty memorandum described a hypothetical case in which eight dreadnoughts might concentrate their fire on a single enemy vessel.[87] The actual effectiveness of concentration of fire was uncertain, however, because when several ships fired on one at longer ranges, it was difficult to distinguish the splashes made by one ship from another, which reduced the combined hitting rate to below that which might have been achieved by the same number of ships firing at separate targets.[88]

On the assumption that big-gun laying would be solved by either new mountings or the perfection of the director system, the main long-range gunnery questions facing the Admiralty by 1906 follow. Naval artillery had to be made effective at distances that were even greater than that of existing battle practice so that British warships could make hits at ranges that were beyond those of the coming generation of torpedoes. To accomplish this, some means of generating ranges with greater precision and reliability was needed in order to enable sights to be set accurately when ranges were long and changing, and possibly changing at a rate that was itself changing. And accurate hitting had to be dependent to the least possible degree on spotting shell splashes so as to improve the ability of several ships to fire on one, which would enable a British fleet to win battles decisively through concentration-of-fire tactics.

Harding specified the technical requirements of a system of fire control that would address the sight-setting aspects of the just-described gunnery agenda in a Naval Ordnance Department study of the Russo-Japanese War printed in 1906.[89] In this work, Harding argued that consistent hitting at long range could only be achieved through "an accurate and reliable range-finding system," "accurate determination of the rate of change of range," knowledge of the various corrections required to convert geometric ranges into gun range, and the automatic integration of the generation of geometric range and application of corrections. He noted even then that there would be "residual error" caused by unknowable ballistical factors.[90] This he conceded would have to be corrected through observation of the fall of shot, but the

elimination of the other unknowns by machine action meant that the quantity involved would be small and constant.[91]

Harding's description of an ideal fire-control system probably owed a great deal
to discussions with Arthur Hungerford Pollen that took place between 1904 and
1906.[92] From 1900 to 1906, Pollen had subjected the fire-control problem to systematic analysis and carried out a number of experiments with Admiralty assistance. He
had initially concentrated his efforts on the development of a two-observer range-
finder. The two-observer approach was inherently more accurate than that of the
single-observer system of Barr and Stroud, but was more cumbersome and vulnerable to disruption by enemy fire. For these reasons Pollen abandoned it in mid-1906
upon learning of the existence of the new-model, 9-foot-base Barr and Stroud single-
observer instrument. Pollen's willingness to accept inferior range-taking was based
on his belief that the inaccuracies of data obtained with a single-observer instrument
would be largely eliminated when the ranges were plotted and then meaned. The plot
would thus provide the target range as well as the target course and speed.

In 1904, Pollen had written a pamphlet in which he described a change-of-range
machine that combined the functions of the dumaresq, Vickers clock, and range
transmitter. In addition to eliminating two manual transfers of data, the device was to
be capable of taking account of varying change-of-range rates without being dependent upon continuous observation of the target or periodic corrections of the rate.
And Pollen suggested that automatic range generation be further enhanced by the
addition of automatic calculation of the deflection based on the mechanical combination of generated ranges and corrections for drift, target, and own-ship course and
speed, time of flight of the projectile, and wind.[93] In 1905, Pollen further refined his
proposals for the calculation of deflection in another pamphlet.[94] And finally, Admi-
ralty-supported experiments in 1905 and 1906 with his range-finder and plotter
prompted him to develop the design of a gyroscopically stabilized range-finder and
bearing-indicator mounting that would increase the speed and accuracy of range and
bearing taking.

By mid-1906 Pollen was convinced that further government backing would allow
him to develop his mounting, plotter, and change-of-range and deflection machines,
which could serve as the basis for a complete sight-setting system that would enable
British warships to shoot accurately at 10,000 yards in spite of high or varying
change-of-range rates.[95] In principle, the main advantages of such a system over
what existed were two. In the first place, it obtained essential gunnery data by exact
measurement rather than through approximation and guesswork. And secondly, it
substituted machine for manual functions to the greatest possible degree in order to
increase the accuracy of calculation, augment reliability through the elimination of
many sources of human error, and ensure rapidity of operation that was essential to
achieve usable results.

If perfected, Pollen's scheme would make it possible for British ships to hit even
when visibility of the target was intermittent and rate conditions were difficult, increase the effective range of naval artillery to beyond that of the torpedo, and enable

accurate sight-setting to occur with minimal assistance from spotting, which would allow several ships to fire on a single target without increasing the rate of missing and mitigate the negative effects of having to rely on a problematical means of correction. There was strong feeling within the Naval Ordnance Department, however, that combining the functions of the dumaresq and Vickers clock would result in "complication and expense, both of which are undesirable."[96] Harding, on the other hand, was an enthusiastic supporter of Pollen's ideas, which were in line with his own predisposition to favor the use of instruments. In addition, Pollen won the support of Scott, who probably recognized that inconsistent hitting when ranges were longer than he had originally assumed had undercut his argument that sight-setting be based on observation of fire alone.[97]

Even the prospect of solving the Royal Navy's main outstanding tactical problems, and the agreement of previously opposed fire-control factions, was not enough to overcome the objections of certain Admiralty officials to Pollen's financial terms. In September 1906, however, Fisher—in his capacity as First Sea Lord—intervened on Pollen's behalf because he believed that the perfection of the inventor's ideas would give the Royal Navy a monopoly on long-range firing that would enable British ships to hit enemy vessels before they could be hit in return. Immunity to effective enemy gunfire so obtained would mean that thick armor would serve little purpose, and that the displacement devoted to it could be better used for higher speed and a heavier armament. The adoption of the Pollen system, Fisher was thus convinced, would enhance the viability of his pet scheme to replace battleships—including even those of the dreadnought type—with fast, heavily armed, but lightly protected battle cruisers.[98]

At Fisher's insistence, a compromise was struck, whereby Pollen was promised sufficient funds and technical cooperation to build prototypes of his observing and plotting instruments. If these performed successfully in trials, Pollen would win a lucrative production contract, and the stage would be set for further support to perfect the balance of the system. Insofar as the Admiralty's interests were concerned, the development of even the partial system would mean that a dumaresq could be set accurately for target course, speed, and bearing, and thus give reliable indications of the change-of-range rate, which could then be put on the Vickers clock. Although this left many problems unsolved, the result would still be a substantial advance. Pollen's instruments were ready for trial by the end of 1907. By this time, however, changes in financial conditions and command had drastically shifted the political balance within the Royal Navy against the civilian inventor.

Financial Crisis and Technical Reaction, 1907–1912

The agreement of September 1906 gave Pollen £6,500 in advance to build trials instruments, and provided for the payment to the inventor of £100,000 for the rights to his system and the installation of production models of his gear in forty ships in the event of the trials being a complete success, the stringent terms of which were also defined in advance.[99] The Admiralty outlay of cash in advance for the development of trials instruments in 1905 and 1906, which made the government a co-investor, was unprecedented, and the size of Pollen's potential remuneration was nearly as much as that spent on all fire-control instruments in fiscal year 1905–6,[100] both of which may be taken as indicators of the value placed on Pollen's ideas by the Admiralty. In November 1907, however, the navy's leaders became aware that unanticipated expenditure in other areas would drive up the next year's estimate by £1,000,000, an increase that the cabinet categorically refused to accept. In order to preserve the planned capital-ship-building program, the Admiralty was compelled to make drastic cuts elsewhere in the 1908–9 estimate.[101]

Of the sum obligated to Pollen, £50,000 was to have been spent in fiscal year 1907–8. But in November 1907, the Admiralty decided, in advance of the trials, that this sum would not be paid,[102] which along with other smaller cuts reduced spending on fire-control instruments for the fiscal year by more than a third to £112,500. Although it is difficult at this point to say for certain, there can be little doubt that these retroactive cuts were at least in part connected to the Admiralty's desperate efforts to economize in the face of the determination of the Liberal government to hold the line on—if not roll back—naval spending. Expenditure on fire control for the next four years (1908–9 to 1911–12) would settle to the level of £75,000–90,000 per annum, or well below the peak of £122,500 reached in 1906–7, and half or little more than half the amount that had been planned for 1907–8.[103]

The onset of what turned out to be some five years of fiscal austerity with respect to fire-control equipment coincided with an important change in leadership with regard to naval gunnery. In September 1907, Jellicoe completed his term as director of naval ordnance, and was replaced by Capt. Reginald Bacon. Jellicoe had been skeptical of the mechanical practicability of Pollen's proposals, but was willing to risk limited amounts of money on experiment because he believed that success, if achieved, would provide the navy with a tactically valuable advance in gunnery capability.[104] Bacon, on the other hand, seems to have been troubled by the novel business arrangements set up between Pollen and the Admiralty, was opposed in principle to the mechanization of fire control on the grounds that it was inherently unreliable, and furthermore was convinced that the capability that mechanization offered was not essential to the achievement of effective hitting because comparable results could be obtained by a combination of existing manual methods of range generation and much improved spotting.

Bacon's distrust of fire-control instruments and faith in spotting probably owed something to his experience as captain of the *Dreadnought*. During gunnery experiments with this ship in early 1907, Bacon had found the new-model Barr and Stroud 9-foot-base range-finder to be unsatisfactory, and that because of good-loading drill as well as the larger number of big guns, four-gun salvos could be fired at 15-second intervals, which was triple the speed of earlier battleships.[105] In theory, such rapid fire greatly increased the value of spotting when the range was changing because the shift in the relative positions of the target and firing ship between salvos was reduced by a third. Successful firing by observing the fall of shot, however, had been possible because ranges were short (2,400 to 5,000 yards), the targets stationary, speeds low, and the sea calm, the last factor allowing effective salvo firing even though the big guns were laid at fixed angles of elevation and train.[106]

Bacon had reason to believe, on the other hand, that advances in gunnery materiel and technique would enable any all-big-gun capital ship to replicate the successes of *Dreadnought's* experimental cruise under more difficult conditions. By the end of 1907, the evaluation of fire effect at long ranges through the observation of the fall of shot had been further enhanced by the adoption of new spotting rules that did not require precise measurement of the distance between splashes and the target.[107] In late 1906, improved mechanisms for the control of hydraulically powered gun trains were tested successfully,[108] with satisfactory upgraded control of hydraulically powered elevation established by early 1908.[109] The mountings of dreadnought and the electric gear of Invincible, which proved to be too slow in operation, and those of some pre-dreadnoughts, were modified accordingly, while later capital ships were equipped with the new controls from the start.[110]

With these changes, big guns could be laid for elevation by continuous-aim as well, or even better, than medium-caliber quick-firers.[111] And although the problem of control over train was not solved completely, it became much easier for big guns to fire effective salvos at ranges of 8,000 yards or more. Rapid fire, the new rules, and the new mountings appear to have convinced Bacon that it would even be possible to correct variable rate error through observing the fall of shot.[112] And in 1906 and 1907, experiments indicated that at least two ships could fire on one with little if any diminution in hitting efficiency even when controlling fire by spotting, provided that salvos were fired alternately and their individual effects identified by keeping track of their times of flight.[113] In late 1907, Bacon was thus probably convinced that imprecise estimates of the change-of-range rate could be corrected through observation of fire in spite of the fact that the Royal Navy had yet had no experience in firing at moving targets,[114] and that concentration of fire could be accomplished even when spotting was required to correct sight-setting.

Bacon's chief advisor on gunnery matters during *Dreadnought's* experimental cruise was Lt. Frederic Dreyer, whose views on gunnery were similar. In April 1907, Dreyer displaced Harding, while the rotation of Scott to sea duty at the same time removed another major supporter of Pollen. In November 1907, Bacon asked Adm. Sir Arthur Knyvet Wilson, Dreyer's senior service patron, to umpire the upcoming trials of the Pollen system. Wilson agreed, and was given Dreyer as his assistant. The two officers then concocted an inexpensive manual method of observing and plotting ranges and bearings as an alternative. In preliminary tests in December 1907, Pollen's mounting and plotter demonstrated that they were capable of fulfilling the agreed-upon conditions of complete success. In the official trials of January 1908, which were conducted with the cruiser *Ariadne* and battleship *Vengeance*, however, artificially favorable conditions enabled the manual methods contrived by Wilson and Dreyer to produce acceptable results; and on these grounds Wilson recommended the adoption of his own system and the rejection of that of Pollen, courses of action that were endorsed by Bacon.[115]

The Pollen system was not the only victim of Bacon's faith in manual methods of gunnery. At about this time, Admiralty support for both gyroscope and director-assisted gun-laying, which were open to the same sorts of criticism about mechanical complication as Pollen's approach, was in the former instance canceled[116] and in the latter suspended.[117] Bacon and Dreyer were also convinced that other steps could be taken to give a British battle fleet a decisive gunnery advantage. In the spring of 1908, Bacon, apparently at Dreyer's suggestion, proposed that the caliber of British naval big guns be increased from 12 to 13.5 inches.[118] While there were a number of factors favoring such a move, the preeminent argument appears to have been based upon the belief that because dependence on spotting was likely to degrade the effectiveness of combining the fire of two or more ships on one, "concentration of fire without waste of energy is nowhere to be found except by concentrating intensity of fire in units."[119] The immediate implementation of such proposals, however, was impracticable because they entailed substantial increases in expenditure.

Bacon succeeded in gaining Fisher's backing for his reversal of previous gunnery policy by convincing his chief that Wilson and Dreyer had invented a cheaper viable alternative. But Pollen was encouraged to continue his work through a substantial financial settlement from the Admiralty and the support of dissenting navy officials. In late 1908, Pollen further strengthened his financial resources through private investment by founding the Argo Company. In early 1909, the Argo Company offered the Admiralty proposals for an improved gyroscopically stabilized rangefinder and bearing-indicator mounting, plotter, and automatic clock with associated transmitting mechanisms and automatic sights. Jellicoe's return to the Admiralty as Third Sea Lord and Controller—and thus Bacon's immediate superior—in October 1908 had assured Pollen a fair hearing, and secret communications with Fisher through an intermediary caused the First Sea Lord to renew his support for mechanized methods of fire control.[120] In April, as a consequence, the Argo Company was rewarded with an order for trials instruments.[121]

Pollen's first automatic clock—serious design work upon which had begun in 1907[122]—was based on fundamentally different principles than the dumaresq–Vickers clock combination. There were two critical problems that had to be overcome in order to produce a device that was capable of generating the target range without being dependent on continuous manual inputs or periodic corrections. In the first place, variation in the change-of-range rate had to be mechanically generated in order to avoid dependence upon continuous observation of the target, which was unlikely to be achievable in a real engagement. In the second place, the clock drive had to be strong enough to actuate a range transmitter when the rate setting was being varied continuously, which the Vickers clock could not do because of the weakness of the disk-roller drive. In the Argo clock Mark I, the carrying out of these two functions was made possible by a mechanism that in essence replicated the relative motion of the firing ship and target in an arrangement that was powerful enough to drive a range transmitter as well as a pointer on a dial.[123]

Pollen's position was improved by the failure of manual methods of plotting in battle practice. In 1907, battle practice had still been carried out with a stationary target. Firing ranges were from 6,000 to 8,500 yards.[124] In 1908, battle practice was carried out at the same distances but with a moving target.[125] In 1909, battle practice conditions appear to have been the same as in the previous year. Rear Adm. Richard Peirse, who had replaced Scott's successor as Inspector of Target Practice in March 1909, pointed out that when moving targets were substituted for stationary targets in the 1908 battle practice, the result had been a sharp decline in the percentage of hits to shots fired by Royal Navy capital ships, which indicated that dumaresqs were not being set properly for target course and speed.[126] Exact comparisons of battle practice results from year to year are impossible because of the way in which the scores were tabulated, but Pollen estimated the rate of hitting in 1907 to be 30 percent, and that of 1908 and 1909 some 20 percent.[127]

In July 1909, Capt. Constantine Hughes-Onslow of the Naval War College at Portsmouth circulated the first edition of his systematic study of fire control, which provided a clear explanation of the nature and complexity of the gun-laying and sight-setting issues. Hughes-Onslow's monograph presented unequivocally strong arguments in favor of the Pollen system, and together with the disappointing outcome of the 1908 battle practice, appears to have changed Bacon's views on the fire-control issue. The analysis of Hughes-Onslow was corroborated by the reports on the 1909 battle practice by the staff of the Inspector of Target Practice. In September 1909, design difficulties caused the postponement of trials of the Argo clock, while the need for automatic sights was removed by the development of an alternative system by Vickers (about which more later), but in October trials of the Argo mounting and plotter in the cruiser *Natal* were completely successful, and their adoption was recommended by the trials umpire, Capt. Frederick Ogilvy, a long-time associate of Scott.[128]

In December 1909, however, Ogilvy was taken ill and died, and Bacon retired, which was followed in January 1910 by Fisher's resignation and replacement by Wilson, who remained an implacable opponent of the Pollen system. And while Pollen enjoyed the unstinting support of Peirse, the Inspector of Target Practice, his relations with Jellicoe, a long-time admirer of Dreyer, deteriorated. Although a production order for the Argo mounting was made in April 1910, the plotter and clock were rejected by the Ordnance department after further trials in the *Natal* during June 1910, the favorable report of the superintending committee notwithstanding.[129] Technical shortcomings of the Argo equipment were undoubtedly a factor, but their adoption was also strongly opposed by Capt. A. G. H. W. Moore, Bacon's successor, who like Bacon, Wilson, Dreyer, and others was incensed by the idea that Pollen would profit handsomely from the perfection with Admiralty assistance of his as yet undeveloped proposals, and in any case was convinced that the fire-control practices and instruments in hand were good enough.[130]

Moore's optimism about the adequacy of existing service methods was in large part based on his confidence in a report by Adm. W. H. May on certain tactical exercises of the Home Fleet. May maintained that estimates of target course and speed by eye were accurate enough for fire control.[131] His conclusions, however, were vigorously challenged by Peirse,[132] and a naval officer informed Pollen that in the Home Fleet exercises guesses about speed had been "generally from 15 percent to 30 percent wrong, and the course error was said to average about 10 degrees."[133] In the absence of an effective means of plotting and, given the inaccuracy of guess-work, the staff of the Inspector of Target Practice in 1909 had recommended that corrections for the rate settings of the Vickers clock be obtained by observation of the fall of shot, an expedient that had been tried experimentally by a few ships before 1908, and used with some success in the 1909 battle practice.[134]

Obtaining rates by observing splashes, however, was bound to be a problemati-cal expedient. In theory, attaining the highest probability of hitting depended upon the achievement of salvos that were correct for line and whose constituent projec-tiles were falling both short and over in a "straddle" of the target. In practice, overs were impossible to observe when shots were correct for line because the target blocked the view. Knowledge of the accuracy of a salvo thus depended upon seeing splashes that were short. In addition, at ranges of 10,000 yards or more, the image of the ship, and its immediate foreground and background were visually so close to-gether[135] that in anything other than a dead calm it was virtually impossible to distin-guish between shells that fell short and those that fell over unless the splash—because it was both short and correct for line—blocked the view of part of the target. If wave action was even moderate and visibility less than perfect, the range at which shorts and overs could be distinguished from one another if splashes were not correct for line could be not much over 7,000 yards.[136]

Accurate spotting for range or rate under average conditions of wind and wave in battle practices carried out at 6,000 to 8,500 yards, in other words, required salvos to be correct for line for much if not most of the time. This was, however, difficult to accomplish because of shortcomings in gun-laying. While the new big-gun mountings had enabled guns to follow the target through a ship's rolling motion, control over gun training—a much more difficult task because it involved the back-and-forth movement of not only the guns but the entire turret—was not good enough to enable gun-layers to compensate fully for yaw.[137] In addition, when guns were not trained exactly across the broadside—that is, when they were pointed either some-what ahead or astern—the canting of the trunnions with respect to the horizontal plane caused the actual training of the gun to be different than that indicated by the sight.[138]

The foregoing may explain why in the 1909 battle practice, more than half the first shots fired were incorrect for line.[139] And although cross-roll error and unpre-dictable ballistical error[140] were more or less constant and thus could be corrected on the basis of observing the fall of a single salvo, there were other factors that worked against the reliability of spotting. The failure of a gun-layer to fire because he could not see the target because of momentary poor visibility, or to fire even when the gun was incorrectly laid and by so doing produce a wild shot, were not uncommon occurrences, whose effect, especially in the latter case, was "to hopelessly mislead the Control Officer."[141] In general, the fact that effective spotting assumed that all guns were properly laid at the moment of firing and that visibility would be clear and unobstructed for both the gun-layer and the observer—which did not always occur even under the easy conditions of battle practice—raised serious questions in the minds of most naval officers about the practicability of observation of the fall of shot as the primary source of sight-setting information. And some were also concerned about the incompatibility of spotting and efficient concentration of fire.[142]

In 1910, the manual methods of plotting that had been advocated by Wilson and Dreyer were categorically rejected by the same committee that tested the Argo plotter and clock.[143] Efforts to control fire in the absence of an effective system of plotting ranges and bearings by clock tuning—that is, a process of comparing clock-generated ranges with observed ranges in order to find a rate which could then be set on the master clock, whose generated ranges were sent to the guns—and spotting for rate were unsatisfactory.[144] In 1911, the range of battle practice appears to have been increased to perhaps 9,000 yards or more, and the hitting rate of all-big-gun capital consequently fell to 13 percent.[145] Analysis of the 1911 battle practice rankings shows that only four out of the twelve all-big-gun ships in service placed in the top ten scoring units in battle practice in spite of the fact that all-big-gun capital ships were supposed to be superior to pre-dreadnoughts in terms of spotting.[146] That same year, the Admiralty convened a special conference to discuss the poor results being obtained in both battle practice and the gun-layer's test.[147]

The unsettled situation with regard to fire control, and the great improvement in the financial circumstances of the navy following the liberal financial revolution of 1909,[148] which reopened the door to greater spending on suitable gunnery instruments, undoubtedly encouraged Pollen to continue the development of his system. And confidence in his ideas on the part of a number of wealthy naval officers and businessmen enabled him to raise private funds that sustained his design efforts. Obviating navy investment of capital in pretrials also made Pollen's business relationship with the Admiralty follow conventional practice more closely. In the spring of 1912, after nearly two years of additional hard work, he offered the Admiralty a prototype clock that, like his instrument of 1910, combined the functions of a dumaresq, Vickers clock, and transmitter in a single device, but did so with greater accuracy and flexibility of operation. Pollen also informed the navy that he would soon have an improved plotting machine and a range-finder much superior to the Barr and Stroud instrument.[149]

The new model Argo clock was based on a more sophisticated disc-roller drive that was capable of activating a range transmitter, even while the rate setting was being altered, without introducing inaccuracy. Moreover, by a remarkably ingenious arrangement of three such devices, the production prototype, which was known as the Argo clock Mark IV, also generated bearings, which were mechanically fed to the equivalent of a dumaresq that in turn controlled the rate setting of the disc-roller range generator. The clock thus did not require continuous bearing observation, as did a conventional dumaresq, in order to take account of continuous variation in the change-of-range rate. The new clock could also indicate ranges through the turning motion of the firing ship, which the Argo clock Mark I could not. These improvements made it far superior to the dumaresq–Vickers clock combination.[150] By the time the Argo clock Mark IV had arrived, however, Admiralty tactical requirements had changed, and the strength of Pollen's opponents had increased by the adoption of mechanized fire-control methods of their own.

The Quest for Reach Abandoned, 1912

The Royal Navy, it will be recalled, had favored early hitting because of the inherent vulnerability of fire control to disruption by damage inflicted by enemy fire and the need of British battle cruisers when engaging battleships to hit before they could be hit in return on account of their inferior protection. Concern over the former appears to have been much reduced by the transfer of fire-control instruments and range finders from the main-top to positions below that could be protected by armor.[151] The retirement of Fisher as First Sea Lord meant that the scheme to replace the battleship with the battle cruiser languished, and with it some of the demand for a fire-control system that could function at ranges that were always supposed to be greater than those at which an enemy warship could make hits. There was also the question of outranging the torpedo, about which more will soon be said. But it is first necessary to examine the reasons why the Admiralty came to the conclusion that an engagement with the German fleet, which was considered to be the Royal Navy's most likely opponent in the near future, would take place at relatively close range.

As early as in December 1908, Bacon had maintained that visibility in the North Sea, where any major battle between the British and German navies was bound to occur, was generally so poor that fighting that took place there would probably be at short distances. This was yet another reason, he had argued, why the Royal Navy could dispense with the development of mechanized methods of fire control.[152] In 1909, experienced naval officers reported that visibility in the North Sea was less than 8,000 yards for seven days out of ten.[153] In April 1910, Adm. W. H. May reported that in the recent tactical exercises of the Home Fleet, which had been held in foggy weather in the North Sea, visibility had been no more than 4,000 to 8,000 yards.[154] Pollen's supporters could argue in reply that there was no guarantee that the Royal Navy would always fight when vision was restricted,[155] but the proposition that battle would most likely take place at shorter rather than longer ranges was strengthened by intelligence assessments of German tactical intentions.

In 1909, Oscar Parkes, a naval artist and journalist, had photographed the first German dreadnought battleships as they were being completed. The pictures revealed that the German vessels, unlike British all-big-gun ships, were equipped with medium-caliber quick-firing guns in addition to an antitorpedo-craft battery of lighter pieces.[156] This information was corroborated by other intelligence sources that reported that the caliber of the secondary battery was even bigger than first assumed.[157] The adoption of medium-caliber quick-firers of relatively large size in addition to the main battery of big guns suggested that the Germans would seek to engage at shorter ranges than were being contemplated by the Royal Navy. By 1910, the belief that the Germans had abandoned long-range gunnery appears to have been common knowledge in British naval circles.[158] A knowledgeable British naval officer later recalled that "so firm was the conviction" that the Germans would fight at close range that "our whole tactics were based upon it."[159]

The self-confident Royal Navy undoubtedly welcomed the German intention to fight at short range, because such a course, given British determination to stand fast and fight, would almost certainly result in a decisive battle rather than an inconclusive skirmish. From the standpoint of gun-laying, shorter range engagements increased allowable margins of error, which may explain why no steps were taken to address certain gun-laying problems, such as cross-roll error, that had been described in detail by Hughes-Onslow in his 1909 report.[160] Observation of the fall of shot was also an easier task at shorter ranges, and this may have been the motivation behind the Ordnance Department's replacement of both instrumental correction for constant ballistical error and even time-of-flight offsets by spotting.[161] The belief that battle ranges would be lower than previously expected also affected attitudes towards the sight-setting issue.

For many years, Pollen had justified his system on the grounds that it would make fire effective at 10,000 yards or even more. But in probable response to the argument that the Germans intended to fight at far shorter distances, Pollen in December 1910 conceded that achieving the highest possible rates of hitting at ranges of 8,000 yards or less was preferable to the practice of cultivating firing at ranges of 8,000 to 11,000 yards.[162] He had good grounds to argue that the conditions of short-range gunnery favored the adoption of his mechanized system because it was much more capable than any manual system of plotting and range generation for dealing with variation in the change-of-range rate, which May's exercises had recently shown would occur more frequently when ranges were short than when ranges were long.[163] But on the other hand, lower ranges meant that margins for error in terms of sight-setting could be much greater and continuous visibility of the target more likely, which gave Dreyer the opportunity to make a convincing case for a simpler and therefore less capable, but far cheaper, system of mechanized fire control.

In 1908, Dreyer had unequivocally repudiated mechanized plotting,[164] but the subsequent failure of his manual methods compelled him to change his views. He thus developed a mechanical method of plotting ranges and bearings against time separately. The range rate alone could be used to set a Vickers clock directly, or the two rates could be put on a modified dumaresq to obtain the target course and speed, which could then be applied to a conventional dumaresq along with the target bearing to obtain a succession of range rates.[165] Dreyer built an improvised rate plotter on his own while serving at sea in 1909 and, after achieving promising results in battle practice, submitted the scheme to the Ordnance Department in 1910. Moore supported Dreyer's efforts, which he probably regarded as much as an antidote to Pollen as a necessary aid to gunnery.[166] By early 1911, Dreyer had combined his plotter, a dumaresq, the Vickers clock, and a deflection calculator and transmitter, on a single metal frame, the collection being known as the Dreyer Table.

Dreyer's plotter was mechanically much simpler than Pollen's, but its output was inherently less accurate, especially when the change-of-range rate was varying. The manner in which the two rates were used in combination with the dumaresq to obtain the target course and speed was also a source of error whose magnitude was bound to increase with variation in the change-of-range rate. The correct indication of continuous variation in the change-of-range rate by the dumaresq was impossible to achieve when visibility was interrupted for any length of time. The manual transfer of rates to the clock, manual application of various corrections to clock-generated ranges, and the manual movement of data from the clock to the transmitter, were also prone to delay and faulty execution because of human mistakes.[167] And such proneness to process error made it much more difficult to determine quickly whether differences between clock-generated ranges and observed ranges were the result of a faulty setting for rate or systemic inefficiency, which in turn meant that it was harder to decide whether to apply a correction for rate or range.

These shortcomings were to a degree offset by advances in other areas. Manual changes in sight-setting—like manual transfers of data between the dumaresq, Vickers clock, and transmitter—had been a serious source of delay and error. By 1912, the general adoption of the Vickers follow-the-pointer automatic sight-setting system had eliminated the problem.[168] Another major difficulty had been posed by inadvertent switching of fire from one ship to another when poor visibility momentarily prevented gun-layers from being able to distinguish the correct place of their target relative to its consorts in a line, which was easy to do because at long ranges the difference in the angles of train for adjacent enemy ships was extremely small. This potential difficulty was overcome with the adoption of the Evershed system of target indication, whose transmission of precise instructions for target bearing ensured that all guns were trained on the right target.[169]

But the practicability of the Dreyer Table was primarily a function of the use of several range-finders to increase the numbers of ranges plotted, an approach that had been recommended by Pollen as early as in 1907.[170] Averaging ranges from different range-finders to a great degree eliminated errors that were inherent to range-taking by the same operator with a single Barr and Stroud instrument,[171] and ranges so obtained were accurate enough to be applied to the Vickers clock and so compensated for any mistakes in the settings for rate. The practice of using plotted range-finder observations to keep the Vickers clock tuned to "the mean range-finder range of the moment" was known as range-finder control. Ironically, its effectiveness was probably much increased in 1912 when the fleet began to receive the first Pollen gyroscopically stabilized range-finder and bearing-indicator mountings, which more than doubled the speed of range-taking.[172]

With improved range-finder control, and the reduction in battle practice range in 1912 from 9,000 yards or more to 7,000 to 8,000 yards, which was coupled to a substantial increase in the size of the target,[173] the role of spotting the fall of shot was greatly reduced. This meant that firing could be independent rather than in salvos, and therefore much more rapid.[174] Accuracy improved also. Specific information about the fleet average percentages of hits to shots fired in battle practice is not available for the years from 1912 onward, but in the 1912 battle practice of the Home Fleet, the average of the top eight all-big-gun ships was nearly 40 percent, or three times that of all-big-gun ships in 1911.[175] Another indicator of improvement may be found in the fact that in 1913, 60 percent of the dreadnoughts in service were to be found in the top ten scoring units, and nine out of the top ten units were all-big-gun capital ships, as compared with the respective figures of 50 percent and eight out of ten in 1912, and 33 percent and four out of ten in 1911.[176]

In addition to accurate rapid fire, the Admiralty could count on a British fleet being able to hit harder than their German opponents because the large increases in funding for the navy from 1909 had made it possible to act on the Bacon and Dreyer initiative of 1908 to increase big-gun caliber. By 1912, the Royal Navy had built or was building no fewer than sixteen battleships and battle cruisers armed with 13.5-inch guns, and had authorized the construction of battleships that were to be armed with 15-inch guns. The 13.5-inch gun fired a projectile that was more than 25 percent heavier than that of the 12-inch guns that equipped the latest German capital ships, while the shells of the 15-inch gun weighed nearly twice as much.[177] The British possession of a substantial superiority in weight of fire meant that the inability to hit before being hit in return would matter less, because even a simultaneous exchange of fire and the assumption of equal marksmanship would still constitute highly unfavorable circumstances for the Germans.

Also by 1912, the potential accuracy and consistency of gun-laying, and therefore the efficiency of spotting,[178] under less than ideal conditions was markedly improved by the adoption of director firing. By 1909, experience had demonstrated that big guns on the latest model mountings could not be laid by continuous-aim with sufficient accuracy when waves were such as to cause even moderate rolling and yawing.[179] In addition, the vision of gunners, who were situated near the waterline and at the same level as other gun positions, was often obscured by smoke and sea spray, preventing them from aiming their weapons. Jellicoe's return to the Admiralty in October 1908 had aided Scott as well as Pollen, and Bacon's withdrawal of support for director-firing experiments was reversed.[180]

After two unsuccessful trials of crude experimental equipment in the battleships *Bellerophon* and *Neptune* in 1910 and 1911, respectively, tests of much more advanced gear on the battleship *Thunderer* in 1912 revealed that a ship equipped with a director was capable of making many more hits than one that was not when water conditions were less than ideal.[181] Director-controlled salvos, moreover, were much easier to spot than rapid independent shooting when two or more ships fired on one, and thus became the basis of Royal Navy concentration-of-fire tactics.[182] In addition, the smaller caliber of the main armament of contemporary German ships, and the less efficient disposition of their turrets, meant that a concentration of two 13.5- or 15-inch-gun-armed British capital ships against one German could be expected to produce at least a 2.5-to-1, and as much as a 3-to-1, superiority in the weight of fire, which equaled, if not bettered, the figures achieved by Nelson's ships at Trafalgar.

This left the problem of the torpedo threat, which showed every sign of getting worse. By 1909, the Royal Navy was receiving "wet heater" (steam) 18-inch torpedoes that were faster at 6,500 yards than the best cold torpedoes had managed at 2,000 yards and that ran at diminishing speed for an additional 1,000 yards.[183] By 1912, the increase in the size of the largest torpedoes from a diameter of 18 to 21 inches had extended the effective range of the best underwater weapons at high speed from 6,500 yards to over 10,000 yards. In addition, the advent of gyroscopes whose orientation was adjustable meant that torpedoes could be set to run on a course that was different from the angle of train of the launching tube (angled fire), which multiplied the number of circumstances in which torpedoes would be effective. And finally, the introduction of torpedo fire-control instruments further improved the chances of hitting at longer ranges.[184]

For these reasons, Bacon in particular is known to have concluded that even a sophisticated fire-control system such as that offered by Pollen would not enable guns to outrange the torpedo. As early as in December 1908, reports that an experimental 21-inch torpedo had a range of 12,000 yards at full speed prompted the then Director of Naval Ordnance to give serious consideration to the proposition that extending the reach of naval guns to beyond that of the torpedo was impracticable.[185] In 1910, Bacon declared in public that the advent of long-range torpedoes had made the line of battle obsolete.[186] Most British officers, on the other hand, remained convinced that rapid heavy hitting at close range would decide the outcome of an engagement, the longer range of torpedoes notwithstanding, but, in any case, the fact that the Germans were thought to be intent on fighting at 10,000 yards or less made increasing gun range to beyond that of the torpedo pointless.

To sum up, in 1912, the Admiralty had reason to believe that Dreyer's less than precisely accurate instruments could serve as the basis for gunnery that was capable of achieving decisive results in battle; that, even when spotting was necessary, British ships would be able to develop an effective concentration of fire; and that further investment in fire-control improvement for the purposes of avoiding the torpedo threat did not matter. The cost of Dreyer's first model table, in addition, was as little as a tenth and no more than a fifth that of Pollen's equivalent.[187] In light of these figures, even Pollen's old supporters, such as Jellicoe and Prince Louis of Battenberg, changed sides and supported Dreyer, which was enough to shift the balance of power decisively against Pollen allies such as Peirse and Scott. The Admiralty thus rejected his plotter without a trial, and also the new-model range-finder, which though significantly superior to the Barr and Stroud instrument, was three times its price. And while the Argo clock Mark IV performed successfully in tests carried out on the dreadnought battleship *Orion*, only six were ordered.[188]

Ships equipped with these instruments appear to have outperformed those with the Dreyer gear, but the Admiralty did not purchase additional units. Dreyer developed an improved table, which introduced a device that resembled the Argo clock

Mark IV but whose design was crude in comparison and thus much less satisfactory in operation. They were ordered for new ships only, however, while the bulk of the fleet soldiered on with the early version of the Dreyer Table. A Barr and Stroud 15-foot-base range-finder, which was much superior to the 9-foot-base instrument when ranges were greater than 10,000 yards, had been developed by 1909, but was not ordered in quantity until 1914.[189] Improving the ranging of projectiles by altering their shape was undertaken to a limited degree beginning in 1909, but in spite of what seems to have been a good understanding of the ballistic issues, the full potential of this new development was not exploited for more than a decade.[190] And the Admiralty's faith in independent fire at close range may partly explain why only three ships were equipped with directors by the end of 1913, and only eight by the outbreak of the war.[191]

Even the Royal Navy's procurement of less than the best possible fire-control equipment, however, was accomplished only after significantly greater expenditure. In fiscal year 1912–13, spending on gunnery instruments—which would have included the Pollen mountings and a few clocks, Dreyer Tables, and Scott director gear—may have risen from £75,000 of 1911–12 to over £120,000. In 1913–14, expenditure on fire-control equipment appears to have exceeded £400,000, or more than quadruple the amount spent by Bacon for the same purposes in 1908–9 and 1909–10.[192] Spending of such magnitude would have been unthinkable before the major changes in tax structure made in 1909, which had greatly improved the financial position of the government in general and the navy in particular. Even so, the increase of 1912–13 came only after a record budget surplus in 1911–12, while the reduction in projected spending on fire-control instruments in 1914–15 to less than half the figure of the previous year was probably a response to Treasury calls for the most stringent economy.[193]

Epilogue, 1913–1919

The Admiralty's belief that a battle against the German fleet could be won by rapid fire at relatively close range and in spite of the torpedo threat was not shared by the commanders of the fleet. Skepticism with regard to the likely effectiveness of the fire-control methods promulgated by the Ordnance Department under the more difficult conditions that could arise in a real battle if ranges were kept open to minimize the danger from torpedoes seems to have prompted the Battle Cruiser Squadron to carry out a high-speed firing trial at targets that were 12,000 yards distant in the spring of 1913, which exposed serious weaknesses in gunnery.[194] These were followed in November by a shoot by warships of the Home Fleet at the obsolete battleship *Empress of India*. Sea conditions, however, were calm and visibility clear, ranges less than 10,000 yards, and the target stationary because of the inadequacy of provisions for towing. As a consequence, hitting was frequent and the resulting damage so severe that the target vessel sank before planned longer range firings could take place.[195]

The lack of long-range firing during the *Empress of India* trials disturbed Adm. Sir George Callaghan, the commander-in-chief of the Home Fleet, who informed the Director of Naval Ordnance on 8 December 1913 that "although everyone is agreed as to the great desirability of *hitting first* we have little to guide us as to the range at which we can fire with good prospect of hitting."[196] In early November 1913, just before the trials, Callaghan had warned his command that in battle, range-finder control would most likely be compromised by human error in range-finding, disrupted visibility, and ranges that were too great for the existing 9-foot-base Barr and Stroud range-finder to measure accurately. In the event of such circumstances arising, Callaghan insisted that the fleet be prepared to employ rate control via dumaresq and Vickers clock supported by rate correction through observation of the fall of shot.[197]

Given both the inexactitude of estimating target course and speed, and the difficulties of spotting, particularly when ranges were much in excess of 10,000 yards, Callaghan's alternative amounted to little more than a counsel of despair.[198] In January 1914, he ordered the ships of his fleet to carry out fire-control exercises with a moving target at a maximum range of 16,000 yards, which was probably double that of normal battle practice.[199] In addition, there were instructions for changes in both the target course and speed, and alterations in the firing ship course, including a course reversal that would have posed the problem of a high and varying change-of-range rate.[200] Callaghan may have called for such procedures in the hope of altering the Admiralty's attitude through a demonstration of the ineffectiveness of the officially prescribed method of fire control. Regrettably, however, there is no record of the results achieved by the Home Fleet.[201]

The Admiralty, for their part, were not unaware that existing provisions for both range-finding and spotting were likely to be disrupted in a real engagement by poor visibility conditions. On 11 November 1913, and in the wake of the *Empress of India* trials, a special Admiralty conference went so far as to consider seriously both the adoption of a spotting position at the bow—where it was hoped observation would be unimpeded by smoke from guns and funnels—and what appears to have been even the resuscitation of two-observer, and thus more accurate, range-finding.[202] On the other hand, Admiralty concerns about the shortcomings of Royal Navy fire control may have been mitigated by intelligence reports that indicated that German range clocks were incapable of dealing with a high and varying change-of-range rate,[203] and that German fleet firing practice took place at 8,000 yards or less, with poor results when ranges were above 10,000 yards and the change-of-range rate was high.[204]

During World War I, the nature of fleet engagements between the British and German fleets in the North Sea differed radically from prewar expectations.[205] The intentions of the Germans to fight a close-range action had been based upon the assumption that the British would attempt to impose a close blockade with its battle fleet, and that losses inflicted on this force by fast surface torpedo craft and submarines would eliminate the Royal Navy's numerical superiority in capital ships. When this did not occur because of the British strategy of distant blockade, which was meant to prevent such a situation from arising, the German fleet sensibly attempted to avoid action.[206] On the other hand, misleading intelligence reports of the existence of German torpedoes with a range at high speed of 15,000 yards appear to have magnified British fear of the danger posed by torpedoes in a fleet action,[207] which led to instructions in the spring of 1916 that the German fleet would not be engaged at less than 14,000 yards until gunfire had decided the issue.[208]

In the largest naval battle of the war at Jutland in May 1916, the gunnery problem was made more difficult than had been the case in peacetime practices by many changes of course on both sides, and the fact that much of the fighting took place at ranges of 14,000 to 18,000 yards. When firing across the bow at retreating targets, cross-roll error compromised gun-laying and thus exacerbated the already considerable difficulties of spotting,[209] while visibility that was often poor and the extended distances made accurate range-taking with the 9-foot-base instruments that equipped most British ships virtually impossible.[210] Without frequent correction of ranges via spotting or instrumental measurement, the plotting and calculating mechanisms of the Dreyer Table could not produce accurate results when confronted by high and changing change of rates. The generally harder conditions, moreover, meant that prewar fire concentration techniques were inadequate.[211] Deployment of the director offered some compensation, but the percentage of hits to rounds fired was less than 3 percent.[212]

In 1919, a special committee of the Grand Fleet assessed the performance of the Dreyer Table. In their final report, the committee pointed out that war experience had shown that observation of the fall of shot would "always be difficult and uncertain, and frequently impossible." Instrumental observation of the target would also be intermittent because of poor visibility. Fire would be opened at the extreme range of the guns (that is, well in excess of 15,000 yards and, therefore, more or less greater than the effective range of existing torpedoes), concentration of fire was "desirable under all conditions to obtain decisive results," and both the firing ship and target would "zigzag to avoid being hit."[213] The committee also maintained that "generally speaking, our present system necessitates too many operations carried out by hand" and that therefore future gear "should aim at making as many operations as possible automatic."[214]

The plotting of ranges and bearings to obtain either separate rates or a measurement of target course and speed were both rejected as unsatisfactory. Information about target course and speed, the report stated, would have to be obtained from a combination of target bearing rate obtained by plotting, measurements of the target inclination, and comparisons of clock-generated range and bearing with observed data. The committee was unequivocally clear about the deficiencies of the late-model Dreyer Table's provision of range generation via automatic dumaresq and clock, which was regarded as inefficient and also incapable of dealing with the longer ranges and higher rate conditions expected in the future. The committee also insisted upon the automatic application of ballistical correction to generated ranges and provision for automatic correction for movement of the enemy ship during the time of flight of the projectile. Given the foregoing, the committee not surprisingly concluded that elements of the Argo clock should be incorporated into the next generation of fire-control table.[215]

The committee's findings were endorsed by Adm. Sir David Beatty, the commander-in-chief of the Grand Fleet, who added that "financial considerations should not be permitted to stand in the way" of the great difficulties that would have to be overcome to produce an efficient fire-control system.[216] Beatty's suggestion that Harold Isherwood, Pollen's chief design engineer, be consulted was in the event not only taken up by the Admiralty, but extended to include D. H. Landstad, Isherwood's assistant at the Argo Company. These two men were to play leading roles in the design of the Admiralty Fire Control Table Mark I, the precursor of a highly successful series of fire-control systems that were to equip all of the Royal Navy's major warships built during the interwar period.[217] The cost of the new system, in the end, came to some ten times the price (after adjustment for inflation) of the comparable prewar Argo instruments, and about forty times the price (after adjustment for inflation) of the late-model Dreyer Table that preceded it.[218]

Wartime experience also resulted in the further automation of director firing through the introduction of gyro-director training gear. This system kept the director sight correctly pointed when the target was not visible by controlling the director train through instrumentally generated bearing rate from the table. The system also controlled firing gyroscopically, which ensured that the guns were always discharged at the correct point in the roll. It further provided gyroscopic compensation for cross-roll and featured improved communications.[219] By the 1930s, these advances, together with the postwar tables and better range-finders finally gave British warships the means of shooting with reasonable accuracy when ranges were long and changing, sea conditions imperfect, and visibility intermittent, bringing the Royal Navy's quest for reach based on optical sensors to its culmination.[220] But within a decade, the advent of the electronic age, more effective aircraft, and the guided missile had reduced these accomplishments to something less than a matter of supreme naval importance.

Conclusions

In the 1920s, Hugh Clausen, a civilian employee of the Admiralty,[221] attempted to write an official pamphlet on the development of British naval fire control. He never completed the task. Clausen later recalled that "the more I studied past history, the murkier and more obscure became the background," and that "it seemed such a dangerous thing to write about that I gave it up."[222] Clausen had operated the Dreyer Table in the dreadnought battleship *Benbow* during World War I, and while involved with the design of the postwar generation of fire-control tables worked with the engineers who had developed the Pollen and Dreyer systems.[223] His inability to produce a history in spite of his gunnery expertise and inside knowledge may stand as testament to the inadequacy of technical and personal perspectives as a basis for understanding the complex story of pre-1914 Royal Navy fire control. On the other hand, careful consideration of the mechanical and human issues, and of tactics, finance, and intelligence assessment as well, is capable of yielding an account that is comprehensible, nonproblematic, and significant.

The Royal Navy's self-conscious search for the ability to shoot further during the early twentieth century was driven by two major imperatives. In the first place, a succession of improvements in the range of the torpedo made several extensions of gun reach essential in order to preserve the tactical viability of the surface capital ship. And in the second place, the high vulnerability to gunfire of the centralized system of fire control that was required for accurate shooting at greater distances, and Fisher's plan to replace the battleship with the less well protected battle cruiser, placed a premium on the ability to hit at greater ranges than the enemy. The demand for increases in gun range was in addition influenced by three lesser factors: the normal visibility conditions in the anticipated main theater of operations; the tactical intentions of prospective opponents, either of which could favor or disfavor the possibility of a long-distance action taking place; and the advantage held by individual British capital ships over foreign counterparts with respect to weight of fire, which if large could serve as a make-weight for a degree of weaknesses in long-range gunnery.

Supply forces were two, technology and finance, which were interrelated. In the early twentieth century, engineering had reached a general state of advancement that was sufficient to allow the invention of a fire-control system that could perform reasonably well when ranges were much greater than those thought to be the maximum practicable in the nineteenth century. Neither the basic principles of design nor the mechanical particulars of such a system, however, were obvious, and thus had to be discovered through a protracted and expensive process of experiment. The Royal Navy, moreover, had to contend with restrictions in funding that were at times severe. To a very great degree, therefore, availability of money was the critical variable regarding what could or could not be accomplished.

The initial stages of the quest for reach were for the most part shaped by the torpedo threat and financial circumstances. During the 1880s and 1890s, the short effective range of torpedoes and the primitive state of gunnery meant that there was neither a large incentive nor the capacity to hit with surface ordnance at longer distances. From 1901 to 1904, the improvement in torpedoes prompted some advance in gunnery technology and technique, but inadequate funding limited progress, and accuracy at extended ranges thus remained unsatisfactory. During the second phase from 1904 to 1907, spending on gunnery was increased substantially, which resulted in major improvements. As a consequence, the rate of hitting at greater distances, albeit under unrealistically easy conditions, was raised significantly. In addition, the loosening of purse strings, recognition that a further large increase in the range of the torpedo was probable, and Fisher's advocacy of the battle cruiser prompted the Admiralty to support the development of advanced methods of mechanized sight-setting and gun-laying.

During the third phase from 1907 to 1909, government budget cuts prompted a sharp decline in spending on gunnery. But the development of the Pollen system in particular, which was crucial to the solution of the sight-setting problem, was disrupted by three other factors that exacerbated unfavorable financial circumstances. First, the novelty of joint state and private development of a weapons system meant that the special relations that were set up between the Admiralty and Pollen in 1906 caused resentment, mistrust, and misunderstanding.[224] Second, Pollen's proposals, because they were complicated and expensive, were seen by some as both beyond comprehension and fiscally extravagant, and thus a phenomenon to be opposed. And third, the direction of service gunnery policy was weakened by the short fixed terms of department chiefs, and the lack of a naval staff system. The net result was that Admiralty support for the work of Pollen and Scott was abandoned and shifted to less expensive manual methods of gun-laying and sight-setting.

In the fourth and final phase, which lasted from 1909 to 1912, the discrediting of manual methods of fire control that had been put forward in 1908 resuscitated Admiralty interest in the development of mechanized methods of sight-setting and gun-laying, while the fundamental restructuring of government finance improved naval financial conditions. On the other hand, Fisher's retirement weakened support for the battle cruiser and associated demand for the development of the capacity to hit before being hit in return. Torpedo range increased to the point that guns outdistancing them through improved fire control seemed doubtful. There was growing certainty that the next war would be fought against Germany in the North Sea where visibility was usually much poorer than in the Mediterranean, while intelligence reports indicated that the German Navy intended to fight at close range. Finally, the adoption of heavy-caliber guns for new British capital ships that were much larger than those of their German counterparts made the achievement of early hitting seem less important.

In 1912, Scott's director was adopted because sophisticated mechanized gun-laying was believed to be essential for effective shooting in bad weather. But at the same time, the Admiralty rejected the Pollen system for Dreyer's cheaper but less capable alternative on the grounds that state-of-the-art sight-setting equipment would not be required to deliver overwhelming firepower in the first few minutes of the close-range naval battle that it believed would be fought against the Germans in the near future, and lowered the range of battle practice.[225] This ended the quest for reach that had been initiated in 1901 and would not be resumed until the war had shown that fleet engagements would most likely take place at distances that were at least twice the magnitude of those that had been anticipated in 1912, and under conditions of visibility that were far more difficult than those of peacetime gunnery practice.

The Admiralty's gunnery decisions of 1912 may have been an understandable response to intelligence assessments and financial circumstances. There can be little doubt, on the other hand, that Pollen's conception of fire control made more sense by identifying what was tactically required and exploiting what was technically feasible.

Public recognition of Pollen's achievement even after the war had vindicated his ideas, however, was necessarily resisted by the naval establishment. This was because the gunnery story properly told would almost certainly have brought the Royal Navy's prewar leadership under highly unfavorable scrutiny.[226] What was at stake was the professional reputation of a service already sensitive to criticism over unsatisfactory performances at Jutland and in the antisubmarine campaign, and faced with stiff competition from the air force and army for funding in a time of fiscal austerity. Clausen's concern about the political perils of fire control as an historical subject were thus well founded.

The Admiralty's partisans maintained that the great strides made in gunnery before 1914 were proof of the navy's openness to technical advances, and that shortcomings exposed by the war were attributable to unavoidable mistakes caused by the rapidity of change.[227] In actuality, the Royal Navy was neither simply a passive recipient of technological windfall nor a helpless victim of inevitable mechanical vicissitude. Between 1901 and 1912, the Admiralty sought actively to increase the effective range of surface ordnance in order to solve specific operational problems, and the varying fortunes of its efforts to do so were heavily influenced by considerations of finance, tactics, and intelligence. These forces did not predetermine the outcome, but, given weaknesses in Admiralty organization and certain limitations of technical and business perspective, they complicated and confused decision-making. This in turn made it possible for sound analysis with respect to the probable nature of a fleet action and remarkable mechanical ingenuity, to be overwhelmed in the end by tactical wishful thinking, overreliance on intelligence, and false economy.

The implications of such conclusions for the study of military technological change in the twentieth century are several. In the first place, they would suggest that social and cultural conservatism were not, as some have argued, the primary obstacles to advances in naval armaments.[228] Secondly, in its early stage of development, the phenomenon of command technology identified by William H. McNeill was highly susceptible to disruption for a variety of reasons, and may still well be.[229] And third, while the definition and disruption of command by nontechnological factors in many ways supports the arguments of the "social construction" school of historians of technology,[230] because the actions and interactions of many individuals were also of crucial importance, the role of contingency must be given its due.

Furthermore, the story of the Royal Navy's quest for reach in the early twentieth century is relevant to the consideration of broader topics. First, the development of sophisticated mechanized means of fire control may be seen as one of the most important early manifestations of the control information revolution. Second, the cooperative relationship of Pollen and the Admiralty can be regarded as an early attempt to create the kind of state and private partnership in defense procurement that would later become characteristic of the post–World War II military-industrial complex. And finally, from the methodological standpoint, the complexity of analytical technique and quantity of documentary data required to deal adequately with the history of British naval gunnery indicates, perhaps, that coming to terms with other cases of twentieth-century government problem-solving via technological innovation, must involve far more new thought and painstaking labor than previously has been supposed.

Notes

1. Charles H. Fairbanks, Jr., "The Origins of the *Dreadnought* Revolution: A Historiographical Essay," *International History Review* 13 (May 1991): 248 (emphasis in original).

2. Peter Padfield, *Guns at Sea* (London: Hugh Evelyn, 1973), 228–30.

3. Jon Tetsuro Sumida, ed., *The Pollen Papers: The Privately Circulated Printed Works of Arthur Hungerford Pollen 1901–1914*, Publications of the Navy Records Society, vol. 124 (London: Allen & Unwin, 1984), and idem, *In Defence of Naval Supremacy: Finance, Technology and British Naval Policy, 1889–1914* (Boston: Unwin Hyman, 1989; pa. ed., London and New York: Routledge, 1993).

4. The author owes much to the late Ben Clymer, formerly an engineer with the Ford Instrument Company; William Newell, formerly chief design engineer and later president of the Ford Instrument Company; and Stephen Skelley, currently the U.S. Navy's foremost expert on surface fire control, for their clarification of several difficult fire-control engineering problems. The author is especially indebted to Terry Lindell, an independent computer consultant and pioneer student of submarine torpedo fire control, for much technical advice and illuminating general discussion. And several historians generously shared their research findings with me; their contributions have been acknowledged in each case in later notes.

5. Padfield, *Guns at Sea,* 204.

6. For a detailed explanation of these major and other minor forms of ship motion, see S. N. Blagoveshchensky, *Theory of Ship Motions,* 2 vols.; trans. Theodor and Leonilla Strelkoff, and Louis Landweber (New York: Dover, 1962; first published in Russian, 1954).

7. Capt. Constantine Hughes-Onslow, *Fire Control,* sec. 4:2 (Portsmouth: Royal Naval War College, c. May 1909, with additions in June and August 1909), Pollen Papers [private collection, Anthony Pollen; hereafter cited as Anthony Pollen].

8. Hughes-Onslow, *Fire Control,* sec. 4:1.

9. Margins of error were expressed in terms of what was known as danger space, for an explanation fo which see Sumida, *Pollen Papers,* 360. For the purposes of this paper, the plus-or-minus margin for error was considered to be half the danger space (which assumes that the point of aim was half the 30-foot height of the vertical target offered by a ship's hull, or in other words, the middle of the danger space). For the danger space of a 12-inch gun (Mark X) at various ranges, see Great Britain, Admiralty, Gunnery Branch, *Range Tables for His Majesty's Fleet, 1906* (1906), Naval Library, Ministry of Defence, London.

10. Edwyn Gray, *The Devil's Device: Robert Whitehead and the History of the Torpedo,* rev. ed. (Annapolis: Naval Institute, 1991), 253, 259.

11. Admiralty, *First and Second Reports of the Torpedo Design Committee* (Oct. 1904), 134–35, Adm. 1/7759, Public Record Office, Kew. The author is indebted to Dr. Nicholas Lambert for bringing this document to his attention.

12. "Notes on the imperative necessity of possessing powerful fast armoured cruisers and their qualifications" (c. Feb. 1902), typescript memorandum, FISR 5/9, F.P. 4198, Fisher Papers, Roskill Archive Centre, Churchill College, Cambridge (hereafter cited as Fisher Papers).

13. Arthur J. Marder, ed., *Fear God and Dread Nought: The Correspondence of Admiral of the Fleet Lord Fisher of Kilverstone,* 3 vols. (London: Jonathan Cape, 1952–59), 1:251–53.

14. Admiralty, Intelligence Department, *Tactical Exercises, 1903 (Mediterranean, Home and Channel Fleets, and Cruiser Squadron)* (Feb. 1904), 16, 25–26, Naval Library.

15. Adm. Sir John Fisher, *Extracts from Confidential Papers: Mediterranean Fleet, 1899–1902*

(15 Oct. 1902), 3, 32, 90, FISR 8/1, F.P. 4702, Fisher Papers; memoranda and minute of 23 Mar. 1903, Director of Naval Ordnance (hereafter cited as DNO), "Revised Regulations for Heavy Gun Prize Firing" (1 Mar. 1904), Adm. 1/7756, PRO; Adm. Sir Frederic Dreyer, "A Brief History of the Development of Fire Control in the Royal Navy," typescript mimeograph copy (n.d., but post–World War I and probably for the Royal Commission on Awards to Inventors proceedings of 1925), courtesy of Anthony Pollen, and Adm. Sir R. H. Bacon, *The Life of Lord Fisher of Kilverstone*, 2 vols. (London: Hodder & Stoughton, 1929), 1:251–52. For the torpedo threat as a spur to the development of long-range gunnery at this time, see also Adm. Sir Percy Scott, *Fifty Years in the Royal Navy* (New York: George H. Doran, 1919), 162.

16. *Principal Questions Dealt with by Director of Naval Ordnance* (Dec. 1903), 145, Naval Library, and Scott, *Fifty Years*, 163. For Fisher's urging of the Admiralty to development long-range firing, see Fisher to Selborne, 9 Aug. 1901, Marder, ed., *Fear God*, 1:206.

17. Fisher, *Extracts from Confidential Papers*, 26.

18. Scott, *Fifty Years*, 163–65. In 1902, Capt. C. F. King Hall devised a point system to indicate the effectiveness of naval gunnery at various ranges. According to his scale, the results of firing at 2,000 yards were half as effective as firing at 1,000 yards, one-third as effective when firing at 3,000 yards, and one-fourth as effective when firing at 4,000 yards; see Admiralty, Intelligence Department, *Battle Fleet Actions: Notes by Captain C. F. King Hall, R.N., with Remarks by Captain H. J. May, C.B., R.N., Royal Naval College* (Oct. 1902), Adm. 1/7617, PRO. See also similar scale given in Fisher, *Extracts from Confidential Papers*, 12.

19. "Abstract of Battle Practice Results, 1904," in Rear Adm. Percy Scott, "Gunnery Lecture, No. IV" (28 Feb. 1905), in Percy Scott, *Gunnery* (June 1905), 56, WH7/65, Arnold White Papers, National Maritime Museum, Greenwich, and Capt. William S. Sims, "Report of Gunnery Information obtained during a visit to England, British Battle Practice, Fire-Control, Range-Indicators, Range- and Deflection-Transmitters, and Range- and Deflection-Receivers, etc." (June 1905), Sims Papers, Library of Congress. This document was brought to the author's attention by Christopher Havern, a graduate student at the University of Maryland, who is currently working on the gunnery question in the U.S. Navy. The hitting rate of 30 percent characteristic of the prize firings appears to have set the Royal Navy's minimum standard of gunnery achievement in the early twentieth century; see Admiralty, Gunnery Branch, *Interim Report of the Mediterranean Committee on Control of Fire, &c.* (Mar. 1904), 3, Adm. 1/7756, and idem, *Report of Committee on Control of Fire in Actions, &c.* (Apr. 1904), 7, Adm. 1/7758, PRO.

20. Peter Padfield, *Aim Straight: A Biography of Admiral Sir Percy Scott* (London: Hodder & Stoughton, 1966), 79–89.

21. The projectile of a 12-inch gun made a splash 200 feet high; see Hughes-Onslow, *Fire Control*, sec. 4:15.

22. Scott to Fisher, "Remarks on existing Prize Firing Regulations in comparison with scheme proposed by the Captain of His Majesty's Ship 'Excellent,'" 11 Nov. 1903, DNO, "Revised Regulations for Heavy Gun Firing" (1 Mar. 1904), Adm. 1/7756, PRO.

23. Percy Scott, *H.M.S. "Dreadnought"* (n.d., probably 1905), RIC 19, Richmond Papers, National Maritime Museum, Greenwich.

24. *Controller of Navy's Portion of the Vote*, projected estimates for 1900–1901 to 1906–7, Naval Library. Expenditure on telescopic sights apparently began in 1900–1901 at £5,075, and peaked in 1903–4 at £20,000, ending in 1906–7 at £8,000. See also DNO, "New Pattern Telescopic Sights for Barbette & Turret Guns" (15 Apr. 1903), Adm. 1/7686, PRO.

25. Scott, *Fifty Years*, 184–85.

26. Hughes-Onslow, "Introductory Remarks," *Fire Control*, 4.

27. Sumida, *In Defence*, 73.

28. For a diagrammatic explanation of the phenomenon of change of range, see Sumida, ed., *Pollen Papers*, 367–69.

29. British Patent 17,719/1904, and "Dumaresq Instrument Designs & Patents: Notes as to

History" (31 Jan. 1916), Adm. 1/8464/181. The latter gives only the history of the dumaresq from the date of patent. For its invention in 1901, see Dreyer, "Brief History of the Development of Fire Control."

30. Report by Lt. A. V. Vyvyan, in R. A. Burt, *British Battleships, 1889–1904* (Annapolis. Md.: Naval Institute Press, 1988), 50–51, and Scott, *H.M.S. "Dreadnought."* Vyvyan described an experimental clock device invented by Lt. Fawcet Wray that when set with a rate from the dumaresq indicated the amount of range change at 2-second intervals, which could be used to guide changes in sight-setting.

31. For the shortcomings and later improvement of voice pipes, see "Committee on the use of voice pipes for fire control purposes," report of (8 July 1910), *Important Questions Dealt with by D.N.O.; Copies, Precis, etc.,* vol. 1, *1912,* Naval Library.

32. Report by Vyvyan, Aug. 1902, quoted in Burt, *British Battleships,* 50, Fisher, *Extracts from Confidential Papers,* 15, and Rapidan (Edward Harding), *The Tactical Employment of Naval Artillery* (London: Offices of Engineering, 1903). According to Harding, the masthead spotter's observations supplemented rather than in all cases replaced those of individual gunners. For the varieties of fire control in 1903, see DNO, "Communications and Control of Gun Fire in Action; Correspondence and Reports relating to various schemes" (31 Mar. 1904), Adm. 1/7756, PRO.

33. Admiralty, Gunnery Branch, *Report of Battle Firing Carried Out by H.M. Ships "Vengeance" and "Eclipse," 1903* (Jan. 1904), 13, Adm. 1/7756, PRO, and "Fire Control and Long-Range Firing: An Essay to Define Certain Principia of Gunnery, and to Suggest Means for their Application" (Dec. 1904), in Sumida, ed., *Pollen Papers,* 32–33, and Hughes-Onslow, *Fire Control,* sec. 1:8.

34. Percy Scott claimed to have invented the salvo system in December 1903; see Scott, *Fifty Years,* 180–81. Edward Harding's important article on gunnery of 1903, however, seems to refer to the practice when he mentions "simultaneous" as distinct from "rapid," "independent," and "slow" fire; see *Tactical Employment of Naval Artillery,* 13. For its evolution in the Mediterranean Fleet between 1898 and 1904, see Bacon, *Life of Lord Fisher,* 1:251–52. It was described by Vyvyan in his August 1902 report; see Burt, *British Battleships,* 46, 50.

35. For errors of guns, see Philip R. Alger, *The Groundwork of Practical Naval Gunnery: A Study of the Principles and Practice of Exterior Ballistics, as Applied to Naval Gunnery,* 2d ed. rev. (Annapolis, Md.: United States Naval Institute, 1917), chap. 20.

36. Jellicoe Memoranda and Minute (17 Aug. 1905), DNO, "Director for Turret Firing" (12 Feb. 1907), Adm. 1/7955, PRO.

37. "Bracketting" was mentioned by Harding in 1903 (see *Tactical Employment of Naval Artillery,* 18), and was described as a technique to be used in conjunction with conventional spotting. It was also employed in an experimental firing practice on the China station in 1903; see *Report of Battle Firing,* 5. For the use of bracketing and conventional spotting together in battle practice, and the recommendation that bracketing alone was preferable when ranges were very long, see "A.C.: A Postscript" (mid-1905), in Sumida, ed., *Pollen Papers,* 61–62.

38. *Report of Battle Firing,* 8, 11, and Hughes-Onslow, *Fire Control,* sec. 1:11, 14.

39. *Tactical Employment of Naval Artillery,* 38. For Harding's authorship of this article, see Sumida, *In Defence,* 48.

40. *Tactical Employment of Naval Artillery,* 18.

41. Admiralty, Gunnery Branch, *Joint Report of the Mediterranean and Channel Committees on Methods of Controlling Gun Fire in Action* (June 1904), Adm. 1/7758, PRO, quoted in Arthur J. Marder, *The Anatomy of British Sea Power: A History of British Naval Policy in the Pre-Dreadnought Era, 1880–1905* (New York: Knopf, 1940), 522–23. See also Admiralty, Gunnery Branch, *Report of Committee on Control of Fire in Action, &c.* (Apr. 1904), Adm. 1/7758; idem, *Report of the Mediterranean Committee on suggestions made by the Vice-Admiral Commanding Channel Fleet relative to proposed Systems of Firing* (May 1904), Adm. 1/7759; idem, *Report of Committee ("H.M.S. Victorious") on Control of Fire in Action, &c.* (June 1904), Adm. 1/7758; and idem, *Joint Report of the Mediterranean and Channel Committees relative to the Systems of Firing and Allocation of Ammunition* (June 1904), Adm. 1/7759.

42. Admiralty, Gunnery Branch, *Joint Report of the Mediterranean and Channel Committees on Methods of Controlling Gun Fire in Action* (June 1904), Adm. 1/7758, PRO, quoted in Marder, *Anatomy,* 522–23. See also Admiralty, Gunnery Branch, *Report of Committee on Control of Fire in Action, &c.* (Apr. 1904), Adm. 1/7758; idem, *Report of the Mediterranean Committee on suggestions made by the Vice-Admiral Commanding Channel Fleet relative to proposed Systems of Firing* (May 1904), Adm. 1/7759; idem, *Report of Committee ("H.M.S. Victorious") on Control of Fire in Action, &c.* (June 1904), Adm. 1/7758; and idem, *Joint Report of the Mediterranean and Channel Committees relative to the Systems of Firing and Allocation of Ammunition* (June 1904), Adm. 1/7759.

43. *Interim Report of the Mediterranean Committee on Control of Fire,* 45.

44. Marder, *Anatomy,* 523–24.

45. "Control of Fire." See also Padfield, *Aim Straight,* 140. For Scott's belief that extreme long range was 7,000 to 8,000 yards, see Percy Scott, "Proposed Method of fighting the guns of the most powerful ship in the world, namely H.M.S. 'Dreadnought,'" typescript memorandum (1905), Adm. 1/7955, PRO.

46. Kerr to Selborne, 29 May 1904, Selborne 41, f. 162, and Kerr to Selborne, 1 June 1904, Selborne 41, f. 166, Selborne Papers, Bodleian Library, Oxford (notes courtesy of Dr. Nicholas Lambert).

47. "Remarks of the Captain of H.M.S. 'Excellent' on the Joint Report of the Mediterranean and Channel Committees on Fire Control," 2 July 1904, 12, Adm. 1/7758, PRO. These were also Scott's views in 1905; see Rear-Admiral Percy Scott, "Gunnery Lecture, No. IV" (28 Feb. 1905), in Scott, *Gunnery,* and also Scott, *H.M.S. "Dreadnought."*

48. "Gunnery Lecture, No. III" (Oct. 1904), in Scott, *Gunnery,* 37–39, and Council Office, "Gunnery Training at the Three Home Ports: Report on by Captain Percy Scott" (29 Nov. 1904), Adm. 1/8005, PRO. See also Padfield, *Aim Straight,* 138–39.

49. Admiralty, Gunnery Branch, *Fire Control: A Summary of the Present Position of the Subject* (Oct. 1904), Naval Library. For the authorship of the Admiralty summary by Harding, see Sumida, *In Defence,* 48–49.

50. *Fire Control: A Summary,* 8–9.

51. Scott, *Fifty Years,* 163–65.

52. *Controller of Navy's Portion of the Vote,* projected estimates for 1899–1900 to 1906–7, Naval Library.

53. Fisher to Selborne, 19 Nov. 1904, and Selborne's reply, Selborne Papers, Bodleian Library, Oxford (notes courtesy of Dr. Nicholas Lambert).

54. *Controller of Navy's Portion of the Vote.* For Admiralty parsimony as a factor influencing the failure to procure gunnery instruments, see Scott, *Fifty Years,* 164. For the Admiralty's recognition that satisfactory instruments would require large expenditure, see Sims, "Report of Gunnery Information Obtained During a Visit to England," 29. For Percy Scott's conviction of June 1905 that recent changes had cleared the way for rapid and major change, see Scott, *Gunnery,* preface. For the very high costs of electrical communication equipment, which probably accounted for most of the spending increase, see "Fire Control Instruments—Types to be adopted for future use," in Admiralty, Gunnery Branch, *Record of the Principal Questions Dealt with by Director of Naval Ordnance, January to December 1907,* 44–45, Naval Library.

55. Admiralty, Gunnery Branch, *Paper Prepared by the Director of Naval Ordnance and Torpedoes for the Information of his Successor* (hereafter cited as *DNO Paper*) (Feb. 1905), 17–19, and ibid., (July 1907), 17–18, Naval Library; and Scott, *Fifty Years,* 192. The author is indebted to Capt. G. StM. Mills, CBE, RN, for his presentation of a set of *DNO Paper* for the years 1895–1912 via Comdr. (RAN) James Goldrick to the Naval Library.

56. DNO, "Calibration of Guns: Report of Committee &c." (20 July 1905), Adm. 1/7835, PRO; Admiralty, Gunnery Branch, *Extracts from the Report of the Calibration Committee* (Sept. 1905), Naval Library; Scott, *Fifty Years,* 184–85, 192; and Padfield, *Aim Straight,* 150–51.

57. *Range Tables for His Majesty's Fleet 1906.*

58. Sir John Fisher, "The Guns of the Battleship and Armoured Cruiser" (14 May 1904), in P. K. Kemp, *The Papers of Admiral Sir John Fisher*, 2 vols (London: Navy Records Society, 1960–64), 1:32–34, and Fred T. Jane, *All the World's Fighting Ships (Naval Encyclopedia and Year Book)* (London: Sampson Low, Marston, 1904), 380–82. For a detailed examination of the relative destructive effect of big guns, intermediate-caliber guns, and medium-caliber quick-firers, see "The Effect of Shell from different Calibres of Guns on the latest type of Krupp Cemented (K.C.) Armour," in Admiralty, *The One Calibre Big Gun Armament for Ships* (June 1908), 7–9, FISR 8/31, F.P. 4881, Fisher Papers.

59. Report of Capt. William Pakenham, in Admiralty, Intelligence Department, *The Russo-Japanese War: Reports from Naval Attaches, &c., Vol. II* (July 1905), Naval Library. For the effect of Pakenham's reports on the Admiralty, see Marder, *Anatomy*, 530–32. Pakenham maintained that hitting was possible at 21,872 yards, reasonable at 15,310 yards, and decisive at 10,930 yards; see *Russo-Japanese War: Reports from Naval Attaches, &c., Vol. II*, 162, and Marder, *Anatomy*, 531. These were preposterous figures, however, given the primitive methods of fire control on both sides. For the Admiralty's considered judgment of this issue, see Admiralty, Naval Ordnance Department [Capt. (R.M.A.) Edward Harding], *The Russo-Japanese War from the Point of View of Naval Gunnery* (1906), 1, Naval War College Library, Greenwich. The author is indebted to Mr. Eric Grove for bringing this important document to his attention.

60. Marder, *Anatomy*, 532.

61. "Report No. 1 of the Committee on the Electrical Equipment of Warships" (26 July 1902), in Admiralty, *Reports of the Committee on the Electrical Equipment of H.M. Ships* (Dec. 1903), 5–6, and *DNO Paper* (Feb. 1905), 12, both in Naval Library.

62. Jellicoe, "Director for Turret Firing" (12 Feb. 1907).

63. Director firing was proposed for long-range firing as early as in 1902; see Vyvyan's report in Burt, *British Battleships*, 50–51.

64. Jellicoe, "Director for Turret Firing" (12 Feb. 1907). It needs to be remembered that director firing was proposed at a time when continuous-aim laying for big guns was still a matter of some doubt, and that therefore it was at first considered to be an alternative to the latter. But as early as in 1908, some thought appears to have been given to the possibility of combining director firing and continuous-aim; see "Director Firing: Extracts from a Long-Course Lecture by Lieutenant L. V. Wells, R.N.," in Admiralty, Gunnery Branch, *Half-Yearly Summary of Progress in Gunnery* (July 1908), 42, Naval Library. For Jellicoe's support of director-firing experiment through the summer of 1907, see *Half-Yearly Summary of Progress in Gunnery* (July 1907), 34–35.

65. Scott, *H.M.S. "Dreadnought."*

66. This was apparently a "well-known" technique; see Viscount Hythe, *The Naval Annual 1913* (Portsmouth: Griffin, 1913), 320. For the engineering theory of disc-roller integrators, see A. B. Clymer, "Mechanical Integrators" (unpublished M.S. thesis, Ohio State University, 1946), chap. 1, and U.S. Navy Department, Bureau of Ordnance, *Basic Fire Control Mechanisms* (Sept. 1944), 114–35.

67. "Remarks on Long Range Hitting" (15 Dec. 1903), in Scott, *Gunnery*, 27.

68. For the practice of setting the clock by range rates derived from the timing of rangefinder observations, and the fact that the rate markings of the dumaresq were not compatible with those of the Vickers clock, see "Remarks on Long Range Hitting," in Scott, *Gunnery*, 27, and *Half-Yearly Summary of Progress in Gunnery* (July 1906), 22–23, 39. Dumaresq's provisional patent specification of August 1904 does state that his device could be used to set machines such as the Vickers clock; see British Patent 17,719/1904.

69. Michael Moss and Iain Russell, *Range and Vision: The First Hundred Years of Barr & Stroud* (Edinburgh: Mainstream, 1988), 55. For a comparison of the accuracy of 9-foot as opposed to 4.5-foot-base Barr and Stroud range-finders from 1,000 to 10,000 yards, see *Half-Yearly Summary of Progress in Gunnery* (July 1906), 16.

70. Sumida, *In Defence*, chap. 3.

71. "Gunnery Lecture, No. IV" (28 Feb. 1905) in Scott, *Gunnery,* plates 1 and 2; Sims, "Report of Gunnery Information obtained during a visit to England," 12–13; and *Half-Yearly Summary of Progress in Gunnery* (July 1906), 22–23.

72. Admiralty, "Application of Heat to compressed air for torpedoes; Consideration of commercial offer" (13 Mar. 1903) Adm. 1/7657, PRO.

73. *First and Second Reports of the Torpedo Design Committee.*

74. Selborne to Fisher, 14 May 1904, Inclosure 1, in Kemp, *Papers of Admiral Sir John Fisher,* 1:xxi.

75. Sir John Fisher, "Types of Fighting Vessels," ibid., 1:74.

76. In the fall of 1906, the Admiralty insisted that Arthur Hungerford Pollen's proposed fire-control system perform accurately at 8,000 yards; see Sumida, *In Defence,* 99.

77. DNO, "Revised Rules for Battle Practice 1906" (29 Jan. 1906), Adm. 1/7896, PRO.

78. Admiralty, *H.M. Ships "Dreadnought" and "Invincible"* (June 1906), 9, FISR 8/8, F.P. 4718, Fisher Papers.

79. Arthur Hungerford Pollen, "The Jupiter Letters: Extracts from Letters Addressed to Various Correspondents in the Royal Navy Principally from HMS *Jupiter*" (May 1906), in Sumida, ed., *Pollen Papers,* 82, and Padfield, *Guns at Sea,* 230.

80. In 1909, the standard amount of time between rate-resetting was 1 minute, although resettings at ½- or ¼-minute-intervals were recommended "if very great accuracy" was obtainable, for which see Admiralty, Gunnery Branch, *Information regarding Fire Control, Range-Finding, and Plotting* (May 1909), 16, Naval Library.

81. This information was provided to the author by Mr. John Brooks, who is currently studying certain technical aspects of early twentieth-century British fire control.

82. For Pollen's contention that "the drive of the Vickers clock does not permit of continuous variation in rate" and that "the rate can be altered from time to time, but it cannot be subject to continuous variation," see Pollen to Peirse, 24 Nov. 1912, Pollen Papers [Anthony Pollen]. For the repetition of this argument by Wilfrid Greene before the Royal Commission on Awards to Inventors, which was not challenged by Admiralty counsel, see *Minutes of Proceedings before the Royal Commission on Awards to Inventors,* 4 (13 July 1925): 13, Pollen Papers [Anthony Pollen]. These statements by Pollen and his representative may have been made with reference to a hypothetical situation in which the disc-roller drive was made to activate a transmitter, although there is no indication that this was the case. And the disc-roller drive of the early production Vickers clock is known to have been mechanically unsatisfactory; see Admiralty, Gunnery Branch, *Fire Control* (May 1908), 5, Naval Library.

83. For a revealing indicator of the dangers of manual transfers of data involving the dumaresq and Vickers clock, see Admiralty, Gunnery Branch, *Fire Control,* 48–49; Hughes-Onslow, *Fire Control,* sec. 3:4; Admiralty, "Invention for a device for applying the Dumaresq to the Range Clock" (25 Oct. 1910), Adm. 1/8131, PRO; and Royle to captain of H.M.S. *Goliath,* 7 July 1911, quoted in report of Vice Adm. Prince Louis Alexander of Battenberg, 24 Aug. 1911, enclosure 2a, Battenberg MB1 T8/42A, Battenberg Papers, University of Southampton (courtesy of Dr. Nicholas Lambert). For the Admiralty's recognition that excessive reliance on manual functions compromised rate-finding, see "Of War and the Rate of Change" (Dec. 1910-Jan. 1911), in Sumida, ed., *Pollen Papers,* 278.

84. Net error was unpredictable because depending upon circumstances, the inaccuracies introduced from the separate sources could cancel or compound to varying degrees. In the former case, net error might be small or nonexistent, in the latter, large enough to prevent hitting from taking place.

85. *Range Tables for His Majesty's Fleet, 1906.*

86. *Tactical Exercises, 1903,* 13, 15.

87. "Remarks by the Board of Admiralty on the attached Memorandum:— ("Notes on comparative Naval Strength")" (July 1906), Adm. 116/3095, PRO. The author is indebted to Dr. Nicholas Lambert for bringing this document to his attention.

88. *Tactical Employment of Naval Artillery,* 18, and Hughes-Onslow, *Fire Control,* sec. 1:14.

89. Harding's name is not given on the document, but the argument and style are unmistakably the same as those of *Tactical Employment of Naval Artillery* and *Fire Control: A Summary.*

90. For the nature of "residual error," see Alger, *Groundwork of Practical Naval Gunnery,* 233–34.

91. *Russo-Japanese War from the Point of View of Naval Gunnery* (1906), 17–18.

92. Sumida, *In Defence,* chap. 3.

93. "Fire Control and Long Range Firing," in Sumida, ed., *Pollen Papers,* 40–41.

94. "A.C.: A Postscript," in Sumida, ed., *Pollen Papers,* 59–61.

95. "The *Jupiter* Letters," in Sumida, ed., *Pollen Papers.* For Pollen's assumption in 1904 that half the danger space was the largest permissible error, see "Fire Control and Long Range Firing," ibid., 34. For Pollen's less exacting standard of accuracy by 1912, which was probably attributable to recognition that salvo spreads of 200 to 300 yards at ranges of 10,000 yards allowed a somewhat greater degree of permissible error, see Boys to Pollen, Nov. 1912, Pollen Papers [Anthony Pollen].

96. This attitude appears to have been based on experience with a clock device that had been invented by Lt. W. W. Fisher; see *Half-Yearly Summary of Progress in Gunnery* (July 1906), 28. The Ordnance Department also favored the division of functions between two machines on the grounds that the breakdown of one would still allow the other to be used to provide some assistance to sight-setting, and because it would allow the clock to be removed from the spotting top and placed in the transmitting station, which could be protected by armor.

97. Scott to Pollen, 31 July 1906, Pollen Papers [Anthony Pollen], and Sumida, *In Defence,* 90.

98. Ibid., chaps. 2–3.

99. Ibid., 99.

100. *Controller of Navy's Portion of the Vote,* 1905–6.

101. Ruddock F. Mackay, *Fisher of Kilverstone* (Oxford: Clarendon, 1973), 387–91.

102. "Notes, Etc., On the *Ariadne* Trials" (Apr. 1909), in Sumida, ed., *Pollen Papers,* 223.

103. *Controller of Navy's Portion of the Vote.*

104. For Jellicoe's belief, in particular, in the importance of being able to hit before the enemy, see Jellicoe, "Considerations of the Design of a Battleship: Paper Prepared by the Director of Naval Ordance," May 1906, in *H.M. Ships "Dreadnought" and "Invincible,"* 17.

105. For reloading times of 30 seconds or less during some of the firings of *Dreadnought's* 12-inch guns on her experimental cruise, see Capt. Reginald Bacon, *Report on Experimental Cruise* (16 Mar. 1907), 97–100, Adm. 116/1059, PRO, and Adm. Sir Frederic C. Dreyer, *The Sea Heritage: A Study of Maritime Warfare* (London: Museum Press, 1955), 57. The average rate of fire of 12-inch guns prior to this time was every 48 seconds; see Fred T. Jane, ed., *Fighting Ships 1905/6* (London: Sampson Low Marston, 1905; reprint, New York: Arco, 1970), 34.

106. Bacon, *Report on Experimental Cruise,* 28, 83–84, 95, 97–100. For Bacon's distrust of all instruments—including range-finders—as expressed to Pollen in December 1908, see Sumida, *In Defence,* 151–52.

107. For faulty spotting as the main cause of missing in battle practice in 1905, see "Remarks by Lieutenant F. C. Dreyer, R.N., on 'Calibration' and 'Battle Practice,' carried out by the Channel Fleet in 1905," *Half-Yearly Summary of Progress in Gunnery* (July 1906), 36–37. For the new rules, see Admiralty, Gunnery Branch, *Fire Control,* 47–48; *One Calibre Big Gun Armament for Ships,* 18; and Hughes-Onslow, *Fire Control,* sec. 1:11. They may be explained as follows. In 1908, 12-inch guns laid to fire to 8,000 yards produced salvo spreads of 224 yards, and half of the constituent projectiles could be counted upon to fall within a quarter of this distance, which was known as the 50-percent zone. With four-gun salvos, therefore, keeping two shots short (overs could not be seen because the target blocked the view) would mean that there was a high likelihood that the third shot would fall within the 144-yard danger space of the target. For Dreyer's contribution to the development of the new rules, see Dreyer vita (n.d., probably 1907), DRYR 1/2, Dreyer Papers, Roskill Archive Centre.

108. *Half-Yearly Summary of Progress in Gunnery* (Jan. 1907), 7–11.

109. The effectiveness of new controls for train was demonstrated in an experimental shoot in the pre-dreadnought battleship *Britannia* in December 1906; see "Experimental Battle Practice carried out by H.M.S. 'Britannia,'" *Principal Questions Dealt with by Director of Naval Ordnance* (1906), 38–39, Naval Library. See also *DNO Paper* (July 1907), 15, and Admiralty, *Letter from Rear-Admiral Sir John Jellicoe (Second-in-Command of Atlantic Fleet, and recently Director of Naval Ordnance) to Sir John Fisher* (Nov. 1907), FISR 8/27, F.P. 4849, Fisher Papers.

110. For the refitting of older mounts, see *DNO Paper* (Jan. 1910), 7–8, and ibid. (June 1912), 6–7. For the defects of *Invincible*'s electric mountings, see John A. Roberts, *Invincible Class* (Greenwich: Conway Maritime Press, 1972), 10. For the mechanical nature of the changes in elevation and training controls, see Admiralty, Gunnery Branch, *Manual of Hydraulics, 1906 and its Several Addenda, Addenda* (1913), 9–19, Naval Library.

111. *One Calibre Big Gun Armament for Ships*, 10–17.

112. For Bacon's recommendation that spotting for rate be tried in battle practice, see "Memorandum by Director of Naval Ordnance on Towing Targets," Admiralty, Navy Estimates Committee, *Report upon Navy Estimates for 1908–9* (Feb. 1908), 114, FISR 8/11, F.P. 4724, Fisher Papers. For experimental attempts to spot for rate prior to 1908, "Remarks by Commander F. C. Dreyer, R.N. on the question of how to best obtain and maintain the gun range in action," see typescript memorandum, 22 July 1910, Adm. 1/8417, PRO. Bacon's thinking may also have been affected by the fact that sight-setting had been made more exact by advances in the application of ballistic corrections to ranges and bearings; see *Russo-Japanese War from the Point of View of Naval Gunnery,* 8, 16; Hughes-Onslow, *Fire Control,* sec. 2:17–20; and Dreyer, "Brief History of the Development of Fire Control."

113. Admiralty, Gunnery Branch, *Concentration of Fire Experiment: Extracts from Reports of the Firing at H.M.S. "Landrail," carried out at Portland, 4th October 1906* (Jan. 1907), and reference to the 1907 firings at the target ship *Hero* in Admiralty, Gunnery Branch, *Report on Firings at H.M.S. "Empress of India," carried out by First Fleet on the 4th November 1913* (Sept. 1914), 2, both in Naval Library.

114. For firing at stationary targets in battle practice through 1907, see Hughes-Onslow, *Fire Control,* sec. 2:17, and "Memorandum by Director of Naval Ordnance on Towing Targets," *Report upon Navy Estimates, 1908–9,* 114.

115. Sumida, *In Defence,* chap. 4.

116. DNO, "Obry (Petravic) gun Firing Apparatus, Trial of & report on" (17 Sept. 1908), Adm. 1/8011, PRO. The Petravic system, according to J. B. Henderson, an Admiralty engineer, may at this time have been unworkable; see *Consideration of the services of Prof. J. B. Henderson in respect of his inventions relating to gyro-compasses and director gear* (Sept. 1916), Adm. 1/8590/111, PRO, and also Dreyer, *Sea Heritage,* 85. On the other hand, the Admiralty's decision not to adopt the gear may have had something to do with the secrecy conditions insisted upon by the inventor; see Hugh Clausen, "Invention and the Navy, The Progress from Ideas to Ironmongery," a lecture given at the Royal Society of Arts to the Institute of Patentees and Inventors (30 Jan. 1970), courtesy of Anthony Pollen. For the deployment of a later variant of the Petravic System in the German navy during World War I, see N. J. M. Campbell, *Battlecruisers: The Design and Development of British and German Battlecruisers of the First World War Era* (Greenwich: Conway Maritime Press, 1978), 49, 57. For the contemplation of gyroscopic stabilization of guns by Henderson and Pollen in 1907, see J. B. Henderson, "Gyroscopic Gun Sights" (16 May 1907) and Jellicoe minute, 30 May 1907, Adm. 1/7936, PRO, and Hughes-Onslow, *Fire Control,* sec. 4:20.

117. Scott, *Fifty Years,* 201.

118. For Bacon's proposals of late 1907 to increase gun caliber from 12 to 13.5 inches, see Sumida, *In Defence,* 161. Bacon may also have believed that the flatter trajectory of the heavier gun would increase allowable margins of error for sight-setting. This appears to have been demonstrated in the *Britannia* trials of December 1906, with which Bacon would undoubtedly

have been familiar; see "Experimental Battle Practice carried out by H.M.S. 'Britannia,'" *Principal Questions* (1906), 38. For Dreyer as the original instigator of the increase in gun caliber, see Dreyer, *Sea Heritage,* 59–60.

119. *One Calibre Big Gun Armament for Ships,* 47. See also Admiralty, *A Discussion of the Relative Merits of the 13.5-inch and the 12-inch Gun as the Armament for Battleships* (Mar. 1908), DRYR 2/1, Dreyer Papers. For Julian Corbett and Frederic Dreyer as the authors of these two papers, see Jon Sumida, "The Historian as Contemporary Analyst: Sir Julian Corbett and Admiral Sir John Fisher," in James Goldrick and John B. Hattendorf, eds., *Mahan is Not Enough: The Proceedings of a Conference on the Works of Sir Julian Corbett and Admiral Sir Herbert Richmond* (Newport: Naval War College Press, 1993), 132–34.

120. For Pollen's discussions with Thomas Evans Crease, Fisher's naval secretary, see Pollen to Craig-Waller, 8 June, 1923, Pollen Papers [Anthony Pollen].

121. Sumida, *In Defence,* chap. 5.

122. Harding to Usborne, 12 Jan. 1913, 4, Pollen Papers [Anthony Pollen].

123. Sumida, *In Defence,* 208.

124. *Half-Yearly Summary of Progress in Gunnery* (Jan. 1907), 48.

125. *Information regarding Fire Control,* 30.

126. Peirse to Moore, 18 May 1910, Admiralty, "Gunnery: Effects on (a) plotting for range, (b) rate of change of bearing & range, (c) possibility of concentration of fire &c. of new developments in Fleet Tactics," Adm. 1/8051, PRO.

127. "Of War and the Rate of Change," in Sumida, ed., *Pollen Papers,* 280.

128. Sumida, *In Defence,* chap. 5.

129. Ibid., chaps. 5–6.

130. Ibid., 204–5. For the mechanical shortcomings of the Argo clock Mark I, which appear to have been relatively minor and therefore correctable, see "The A. C. Range and Bearing Clock, Mark II" (8 May 1911), Pollen Papers [Anthony Pollen].

131. For the memorandum that probably informed Moore's opinion, see May to Secretary of Admiralty, 25 Apr. 1910, "Gunnery," Adm. 1/8051. For the criticism of May's remarks, see Peirse to Moore, 18 May 1910, ibid. In general, testimony by various officers involved in the trials was conflicting. For the complete account of the tactical exercises, see Admiralty, *Notes on Tactical Exercises: Home Fleet, 1909–1911* (Sept. 1911), Naval Library.

132. Peirse to Moore, 18 May 1910, "Gunnery," Adm. 1/8051.

133. "The Quest of a Rate Finder," in Sumida, ed., *Pollen Papers,* 262.

134. "Correction of the Rate by the Fall of Shot," typescript memorandum, comp. Staff of the Inspector of Target Practice, *Battle Practice, 1909,* Lecture II: "Details relative to Transmitting Stations and Plotting Arrangements," 12, Pollen Papers [Anthony Pollen], and *Information Regarding Fire Control,* 12–13. For experimental efforts to spot for rate prior to 1908, see "Remarks by Commander F. C. Dreyer, R.N., on the question of how to best obtain and maintain the gun range in action."

135. "Fire Control and Long-Range Firing," in Sumida, ed., *Pollen Papers,* 32–33, 50.

136. Hughes-Onslow, *Fire Control,* sec. 1:6, and Admiralty, Gunnery Branch, *Spotting Rules* (Nov. 1916), 6. See also Padfield, *Guns at Sea,* 232.

137. Hughes-Onslow, *Fire Control,* sec. 4:5–7; and "Extract from Report on Gunnery Exercises," *Important Questions Dealt with by D.N.O.,* 1:74.

138. Hughes-Onslow, *Fire Control,* sec. 4:7–8.

139. Staff of the Inspector of Target Practice, *Battle Practice, 1909,* Lecture I, 8, Pollen Papers [Anthony Pollen].

140. Alger, *Groundwork of Practical Naval Gunnery,* 233–34.

141. Lt. E. Altham, *Notes on Director Firing,* 19 June 1913, Admiralty, "Notes on Director Firing" (19 June 1913), 2, Adm. 1/8330, PRO. See also "Final Report of the Committee on Director Firing" (15 Nov. 1912), MB1/T22/161, Battenberg Papers, Southampton University. The author is indebted to Dr. Nicholas Lambert for bringing this document to his attention.

142. "Quest of a Rate Finder" (Nov. 1910), in Sumida, ed., *Pollen Papers*, 262, Hughes-Onslow, *Fire Control*, sec. 1:14, and Pollen to Scott, 3 June 1912, 2, Pollen Papers [Anthony Pollen].

143. "Aim Corrector Trials" (31 May 1910), DRAX 3/3, Drax Papers, Roskill Archive Center.

144. "Quest of a Rate Finder," in Sumida, ed., *Pollen Papers*, 260–62. For finding the rate through the use of a "tuning clock" and a description of the procedures used to correct rate error through spotting, see "Details of System of Control" (4 Mar. 1911), David T. Graham-Brown Papers, MS. 18, Historical Library, *H.M.S. Excellent*, Portsmouth. The author is indebted to Dr. Nicholas Lambert for bringing this document to his attention.

145. Pollen to Scott, 3 June 1912, Pollen Papers [Anthony Pollen], and "Final Report of the Committee on Director Firing," 12.

146. Sumida, *In Defence*, 283.

147. Admiralty, "Gunnery in the Royal Navy: Conference at Admiralty Dec. 1911/Jan. 1912; Report and Action" (13 Feb. 1913), Adm. 1/8328, PRO.

148. Sumida, *In Defence*, 188–89.

149. Ibid., chap. 6.

150. Ibid., 208–11. In addition, corrections of the clock bearing based on observation of the target, in conjunction with alterations of the clock range and bearing rates suggested by observed ranges and bearings, would automatically correct the course and speed settings, and thus improve the accuracy of subsequent calculation; see Admiralty, Gunnery Branch, *The Argo Range and Bearing Clock, Mark IV* (Feb. 1914), 3, Naval Library. This is not described accurately in the text of Sumida, *In Defence*, 211, but evident in the schematic diagram on p. 214.

151. Experiments in 1908 had confirmed earlier fears that exposed fire-control communications were likely to be disrupted at the beginning of an action; see Admiralty, Gunnery Branch, *Fire Control*, 9. For the shifting of instruments below, see *Information Regarding Fire Control*, 2–3. For the placement of range-finders behind armor on turrets and in a revolving position above the conning tower, and the belief that this reduced or even eliminated the need for mast-top fire-control positions, see "Mastless Ships from a Gunnery Point of View" (1908) in Ship's Cover 243 (*Neptune*), National Maritime Museum, Greenwich, and Oscar Parkes, *British Battleships: A History of Design, Construction and Armament*, new and rev. ed. (London: Seeley Service, 1957; 4th impression, 1971), 533, 538. For the Admiralty's confidence in the survivability of centralized fire control by 1913, see Admiralty Permanent Secretary to H.M. ships, 14 Feb. 1913, *Important Questions Dealt with by D.N.O.*, vol. 2, *1913*.

152. Pollen draft letter to Bacon, 16 Dec. 1908, quoted in Sumida, *In Defence*, 151–52.

153. Hughes-Onslow, *Fire Control*, sec. 3:1. Adm. Sir Charles Beresford maintained that visibility would most likely be no more than 4,000 yards and certainly not 7,000 yards; see *Report and Proceedings of a Sub-Committee of the Committee of Imperial Defence appointed to Inquire into Certain Questions of Naval Policy raised by Lord Charles Beresford* (12 Aug. 12, 1909), printed for the C.I.D. in Oct. 1909, 19, Cab 16/9A, PRO.

154. May to Secretary of the Admiralty, 25 Apr. 1910, "Gunnery," Adm. 1/8051.

155. Hughes-Onslow, *Fire Control*, sec. 3:5.

156. Parkes, *British Battleships*, 507.

157. Report from Berlin to McKenna, "Fundamental Basis of Naval Warfare Adopted by the German Navy" (29 Mar. 1909), MCKN 3/19, McKenna Papers, Roskill Archive Centre.

158. Unknown British naval officer to Pollen, Oct. 1910, "The Quest for a Rate Finder," in Sumida, ed., *Pollen Papers*, 262.

159. Henry G. Thursfield, *Development of Tactics in the Grand Fleet: Three Lectures*, delivered 2, 3, and 7, Feb. 1922, Thursfield Papers, THU 107, National Maritime Museum, Greenwich. For the British prewar expectation that battle ranges would be less than 10,000 yards, see also Jellicoe to Jackson, 28 May 1916, in A. Temple Patterson, ed., *The Jellicoe Papers: Selections from the Private and Official Correspondence of Admiral of the Fleet Earl Jellicoe of Scapa*, 2

vols. (London: Navy Records Society, 1966–69), 1:244, and Winston S. Churchill, *The World Crisis, 1916–1918*, 2 vols. (New York: Scribner's, 1927), 1:131.

160. Hughes-Onslow, *Fire Control* (June 1909), sec. 4:7.

161. The Argo clocks Marks I and II had been designed with mechanical provision for the application of time-of-flight and other ballistic corrections. They were not provided for the production Argo clock Mark IV or the Dreyer Table. For the reliance on spotting correction as a means of compensating for ballistical error by 1914, see Admiralty, Gunnery Branch, *Manual of Gunnery (Volume III) for His Majesty's Fleet, 1915*, 11–12, Naval Library.

162. "The Quest for a Rate Finder," in Sumida, ed., *Pollen Papers*, 279.

163. May to Secretary of the Admiralty, 25 Apr. 1910, "Gunnery," Adm. 1/8051. It is worth pointing out that in an exercise where two fleets met head on, which would have produced a constant but very high change-of-range rate, fire was not considered to be practicable at more than 5,000 yards; see *Notes on Tactical Exercises, Home Fleet, 1909–1911*, 108. See also "Of War and the Rate of Change," in Sumida, ed., *Pollen Papers*, 282.

164. Dreyer to Hughes-Onslow, 18 Oct. 1908, in Sumida, *In Defence*, 151.

165. For Pollen's earlier exploration of this approach, and claims regarding it, see Pollen to Craig-Waller, 8 June 1923, Pollen Papers [Anthony Pollen].

166. For Dreyer's view of his work in this light, see Dreyer letter to unknown naval officer, 14 June 1913, DRYR 4/3, Dreyer Papers. For Moore's treatment of Dreyer's invention as a weapon to be used against Pollen, see Sumida, *In Defence*, 218.

167. For the importance to the Admiralty of automatic transmission of the range directly from the clock, see Inspector of Target Practice to Director of Naval Ordnance, 26 Apr. 1910, DNO, "Fire Control Instrument devised by Asst. Paymaster C. R. F. Noyes R.N., Reports, &c.," Adm. 1/8145, PRO.

168. By 1909, one battle cruiser and two dreadnought battleships had been equipped with follow-the-pointer sights; see *DNO Paper* (Jan. 1910), 14, 17–18. By 1912, all dreadnoughts and later pre-dreadnoughts were so equipped; see ibid. (June 1912), 14. For the shortcomings of manual sight-setting, see A. V. Vyvyan, "Electrically Controlled Sights" (1904), DNO, "Vyvyan-Newitt, Siemens & Vickers Electrically Controlled Gunsights" (2 Mar. 1905), Adm. 1/7832, PRO, and Staff of Inspector of Target Practice, *Battle Practice, 1909*, Lecture I, 8. For the follow-the-pointer system, see Admiralty, "Gun Sighting Apparatus, Improvements in" (16 Mar. 1909), Adm. 1/8046, PRO.

169. *DNO Paper* (June 1912), 12, and *Controller of Navy's Portion of the Vote, 1914–15*.

170. Harding manuscript note (n.d., probably 1925) on Hughes-Onslow, *Fire Control*, Pollen Papers [Anthony Pollen]. Harding noted that Jellicoe had rejected the idea on the grounds that it would have made it easier for the Pollen gear to meet the previously established complete success conditions in the forthcoming trials.

171. For a revealing description of the range-finder error problem when one operator made a succession of observations, see "Proof of Evidence of Vice-Admiral Arthur Craig-Waller, C.B." (n.d., probably June 1923); see Pollen to Craig-Waller, 8 June 1923, Pollen Papers [Anthony Pollen].

172. Sumida, *In Defence*, 249–50.

173. For the reduction of battle practice ranges in 1912, see Battle Practice Chart, H.M.S. *Orion* (1912), Craig-Waller Papers [Michael Craig Waller], and "Final Report of the Committee on Director Firing," 12. During the *Orion* shoot, which took place in May 1912, the speed of the target was 8 knots and that of the firing ship 15, and firing took place at between 8,000 and 7,000 yards at a rate of about 100 yards per minute. Jellicoe maintained that the longest prewar practice range was 9,500 yards; see Adm. Viscount Jellicoe, *The Grand Fleet 1914–1916: Its Creation, Development and Work* (New York: Doran, 1919), 38. Jellicoe may have been referring, however, not to regular battle practice but to the experimental firings at *Empress of India*, which took place at ranges that were as high as 9,800 yards, but involved a stationary target; see *Report on Firings at H.M.S. "Empress of India,"* 11, 13. For the regarding of 9,800 yards as very long range even in late 1913, see Callaghan to DNO, enclosure 1, 8 Dec. 1913, DNO, "Gunnery

Practice at Sea: Sinking of 'HMS Empress of India' (4/11/13)" (8 Dec. 1913), Adm. 1/8346, PRO.

174. In 1908, independent fire had been considered to be appropriate "under certain circumstances" that were not explained; see Admiralty, Gunnery Branch, *Fire Control*, 56. By 1913, with the adoption of range-finder control, independent fire appears to have been viewed as the preferred option if conditions of wind, wave, and visibility allowed continuous observation and accurate continuous-aim gun-laying; see *Report on Firings at H.M.S. "Empress of India,"* 12.

175. Typescript memorandum, "Class I Ships" (1912), giving battle practice scores for the top twenty-two ships, in Craig-Waller papers [Michael Craig Waller]. These scores were interpreted on the assumption that the number of rounds fired corresponded to standard ammunition allowances for battle practice (four rounds per gun); see "Revised Rules for Battle Practice 1906," and Admiralty Permanent Secretary to various fleets (6 Mar. 1913), "Enclosure III: Allocation of the Annual Allowance of Ammunition of the Preliminary Armament of Battleships, Battle Cruisers and Cruisers, to the Various Practices," *Important Questions Dealt with by D.N.O.*, vol. 2.

176. Sumida, *In Defence*, 283.

177. For projectile weights of British and German naval guns, see Surgeon-Lieutenant Oscar Parkes and Maurice Prendergast, eds., *Jane's Fighting Ships 1919* (London: Sampson Low, Marston, 1919; reprint, New York: Arco, 1969).

178. For the effect of the adoption of director gun-laying on spotting, see Altham, *Notes on Director Firing*, esp. 3.

179. Hughes-Onslow, *Fire Control*, sec. 4:16–17; *Battle Practice, 1909*, Lecture I: "General Lessons to be drawn from the Year's Experience at Sea"; and references to the experiences of 1909 and 1910 in "Extract from Report on Gunnery Exercises etc., H.M.S. 'Defence', for the year ending 13th February 1911," *Important Questions Dealt with by D.N.O.*, 1:74.

180. For the lack of support for director firing in 1907–8 and Jellicoe's intervention in 1908, see Scott, *Fifty Years*, 201, 238–39. For the extent of the Admiralty's director development effort by the end of 1909, see *DNO Paper* (Jan. 1910), 17–18.

181. Sumida, *In Defence*, 207, 229, and also Inspector of Target Practice, "Report on Director Firing Trial of H.M.S. 'Neptune' carried out 11th March 1911 at Tetuan" (20 Mar. 1911), in Ships Cover 348 (*Lion/Princess Royal*), f. 143, National Maritime Museum, Greenwich, "Final Report on Director Firing." For the possibility that directors were adopted in part to compensate for the unsatisfactory rolling characteristics of the first 13.5-inch gun-armed battleships, see Jon Sumida, "British Naval Administration and Policy in the Age of Fisher," *Journal of Military History* 54 (Jan. 1990): 15–16.

182. For concentration of fire and necessity of using salvos, see Altham, *Notes on Director Firing*, 3. For the opinion that the adoption of the director would facilitate concentration of fire, see "Firings Carried out at 'Empress of India'," *Important Questions Dealt with by D.N.O.*, vol. 2, and *Report on Firings at H.M.S. "Empress of India,"* 6. For the concentration of fire experiments of 1909 that confirmed the results of the earlier trials of 1906 and 1907 previously described, see Admiralty, Gunnery Branch, *Fleet Fire Control and Concentration of Fire Experiments: Reports from the Atlantic, Home, and Mediterranean Fleets on Preliminary Practices carried out during the Year 1909, with Sub-calibre and Aiming-rifle Ammunition* (Apr. 1910), Naval Library. For Jellicoe's intention in 1914 to use concentration firing in pairs by the four leading ships of the Grand Fleet on the two leading German ships in 1914, see *Extracts from Grand Fleet Battle Orders* (18 Aug. 1914), in Patterson, ed., *Jellicoe Papers*, 1:56.

183. *DNO Paper* (Jan. 1910), 23; Hughes-Onslow, *Fire Control*, sec. 3:3–4; and Admiralty, *Addenda (1911) to Torpedo Manual, Vol. III., 1909* (Feb. 1912), 9, Naval Library.

184. *DNO Paper* (Jan. 1910), 23; DNO, "Torpedo Battle Practice—Introduction of Regulations & Confidential Instructions governing the practice for 1910" (17 Mar. 1910), Adm. 1/8147, PRO; *DNO Paper* (June 1912), mimeograph addendum; typescript copy of memorandum with regard to the torpedo threat in general and angled fire in particular, First Sea Lord to First Lord, 25 Nov. 1912, *Important Questions Dealt with by D.N.O.*, vol. 2; Naval Torpedo School

[H.M.S. *Vernon*], *Thirty-Third Annual Report on the Instruction and Practice of Torpedo Warfare in H.M. Navy, 1912* (21 Feb. 1913), 64; Admiralty, Gunnery Branch, *Pamphlet on Torpedo Control* (Sept. 1917), 7, plate II; all (except where noted) in Naval Library. For the 10,000-yard-plus range of the 19.7-inch caliber G/7 torpedos that equipped most German all-big-gun capital ships during World War I, see *Jane's Fighting Ships 1919*, 514, and Eberhard Rössler, *Die Torpedos der deutschen U-Boote: Entwicklung, Herstellung und Eigenschaften der deutschen Marine-Torpedos* (Herford: Koehlers, 1984), 42–43.

185. Currey to Bacon, 14 Dec. 1908, and Bacon minute, 17 Dec. 1908, in Ship's Cover 224 (*Australia* and *New Zealand*), National Maritime Museum, Greenwich. The author is indebted to Dr. Nicholas Lambert for bringing this document to his attention.

186. Rear Adm. R. H. S. Bacon, "The Battleship of the Future," *Transactions of the Institution of Naval Architects* 52 (1910). In this piece, Bacon estimated that battle would take place at 6,000 yeards. For Pollen's reply to Bacon, and general assessment of the torpedo threat, see "The Gun in Battle," in Sumida, ed., *Pollen Papers*, 302–4.

187. Sumida, *In Defence*, 253–54.

188. Ibid., 235.

189. Ibid., 251.

190. "Outfits of Projectiles" (18 Nov. 1908), Ship's Cover 243 (*Neptune*), and "Extracts from a Long-Course Lecture, by Lieutenant A. Domvile, R.N., on the Latest Types of Projectiles, Their Uses in Action, and Effect," *Half-Yearly Summary of Progress in Gunnery* (Jan. 1909), 82–83, and Admiralty, Technical History Section, *Ammunition for Naval Guns* (May 1920), 39–40, Naval Library. For a general discussion of the ballistical issues, see Peter Hodges, *The Big Gun: Battleship Main Armament 1860–1945* (Annapolis: Naval Institute Press, 1981), 32.

191. Admiralty, Technical History Section, *Fire Control in H.M. Ships*. Series: *The Technical History and Index: A Serial History of Technical Problems dealt with by Admiralty Departments*, pt. 23 (Dec. 1919), 10, Naval Library.

192. Controller of Navy's Portion of the Vote, projected estimates for 1911–12 to 1913–14.

193. Sumida, *In Defence*, 193–94, app., table 1.

194. Ibid., 251–52.

195. *Report on Firings at H.M.S. "Empress of India."*

196. "Gunnery Practice at Sea: Sinking of 'HMS Empress of India.'"

197. *Home Fleets General Orders, No. 13—Fire Control Arrangements*, and *No. 14—Fire Control Organization* (25 Oct. 1913); and "Enclosure No. I to Home Fleets General Order No. 14 of 5th November 1913," DRAX 1/9, Drax Papers. For what appears to be the existence of a strong element of doubt about the efficacy of range-finder control in the Home Fleet, and the fact that it was practiced for the most part in only the Second Battle Squadron, see memorandum, 15 May 1914, "Practices of Ships Fitted with Director Firing," *Important Questions Dealt with by D.N.O., vol. 3, 1914.* For the fact that only the dreadnoughts of the Second Battle Squadron were equipped with the numbers of range-finders required to practice range-finder control successfully, see "Colossus Report on Rangefinders" (20 May 1914), ibid.

198. For warnings about the likelihood and danger of inappropriate application of rate correction, see *Enclosure No. II to Home Fleets General Order No. 14 of 5th November 1913*, 2, DRAX 1/9, Drax Papers, and *Spotting Rules*, 13. For the necessity of spotting for rate correction by default because of the belief that plotting could not produce a viable rate estimate unless range-finding conditions were good, in which case rate control would not be necessary, see *Manual of Gunnery (Volume III.) for His Majesty's Fleet, 1915*, 6. For inconsistent gun-laying as the cause of wide variation in the size of salvo spreads in 1913, and the negative effect of this on spotting, see *Report on Firings at H.M.S. "Empress of India,"* 12. For the problem of applying a correction for rate after a turn by the firing ship or target, see Admiralty, Gunnery Branch, *Reports of the Grand Fleet Dreyer Table Committee, 1918–1919* (Sept. 1919), 8, Adm. 186/241, PRO.

199. *Report on Firings at H.M.S. "Empress of India,"* 11, 13.

200. *Home Fleets General Order, No. 16—Rangefinder, Fire Control, and Torpedo Control Exercises* (26 Jan. 1914), superseding General Order No. 16 of 9 Nov. 1912, DRAX 1/9, Drax Papers. These exercises did not involve shooting, and thus did not require Admiralty preapproval because no additional ammunition expenses would be incurred. Callaghan had been reprimanded by the Admiralty earlier in the year for sanctioning the high-speed experiments of the Battle Cruiser Squadron, which had involved firing; see Sumida, *In Defence,* 252. For the possibility that Callaghan's actions were inspired by Pollen, see Pollen to Peirse, 23 July 1912, Pollen Papers [Anthony Pollen], and compare diagrams and discussion in "Gun in Battle," in Sumida, ed., *Pollen Papers,* 320–22, with Callaghan's fire-control exercises.

201. Record keeping for such exercises was optional, and there was no requirement for any kept to be forwarded to higher authorities; see *Home Fleets General Order, No. 16* (26 Jan. 1914).

202. "Report of Conference on Provision of a Spotting Position right forward for existing & new ships," 11 Nov. 1913, *Important Questions Dealt with by D.N.O,* vol. 2. For the practicability of rangefinding at long ranges as demonstrated in tests carried out by the Barr and Stroud Company in May 1914, and the skepticism of the Admiralty with regards to these findings, see "Colossus Report on Rangefinders."

203. "Intelligence, 1914," 155, Adm. 137/1013, PRO.

204. Admiralty, War Staff, Intelligence Division, *Germany: Results of Firing Practices, 1912–13* (Dec. 1914) Naval Library. This document was printed for restricted circulation in December 1914, but could have been in British hands before the outbreak of war. For other possibly prewar intelligence reports of German firing practice from 1910 to 1913 at 8,000 yards or less, see idem, *German Navy; Part IV: Sections 1, 2, and 4. Gunnery Information; Section 4— Target Practice, Range-Finders, and Control of Fire* (July 1917), C.B. 1182A, Naval Library. For the correctness of the British assessment of prewar German intentions, see Comdr. Georg von Hase, *Kiel and Jutland* (New York: Dutton, 1922), 153. For the deficiencies of German fire-control equipment, which German officers attributed to their prewar tactical assumptions, see Siegfried Breyer, *Battleships and Battle Cruisers 1905–1970,* trans. Alfred Kurti (New York: Doubleday, 1973), 68.

205. For the explicit concession of this point by an official British source, see *Fire Control in H.M. Ships,* 21.

206. For German prewar tactical expectations, see Paul G. Halpern, *A Naval History of World War I* (Annapolis: Naval Institute, 1994), 22–23, 27, 29, 39, 47, 288.

207. For the German H/8 torpedo, which was as capable as British intelligence reported but in the spring of 1916 mounted in only one ship (*Bayern*), see *Jane's Fighting Ships, 1919,* 514, and Rössler, *Die Torpedos der deutschen U-Boote,* 43–44.

208. For the British intention in 1914 to fight at ranges of between 9,000 to 12,000 yards, evidence that this was still the case as late as December 1915, and change in instructions of 1916 because of the torpedo threat, see "Extracts from Grand Fleet Battle Orders," 18 Aug. 1914; Jellicoe to Beatty, 22 Dec. 1915, and Jellicoe to Jackson, 28 May 1916, in Patterson, ed., *Jellicoe Papers,* 1:62, 190, 244. For Jellicoe's fear of the torpedo in 1914, see Jellicoe to Secretary of the Admiralty, 30 Oct. 1914, and Jellicoe to Churchill, 14 July 1914, in Patterson, ed., *Jellicoe Papers,* 1:38–39, 75–77, and Jellicoe, *Grand Fleet,* 50. For Jellicoe's mental exhaustion by the spring of 1916, which may have increased his susceptibility to panic over the report of an improved German torpedo, see Anthony Pollen, *The Great Gunnery Scandal: The Mystery of Jutland* (London: Collins, 1980), 150–51.

209. For "cross-roll" error in World War I, see Clausen to Anthony Pollen, 6 Dec. 1969, courtesy of Anthony Pollen. For the difficulty in particular of laying the 15-inch guns of the battleship *Barham* correctly for line at the battle of Jutland when firing ahead, which would have made spotting for range much more difficult and which was probably attributable at least in part to cross-roll error, see Sumida, *In Defence,* 323n82.

210. For the uselessness of range-finders during the battle cruiser phases of the engagement, see Sumida, *In Defence*, 306.

211. *Fire Control in H.M. Ships*, 23.

212. John Campbell, *Jutland: An Analysis of the Fighting* (Annapolis: Naval Institute Press, 1986), 353.

213. *Reports of the Grand Fleet Dreyer Table Committee*, 14.

214. Ibid., 18.

215. Ibid., 14–15, 18. See also Admiralty, Naval Staff, *Extracts of Gunnery Practices in Grand Fleet, 1914 to 1918: Battleships and Battle Cruisers* (Mar. 1922), and *Grand Fleet Gunnery and Torpedo Memoranda on Naval Actions, 1914–1918* (Apr. 1922), Naval Library.

216. *Reports of the Grand Fleet Dreyer Table Committee*, 18.

217. Sumida, *In Defence*, 315.

218. Ibid., 253, 316.

219. "Low-Angle Control," in John Campbell, *Naval Weapons of World War Two* (Annapolis: Naval Institute Press, 1985), 8–11.

220. Jon Tetsuro Sumida, "The Best Laid Plans: The Development of British Battle Fleet Tactics, 1919–1942," *International History Review* 14 (Nov. 1992). The Royal Navy did not, however, develop a fully automated system of gun-laying as did the United States; see Norman Friedman, *U.S. Navy Weapons: Every Gun, Missile, Mine and Torpedo Used by the U.S. Navy from 1883 to the Present Day* (Annapolis: Naval Institute Press, n.d.), 35–36.

221. Hugh Clausen (1888–1972), chief technical adviser to the DNO during World War II, was employed at the Admiralty from 1920 to 1953; see Clausen to Anthony Pollen, 11 Nov. 1969, courtesy of Anthony Pollen.

222. Clausen to Anthony Pollen, 5 Apr. 1970, courtesy of Anthony Pollen.

223. Clausen to Anthony Pollen, 19 Jan. 1970, and Clausen, "Invention and the Navy," courtesy of Anthony Pollen.

224. For a perceptive examination of this issue on a broader scale, see Gary E. Weir, *Building the Kaiser's Navy: The Imperial Naval Office and German Industry in the von Tirpitz Era, 1890–1919* (Annapolis: Naval Institute Press, 1992).

225. Royal Navy thinking about the nature of battleship fighting in 1914 thus seems to have more closely resembled the delivery of the single large pulse of firepower characteristic of carrier warfare, than the "simultaneous erosive attrition" model often associated with battleships; see Wayne P. Hughes, *Fleet Tactics: Theory and Practice* (Annapolis: Naval Institute Press, 1986), 93.

226. See, for example, Arthur Hungerford Pollen, "Jutland and the 'Unforeseen,'" *Nineteenth Century* 99 (Aug. 1927).

227. Jellicoe, *Grand Fleet*, 66–69; Dreyer, *Sea Heritage*, 85–87; and Dreyer, "Brief History of the Development of Fire Control," which, incidentally, was submitted to Clausen by Dreyer; see Clausen to Anthony Pollen, 5 Apr. 1970. For the presentation of the inevitability of error argument in the standard history of the period, see Arthur J. Marder, *From the Dreadnought to Scapa Flow*, 5 vols. (London: Oxford University Press, 1961–70), 5:313.

228. Elting E. Morison, *Men, Machines, and Modern Times* (Cambridge: M.I.T. Press, 1966), and most recently Robert L. O'Connell, *Sacred Vessels: The Cult of the Battleship and the Rise of the U.S. Navy* (Boulder: Westview, 1991).

229. William H. McNeill, *The Pursuit of Power: Technology, Armed Force, and Society since A.D. 1000* (Chicago: University of Chicago Press, 1982).

230. Wiebe E. Bijker, Thomas P. Hughes, and Trevor Pinch, eds., *The Social Construction of Technological Systems* (Cambridge: M.I.T. Press, 1987).

When Technology and Tactics Fail: Gallipoli 1915

Timothy H. E. Travers

In early 1915 the British and French governments attempted to knock the Turks out of the war by sending a fleet to run the Dardanelles Straits, enter the Sea of Marmora, and bombard Constantinople until the Turkish government fled or surrendered. However, the combined British and French fleet failed to achieve this on 18 March 1915 and in previous attempts. Subsequently, a Mediterranean Expeditionary Force (MEF) of French, British, and Anzacs (Australian and New Zealand Army Corps) was put together to land at Gallipoli, and by controlling the peninsula, enable the fleet safely to pass the straits and bombard Constantinople. The landings took place on 25 April at Helles in the south, and Anzac Cove on the west. The landings succeeded, but the land campaign eventually failed, despite another landing and series of offensives at Suvla Bay and Anzac Beach in early August. The failure of the land campaign has generally been ascribed to lack of preparations, shortage of munitions and men, lack of organization, the strength of the Turks, and the mistakes of various individuals, such as Field Marshal Lord Horatio Kitchener (secretary of state for war), Gen. Sir Ian Hamilton (commander in chief, MEF), Lt. Gen. Sir Aylmee Hunter-Weston (general officer commanding [GOC] 29 Division and then VIII Corps), Lt. Gen. Sir Frederick Stopford (GOC IX Corps), Maj. Gen. Alexander Godley (GOC New Zealand and Australian Division), and so on. However, a study of the capabilities of the technology used at Gallipoli has not received much attention. For example, it is necessary to ask whether the artillery and naval gunnery at Gallipoli was capable of fulfilling its role, because by 1915 artillery was recognized as the key weapon of the war. In fact, it will be the central argument of this essay that a critical factor contributing to the Gallipoli defeat was the basic failure of the artillery, including naval gunnery. In particular it was the technical inability of the artillery (and naval gunnery) to achieve desired ends that created problems, as also occurred on the Western Front in 1915.[1]

It is true that Gallipoli faced shortages in artillery and shells, which contributed to failure, yet this may not have been as critical a problem as the technical shortcomings of artillery and the inexperience of artillery commanders and men. Many commentators have stressed the material shortages at Gallipoli, and clearly Gallipoli was a poor cousin to the Western Front. Thus a postwar report by Brig. Gen. Simpson Baikie, brigadier general artillery (BGRA) to 29 Division in Gallipoli, castigated the War Office Ordnance Department for artillery and shell shortages. Fresh from six months on the Western Front, Baikie arrived at Helles in early May 1915, subsequent to the landings of 25 April. He was surprised to find that no one was paying attention to the consumption of artillery shells; that no arrangements for the supply of shell from England had been made; and that there was a serious deficiency in high-explosive shell (HE), there being little for the howitzers, and none at all for the 18-pounders. This situation did not really improve at Helles for the whole campaign, although at Suvla, in August 1915, greater supplies were available, even if not enough space or gun emplacements could be found. Due to the shortages at Helles, no howitzer was able to fire without Baikie's personal order, and howitzers were kept for infantry attacks in June, July, and August. Eighteen-pounder shells at Helles were also in short supply, all of them shrapnel, except for 640 HE rounds that were used for the attack of 4 June. So acute was the shortage of 18-pounder shell that by 13 July there were only five thousand rounds left at Helles. To counteract this, Baikie withdrew ammunition from all 18-pounders except for a couple of rounds, and placed the bulk of the shells in wagons, ready to move to any danger spot. To add to the problem, when three thousand 18-pounder shells did arrive by trawler on the night of 13 July, they were found to have fuses of a new pattern, but no fuse keys. It took a couple of days for the batteries themselves to manufacture the new keys.[2]

Simpson Baikie went on to detail other artillery difficulties at Helles. There were no spare parts for guns; the four 60-pounders broke down frequently; the eight mountain guns at Helles were antiquated, inaccurate, and only fired 10-pound shells; the 15-pounders were obsolete and inaccurate, having been used in the Boer War, and the bores soon wore out; the sixteen 5-inch howitzers were also obsolete—the gun history sheets of some showed they had been used at Omdurman against the Dervishes in 1898; there were not enough planes or observers to spot for the artillery; and there were only a dozen trench mortars at Helles, which were useless, the only effective tubes being six Japanese trench mortars at Anzac. Some of these problems at Helles were alleviated by borrowing French guns and shells, or using French artillery for offensives, yet this was obviously not a satisfactory solution.[3] The result of these shell shortages was that when offensives were carried out, the preliminary bombardment was considered to be weak in comparison with the Western Front. Hence, while at Neuve Chapelle the preliminary bombardment employed one artillery piece for every four to six yards of enemy front bombarded, on 4 June at Helles, the ratio was one artillery piece for every forty-five yards of enemy front. However, despite all these shortages, it is the case that on 4 June at Helles (Third Battle of Krithia) there were just 92 Turkish guns in defense, while there were 102 Allied guns in attack, plus the guns of the fleet for this offensive. It is true that more howitzers would have helped, but it seems it was not really lack of artillery and shells that caused this offensive ultimately to fail (12,240 British artillery rounds were expended in approximately two hours, quite apart from French and naval shelling), but the inability to apply this firepower accurately and at the appropriate times.[4]

Thus, despite all the difficulties that Simpson Baikie lays out, the suspicion remains that it was not so much the shortages of artillery or shells as the inaccuracy of the artillery, and the poor tactics due to inexperience of artillery-infantry cooperation, which really prevented success at Helles. It was always easier to blame the technology rather than the use made of that technology. It will be useful, therefore, to take a more detailed look at the cooperation of artillery with the infantry, air and navy, including the cooperation of field and heavy artillery; followed by a brief evaluation of the artillery in action, with reference to the results of defensive fire, wire cutting, preliminary bombardment, and counterbattery work. The conclusion will look at the options available at Gallipoli, and evaluate the role of the artillery as a piece of technology.

Looking first, therefore, at artillery-infantry cooperation, reveals that ignorance of modern warfare initially marked the relations of the infantry and the artillery at Gallipoli. At Anzac, Brig. Gen. Charles Cunliffe-Owen of the Royal Artillery frequently made angry entries in his diary about the attitude of the infantry. Hence on 7 May 1915 he wrote:

> The ignorant comments of infantry officers were annoying. They never realized that the field gun trajectory is flat, making it impossible to clear the crest at short ranges, and be able to shell hostile trenches in close proximity to our own. . . . They also think that it is an easy matter to immediately knock out or silence hostile guns. Such necessities as being able to locate hostile guns when well hidden, observation of your own fire, and communications, do not occur to them.

Later in July, Cunliffe-Owen waded in against the Marines: "No Marine officer or man knew anything whatever about shooting on land, and were most unreliable." Then after the disappointments of the early August attacks, the artillery at Anzac was criticized by Major General Godley, commanding the Anzac attack, and Cunliffe-Owen noted: "Godley great nuisance, and most ungrateful unnecessary remarks after all the help given and continual work by artillery."[5]

Cunliffe-Owen therefore set out to educate the infantry. In a letter to the GOC 1st Australian Division, Cunliffe-Owen asked infantry officers not to send alarmist messages to individual artillery units asking for artillery fire. In one incident on 30 May, while supporting an attack from Quinn's Post, "messages were received from all directions saying, for instance, 'Turn all guns on to—.' This sort of message means nothing." wrote Cunliffe-Owen, "The Artillery is most carefully arranged . . . and it is impossible for shell fire to be observed from more than two directions on the same objective. Artillery fire without observation is useless and a waste of ammunition. There is an Artillery officer accessible to every unit who is the officer best able to judge." Cunliffe-Owen then declared that:

> In many cases, when Artillery officers asked to be allowed to slacken their fire, they were requested to 'fire heavily.' Random requests from any part of the line means useless waste of ammunition if complied with, and are due to orders of projected movements in other parts of the line not being known. No ammunition supply can possibly compete with the conditions that have existed yesterday.

Then on 10 June, in a hand-written signal to the 3rd Australian Brigade, Cunliffe-Owen continued his educational efforts:

> I again wish to point out that odd snipers . . . are not an artillery target . . . a request

was made to Capt. Kenyon commanding 2 mountain guns to fire on Turkish snipers which could be plainly seen. . . . As no infantry would fire Capt. Kenyon reached for a rifle & after he had fired the sniping stopped.[6]

Cunliffe-Owen was not the only gunner to notice the lack of infantry appreciation for artillery difficulties, since Col. John Hobbs, Commander Royal Artillery (CRA), 1 Australian Division, tangled with Maj. Gen. William Bridges, GOC 1 Australian Division. As the commander of an infantry division (although once an artillery officer), Bridges believed the artillery should be dedicated to close infantry support, if necessary right in the infantry line, and ordered Hobbs to do this in early May at Anzac. Hobbs was reluctant but complied, at great risk to the guns. It was not a success. But here was a conflict between the concept of artillery as really a trench weapon (Bridges) and artillery in a more independent role (Hobbs). This conflict was actually built into warfare at the time due to the technical nature of the artillery and the variable needs of the infantry. In this case, the conflict became somewhat petty: for example, a senior staff officer told Hobbs to raise the platform of Number 3 gun, while Bridges told him to lower it; or when Hobbs received orders to move his HQ from the Anzac beach to be near the divisional HQ, and then Bridges told Hobbs's brigade major "rather rudely . . . to stay where we were." This reception was typical of divisional HQ wrote Hobbs, who treat my suggestions "with scant courtesy," but later act on them. All of this resulted in an outburst in Hobbs's diary, in late May, when he wrote:

> It grieves me to say that ever since I took up this appointment cold water and little help has been given. Scant praise was ever given but any little fault was always pointed out. At the start of the training at MENA [the Australian training camp in Egypt] I know no confidence was placed in my ability to train the Divisional Artillery. Anything I pointed out was coldly received but in the end generally acted on.

Hobbs concluded that this was because "I am not a professional soldier [he was an architect] so I suppose I am not supposed to know anything . . . compared with the professional soldier." Hobbs may have had problems with the abrasive Bridges, but the root cause of the difficulty was not so much his amateur status as the conceptual gap between artillery and infantry.[7]

This conceptual distance relates to prewar training and ideas, and it is noticeable that during the training at Mena in Egypt before the April landings, 1st Australian Division's schemes tended either to separate out the artillery and infantry elements of the scheme, or to be rather simplistic in regard to artillery. For example, in one scheme of defense, of Brown country versus Grey country, it was simply said that "Grey have all kinds of artillery made by Krupp." Cooperation of artillery and infantry in another scheme consisted of reports of what the artillery were doing rather than real integration. As for artillery training itself, this understandably involved the engineering side of artillery work, practice shoots separate from the infantry, and the old artillery idea of speed in getting the first shot off, anticipating mobile warfare. As one Australian artilleryman wrote home in January 1915: "our time for getting the first shot off was very quick." All of this really relates to simple inexperience, so that the report of another scheme had to explain that infantry commanders could not order artillery brigades and batteries to do something, rather the CRA was the one with command discretion. Yet another scheme of a brigade attack made no mention of the artillery. And in February 1915, battery commanders did not know how to fire off the map. On the other hand, lectures were given by officers with experience on the Western Front, such as Cunliffe-Owen, so there were attempts to be up-to-date. However, it is not surprising that it took the experience of actual warfare to work out the necessity of cooperation, rather than the lip-service cooperation often displayed in training.[8]

At Anzac the story of the artillery was really initial disorganization in which the navy guns, the artillery of each division, the New Zealand heavy artillery, and the mountain guns, all operated separately, with separate HQs and observers. In contrast, at Helles, due to the structure of the landing forces, the artillery was centralized under 29 Division. At Anzac these diverse arrangements resulted in awkward communication and technical problems, especially because the steep cliffs and hills required guns to fire not at their front, but across from the flanks at their neighboring front, and all guns needed forward observation officers (FOOs) with telephone lines. This was a nightmare of cross communications and awkward relationships between units. For example, in early May, the commanding officer of 7th Indian Mountain Artillery Brigade wrote most indignantly to Cunliffe-Owen, stressing his casualties and lack of replacements, the fact that his shrapnel actually had to be boiled in order to prevent it clumping, and the ineffectiveness of his 10-pounder gun and common shell ammunition. Another example occurred when the Turks attacked in force during the night of 18–19 May, and Hobbs reported in his diary what he did not put in his official report:

> At 10:30 Brig. Gen. RA enquired whether 6" Howitzer had fired on the target previously mentioned. The OC replied he had not yet done so having had difficulty in fixing his communication and arranging observation. Fire being necessary he was ordered to put in 3 rounds on previous days target . . . and get communication through PHILLIPS Battery with Major Burgess who was observing for the [Turkish] BRIGHTON Gun and could also observe for him. At 10:55 1st Inf. Bde. reported through Capt. Hoskyns that no fire had yet been directed . . . as requested from NZ No 2 Battery. This was communicated to CRA NZ and 2nd Art. Bde at 11 am.

This communication and observation problem with the heavy New Zealand artillery then became an incident when

> Later it was found that OC NZ No 2 Battery was not sure that he was firing on [Turkish guns] ... and Second Art. Bde. was requested to send an officer to observe 20 rounds which would be fired by NZ No 1 Battery with a view of locating the position [of enemy guns]. To this the OC Second Art. Bde. replied that his FOOs had just returned from the trenches for breakfast. The OC Div. Art. sent instructions that an officer was to carry out the order. A further reply was received from Second Art. Bde. that all his officers were then engaged in the trenches and could not be spared.

It seems that breakfast had triumphed over the needs of the moment! However, Hobbs continued:

> By this time the attack had been repulsed but it was desirable to still bring fire to bear. . . . A further order was sent from OC Div. Art. to OC Second Art. Bde. to send an officer to OC NZ No 1 Battery to point out the target. Although this was done the NZ BC [Battery Commander] was still unable to bring fire to bear on the target. Later in the day Capt. Bidale was dispatched to point out target and see fire registered on it.

Presumably this was eventually done, but Hobbs does not elaborate, except to note that enemy fire cut phone lines so that "communications with observing officers was exceedingly difficult."[9]

Despite this strange episode, efforts were steadily made at Anzac to introduce centralization and stronger cooperation with infantry into the artillery, commencing with Force Order #8 on 30 May 1915. This was a cautious document, with the artillery still making the final decision when requests were made for its use. This was partly institutional conflict, but partly because of organizational and technical problems. Nevertheless, it is very noticeable that from early June an artillery system and structure begins to emerge. For example, an Anzac Corps Signal Memo of 9 June 1915 was issued, detailing how lines were to be laid and maintained, plus cautions such as "A frequent cause of breakdown is that linemen in repairing lines, cut pieces out of other lines."! Then, entries in the diary of the commander of 3rd Australian Artillery Brigade, Col. Charles Rosenthal, show that: on 2 June, all brigade phone lines go to a new HQ station; on 3 June, shoots are now recorded on the divisional artillery map; on 5 June, all signals to brigades and batteries are to be duplicated; on 7 June, the Brigade Forward Observation Station was established, with observers and telephones, and close infantry communication, and all batteries were now to be directed from this station, while the Naval FOO would also observe from the same station. Then on 20 June, Lt. Gen. Sir William Birdwood congratulated Rosenthal as having the most up-to-date Brigade HQ; on 22 June, all brigade phones and lines went underground; on 25 June, HE was used for the first time; on 27 June, new howitzers were registered for the first time; on 15 July, a new map for the area was issued, copied from the Turks, one copy for HQ and each battery; on 19 July, another map of Gallipoli was issued; on 26 July, the 4.7-inch naval gun went into action for the first time, and could reach Turkish long range guns; on 2 August, the first issue of new HE shell took place, 150 rounds per battery; on 15 August, new permanent phone lines were laid; and so on. It is obvious that the artillery were being drawn, perhaps unwillingly, into much closer cooperation with the infantry, although this took some time, and probably relates to the fact that after August the infantry was on the defensive, and there was more artillery available. Thus a series of memos were eventually issued, with an emphasis on artillery cooperation with infantry, both in Helles and Anzac. As an example, on 1 November, Cunliffe-Owen remarks that with the Anzac Division CRA "A scheme was arranged so that each section of infantry line could call on at least 1 battery at once, day or night to shoot in front of their line. Direct wires were laid.

Additional artillery support was given through BGRA of division." The stress in all these memos is on quick and speedy artillery response to the requests of the infantry.[10]

So the evolution of artillery-infantry cooperation, at Anzac, was that of initial disorganization, in April and May, then attempts by the artillery to centralize and organize itself in June, July, and August. Following the August offensives, the artillery was required to respond more quickly to the needs of the infantry, especially by late 1915. Of course, the artillery recognized it was there to serve the infantry, but the gap between ideal and reality needed closing. However, in a number of respects, the artillery was actually technically incapable of responding to the needs of the infantry. This was especially the case with counterbattery work, which it was realized, depended on air observation, since FOOs proved to be limited in their ability to locate enemy guns and spot for fall of shell. The alliance between artillery and air observation, therefore, emerged due to sheer necessity. A typical cry of frustration comes from Colonel Hobbs (CRA 1 Australian Division) at Anzac on 13 May 1915:

> Daily I realize how difficult it is to do what is expected of us. . . . We are dominated by the enemy guns and observation stations from [Hill] 971 on our left and are subject to heavy shell fire at once. Our guns get badly knocked about and many casualties. It is impossible for us to discover enemy batteries—which are hidden in Nullahs and behind ridges which our guns without being able to manoeuvre and with flat trajectories cannot reach.

Hobbs knew what the solution was—air observation. But at Anzac the air was initially under the control of the navy, and the shortage of planes meant that for the first weeks there was no air observation at all. Consequently, Hobbs was frustrated. So on 30 April he asked for help "as the Navy give us no information—& we cannot locate enemy guns." The same situation obtained on 17 May, when Hobbs noted: "Asked as I have done before Gen. Cunliffe-Owen, we cannot get aeroplane observation from the Navy. . . . He says the Navy can't spare them but a plane is coming to [do] reconnaissance for the Howitzer." On 17 June, the refrain was still the same "if we only had assistance from aeroplanes & a little help from the Navy we could do more." As late as 14 October, Hobbs was still complaining:

> The air observation is still most unsatisfactory—notwithstanding that we are now supposed to have plenty of aircraft and arrangements are daily made for spotting—appointments are invariably broken lately from some cause or another—their reports are not always reliable, one gave us enemy guns within 50 yards of our trenches, another a battery of our own guns in action.[11]

In fact, the Anzac BGRA, Cunliffe-Owen, knew of the problem, but could not do anything about it. To put the matter bluntly, the artillery had no experience with wireless transmission (W/T); the air observers had no practice in artillery observation, so that naval and artillery officers had to be hurriedly trained (and the navy was not keen to release officers for this job). Furthermore, air strip signalling, for observers to read while aloft, was in its infancy, and shore W/T stations could not signal to planes. Finally, there were very few planes available, only two squadrons for all of

Gallipoli, including the navy, until September, no planes at Anzac until late May, only five planes at Helles, of which only two were W/T planes, whereas two planes were required daily for shoots because, for geographical reasons, Helles had no forward observation positions. Even at the end of October, there were still only twenty-five pilots and forty-four planes available for the whole of Gallipoli. Other counterbattery problems were the unreliability of maps, so that enemy guns could not be fixed solely by map references; enemy guns were well hidden in scrub; the artillery did not realize that planes took thirty minutes to get to 5,000 feet; the observers' inexperience of reporting fire, so that "Over" and "Short," "Left" and "Right" was used until the Clock code came in after July (here they were several months behind usage on the Western Front); not until Suvla in August 1915 were good land-based receivers available (before this to avoid mutual W/T interference, a signal went out "aeroplane up," and after the shoot "aeroplane down"); also the land W/T sets could only receive over a range of 5,000 yards, so if the plane was spotting further away, it had to come back to within 5,000 yards to send its message. However, one small triumph occurred when the Turks attempted to jam the wireless transmissions on 3 May, allegedly using words such as "Schweinhund" and other compliments, but they were ignored, so the Turks thought the jamming was not working, and temporarily gave up.[12]

Clearly, there were many technical problems that prevented air-artillery cooperation in counterbattery work, and many were the complaints of the artillery. The diary of Col. C. G. Miles, CO Heavy Battery, 2nd Australian Division, gives examples of some useful shoots, but more often the results were as follows: "25 July, put in 1st round at 1920 hours, 'Range correct Right 30 yards' from plane, did so, but then the plane just said it was returning; 12 August, 8 rounds spotted onto map location, but then the plane said 'Pits are empty, am spotting on camp,' and elevated by 150 yards, 'so it is not known what the result was' (it turns out the camp was empty too); 16 August, a shoot at map reference, but no observation received, and plane went home with engine trouble; 18 August, plane did not arrive for shoot"; and so on. From the air observer's point of view, in this case Capt. A. H. K. Jopp (an artillery officer turned observer), the most common problems were that the howitzers fired extremely slowly, that there were breakdowns of the plane engine and the wireless, and that it was very difficult to say, even when the fire was accurate, whether the enemy gun or target was actually destroyed. Judging by an analysis of Captain Jopp's counterbattery shoots, about 40 percent of those that took place were accurate, but a very much smaller percentage would result in actual gun destruction. Curiously, 9th Australian battery commander, Burgess, reported that he did three counterbattery aeroplane shoots at Anzac, but there was obviously no W/T direct communication from the planes, which only sent in observations after the shoot was over. Burgess simply noted "These were found a useful guide."[13]

The air proved very useful for reconnaissance, photographs, and the occasional dropping of bombs, but counterbattery was not a success due to technical difficulties and shortages of planes and pilots. Moreover, according to Wing Commander

Charles Samson (3 Squadron, RNAS), there was no system for allocating planes between navy and land forces: "Each of course naturally thought his own requirements were the most important. I would receive orders to spot for the ships, and then urgent demands for increased cooperation with the Army." Samson would then have a chat with the commanders involved and try to sort out priorities—a method that seems open to the influence of rank. The air service did improve after Col. Frederick Sykes's report of 9 July, which emphasized greater centralization and more planes. But even after the arrival of No. 2 Wing in late August, Cunliffe-Owen was probably still correct when he concluded about counterbattery at Helles that "the aeroplane service was never very good." There were not enough observers, "and frequently when observation shoots were being carried out, the good results were thrown away by the ignorance of the Naval observers in spotting 'properly' for land guns, or suddenly going away with the series unfinished." However, this was not the end of the air-counterbattery difficulties, since, for example, enemy guns could not often be located even from the air, unless they were actually firing, so that a few rounds often had to be fired first in order to get the Turkish artillery to respond. But on many occasions the Turkish batteries either never fired, or ceased fire too quickly to be observed. Then again, accuracy was the key to counterbattery work, yet range tables, that gave absolutely essential details of charges, ranges, elevations and angles of descent, were apparently not sent to the howitzers from the War Office until 6 August 1915. Consequently, much testing had to be done by the heavy guns to evaluate the ballistic properties of the different charges. Then, as previously mentioned, maps were a problem, not just for accuracy, but for agreement on the name of the actual objects on the map. So, for example, in September it was pointed out that Battleship Hill, marked thus by Anzac staff on the new 1/20,000 map, was not known by that name to artillery officers, who called it Hill 700, though there was also the hill BABY 700. Finally, artillery survey was on a learning curve at Gallipoli, as on the Western Front, despite the statement of the postwar Mitchell Report on Gallipoli, which claimed that survey

> was entirely ignored so far as indirect fire was concerned, not only in Gallipoli but on the Belgian coast and in Palestine, until it was proved by surveyors to be essential to accurate fire. It was often stated that the error of the gun was so great that accurate maps or charts were unnecessary.

This was too severe a criticism, but once again illustrates the technical obstacles to achieving results in Gallipoli.[14]

Even within the artillery, there were technical difficulties, especially between field artillery and heavy batteries. An interesting series of messages in mid-July shows the system was limping along at this stage. At 7:37 a.m. on 12 July, the 2nd Australian Field Artillery Brigade (FAB) asked whether the New Zealand 6-inch howitzer would be ready to fire immediately after the field artillery had fired, with the latter trying to provoke the Turkish guns into opening fire, in this case on Mortar Ridge. The reply was "Yes," but at 8:55 a.m. the 2nd FAB FOO reported he could not find the New Zealand FOO to inform him the Turkish Mortar Ridge guns were now firing.

Then at 9:13 another 2nd FAB request for the 6-inch howitzer came in, to silence enemy guns, this time at Turkish Hump. But at 9:35 a.m. the 2nd FAB reported the 6-inch howitzer had not yet fired. Again at 9:47 the 2nd FAB was still asking for howitzer fire, and a note on the message records "Pace [CO 6-inch howitzer gun] did not comply." The next day, a letter from Lt. Col. G. J. Johnston, 2nd FAB CO, complained of slow responses from the howitzers, and that the FOO for the howitzers and 8th Field Battery were using the same observation station, which was impossible. Moreover, the Turks had the initiative and were heavily shelling Anzac trenches, so that a more energetic policy was needed. Meanwhile, the 8th Battery commander complained that a heavy battery was using his observation station and interfering with his fire—let them build their own station, he concluded. Then, also on 13 July, the CRA of 1 Division argued that the New Zealand howitzers were failing to silence the Mortar Ridge guns, while the New Zealand FOO defended himself by saying that his observation station had been abolished, and so he had to communicate through the infantry signal service. In this situation, it is noteworthy that in mid-July, four months after the Gallipoli campaign started, the artillery system was evidently not working very well at Anzac. It was only with the arrival of several heavy batteries in August and September, especially the new 6-inch howitzers on 21 September, that matters improved in accuracy and cooperation.[15]

Turning to naval gunnery and naval cooperation with the infantry, the story becomes a mixture of tragedy and comedy. Comments by the infantry and artillery about each other were often very negative. Hence, after a storm in late June had driven ashore some ammunition boats, Hobbs noted morosely in his diary:

> My opinion of our British Navy in many respects has been considerably discounted. . . . Amateur yachtsmen could handle boat traffic better—& some of their gunnery is rotten. They killed & wounded some of our men on the 28th [June] with their fire from a destroyer trying to hit enemy's trenches.

Another Australian officer, on the staff of 1 Division, remarked in his diary on 29 May:

> From the moment the *Prince of Wales* opened up on Gaba Tepe's guns, on the first Sunday morning, right through the succession of weeks when they opened up on the Olive Grove battery and the Anafarta villages etc., they have seemed to us not able to hit what they aimed at. As the General says, when they hit what they aim at in target practices in peace time, they put it in despatches! "A sea fight can't be the terrible thing we are led to believe," he says. "I can't imagine any boat being so starkly accurate in her gunnery as to hit another boat!"

At Suvla many similar complaints of inaccuracy emerged, for example Brig. Gen. C. de Winton, concerning 14 August: "The ships are now firing heavy shell but do not seem able to find the enemy guns and have no aeroplanes to assist them." Or Brig. Gen. G. Downing, who wrote of 7 August that the navy shells "were dropping dangerously short." And Maj. Gen. Sir Sydenham Smith, who, after reading drafts of the official history, remarked simply that "My own job during the period under review [Suvla in August] was to direct the fire of the supporting ships and my experience in trying to adapt naval fire control and fire discipline to the requirements of the support of infantry on land would make this literature still more harrowing." Meanwhile, the Turks themselves said of naval gunnery during the April landings that "the effect of the British naval artillery was moral without being material."[16]

To be fair to the navy, there were a number of severe technical problems, the most immediate being the inaccuracy of the maps. As Cunliffe-Owen, who initially directed naval fire in support of the Anzac landing from HMS *Queen,* pointed out, the maps, prepared from the 1840's version

> did not agree with the ground, and what was most important for the ships' fire, the compass bearing was two degrees out [meaning an error of 400 yards], so that if a message came saying "Fire in 243 D 4" the ships made most accurate calculations, and the result was that they were 2 degrees off. Owing to the long range and flat trajectory of the ships' guns, it would be dangerous to shoot close to our own troops, and they mostly fired AP [armor piercing] shells.

At an early stage at Anzac, Cunliffe-Owen admitted this inaccuracy: "It must be understood that Naval guns cannot fire on enemy's guns close to our line. It is too dangerous, because all fire from ships is by compass, as the map is not accurate." Moreover, "locating our line in such precipitous country was practically impossible from the sea." Later, Cunliffe-Owen noted that the code most often used by Anzac HQ was "C.O.," which meant "Don't shoot so close to our troops." Consequently, on 25 April when the Helles landings were going in, the air located Turkish guns close to the beach, but the navy took no notice and continued firing too far inland. Wing Commander Samson concluded: "No doubt they did this because they were afraid of hitting our own people." Again, when HMS *Canopus* fired at the infantry objective in support of an attack on 29 April, she had to cease fire at 7:15 p.m., though the attack was not due to start until 7:30. This 15 minutes was simply the margin of error deemed necessary at the time. Again, to be fair to the navy, targets were often very vague, so that for the attack on 19 May at Anzac, ships were asked to fire on "971 Spur," but this could be three miles long, and refer to various spurs. It was not until late May that an air photo map was first available.[17]

Apart from maps and observation problems, at Anzac the navy frequently did not know where the infantry were, partly because communication difficulties were acute. Initially, messages had to go from FOOs to beach by only two flank lines, so observers had to get to these lines, then transmit to the beach, and from the beach to ships. Meanwhile Cunliffe-Owen was on HMS *Queen,* and all signals from *Queen* to other ships were "by flag signal, which takes some time to get to 7 ships." The reverse direction had to follow the same complicated procedure.[18] Not surprisingly, the early messages and gunnery were confused, thus at 9:25 a.m. on 26 April, 3rd AIF Brigade thanked *Queen* for her fire on 224 D "which is doing great execution," but forty-five minutes later at 10:18, 2nd AIF Brigade complained that the navy was shelling its trenches. Then in the afternoon of the same day, *Queen* to Base on 26 April at 2:30 p.m. wirelessed "help me by spotting." An hour later, at 3:37 p.m., the message was the same, *Queen* to Base, "Help me by spotting my fire." On the message is written "Done," and then FOO [via beach station] (W5) to *Queen,* reported her fire "appears correct" for one target, but the gun emplacement fired at could only be located visually and rather crudely: "Close to summit of knoll south west of the 2 huts in [the] saddle." The next day, in the afternoon of 27 April, an observer obviously wondered what the navy was firing at, but the navy reply at 3:00 p.m. was not helpful: "Reply. I was firing at the enemy I saw." The observer evidently then shifted the ship's fire, because at 3:50 p.m., she signaled: "Have now been put on to 224 F." Then on the night of 27 April, at 3:15 p.m., the navy was requested to "play your [search] light on the point of our right flank," presumably to locate Turks creeping up at night. But the only result was an irritated message from the Anzac infantry ten minutes later at 3:25 a.m.: "Keep light off our trenches." This early disorganization at Anzac is very understandable, yet cooperation remained difficult throughout the campaign, as missed shoots and lack of accuracy were noted at various times.[19]

No doubt a hint of arm rivalry comes through when Hobbs concludes that "our observation for the Navy was far from satisfactory, after spotting them on to a Target they would, just as they appeared to be getting effective fire on, clear out without any warning."[20] Probably this was to avoid submarines, and if so, was sensible. Yet some navy-army exchanges late in the campaign sound a strange note of misunderstanding. For example, in November 1915, a casual letter from Capt. Vyell Vyvyan, senior naval officer of 3rd Squadron, to Brudenell White, GSO 1, Anzac, claimed that since the admiral was away, no decisions on a navy fire support plan for Anzac were possible. But Vyvyan enclosed a copy of Sulva's [sic] fire plan, "which appears to my simple mind somewhat over-organized, however, it might suit Sulva's [sic] conditions. . . . If your GHQ are not satisfied with our report and want something in the nature of Sulva's [sic] which I understand they are in love with this copy might be a basis." Brudenell White's reply is not recorded, but the misspelling of Suvla may indicate a certain mental distancing from the situation. Then, a lengthy letter at the beginning of December, from Edgar Grace, the captain of HMS *Grafton,* to Brudenell White, explains why his ship failed to cooperate in a shoot: "in the Navy the 'morning' indicates the period from 4 to 8 am and after 8 am we talk of the forenoon. When I got your signal at 11:15 last night I read it as meaning an early show." Grace continued that he sent a signal at 7:50 a.m., but got no reply until 10:15 a.m., yet this delayed message "could easily have been passed by visual if there were W/T difficulties." In any case, wrote Grace, "Had I known that it was an eleven o'clock shoot I could have told you that I was expecting to be relieved by *Endymion* at 9:30." However, *Endymion* was delayed, and "when finally I started ranging I got an urgent order to take *Endymion's* place supporting the *Lord Nelson.*" Grace then stressed that the Turkish guns at Olive Grove needed to be kept quiet or his ship could not take up "good positions within 6,000 yards or even more anywhere to the North and West of Gaba Tepe. He ranges 10,000 yards and therefore we cannot *stop still* when he is busy. To do really good shooting really we ought to be at anchor especially in any decent breeze." This was reasonable, for ships were vulnerable at anchor, but then Grace added "Further I personally am still quite in the dark as to what was the objective of this attack. We could do so much more if something was indicated as to the nature of the operation." Grace concluded by stating that "W4 [shore] station will not understand when W/T is blocked visual should be *at once* resorted to—and also vice versa."[21]

Grace's letter illustrates the curious gap between the mental worlds of the navy and army, which existed right up to the end of the campaign. It is remarkable that as late as December 1915 the navy and army could not agree on what constituted the morning hour. However, there were also success stories, for example, the attack on Turkish trenches near the Vineyard, at Helles, on 15 November, when land mines, two monitors, another ship, and land artillery, combined to allow 52 Division to succeed. The BGGS of VIII Corps wrote after this attack that "it has been established that cooperation in an attack has become a practical reality and that a system has been established which, with further development, will prove a powerful factor, both in attack and defence." On the other hand, a variety of letters and memos late in the campaign still indicate two different approaches to gunnery and warfare, for instance when HMS *Humber* arranged a shoot with an Anzac observer on 8 December: "Can you spot me on 47 S Q. 1415." Added to this memo is a note by Anzac: "'Com' [communications] opened with *Humber* 1500. Asked if 47 H 9 was correct instead of 47 S Q. She replied 47 S Q but she would only open fire in case [the] Turk gun fired. She did not fire. This entailed some strain on observer & telephone operator. RM." Then again, on 27 November, Captain the Hon. A. Boyle, HMS *Bacchante,* wanted to know whether, in view of the failure throughout 1915 to silence the Turkish Olive Grove guns, a massive bombardment by all the ships firing at the same time, might have a good moral effect, even if they hit nothing. The shoot was arranged for 10 December, but Boyle did not know the result. It seems unlikely to have achieved anything due to the well hidden nature of these reverse-slope guns.[22]

Navy-infantry cooperation at Anzac and Helles was awkward due to inexperience, technical inabilities, and different philosophies, with a patchwork of success and failure toward the end of the campaign. Even more difficult was navy-air cooperation. Here there were complaints against each other. Thus Capt. Charles Dix, RN, wrote after the war that our planes "never succeeded in doing much towards spotting our guns on to those of the Turks." On the other hand, the pilots complained that time and energy was wasted spotting for ships. Planes would go up and make the ready sign. Then they would wait one or two hours for ships to open fire, which would in any case be very slow. These delays could no doubt be overcome, but the real difficulty was, as always, accuracy. The problem for the air observer on a moving target such as troops or other ships, was how to allow for the following corrections: the error of the last shot fired; the delay in firing after the signal was made; the time the signal took to get through; the time of flight of the shell; and the movement of the target. All these obstacles could only be dealt with by estimations, and thus this form of firing was not likely to be accurate. Similarly, long-range targets, even if stationary, were not feasible either, even with spotting. Thus Rear Adm. Cecil Thursby noted that not more than 2 percent of rounds fired at ranges longer than 15,000 yards were effective against enemy guns. Shorter range targets, almost always by indirect fire, were also difficult, and ships had to go through a complicated procedure. First, the military asked for indirect fire on "X" battery. Then the ship's position was accurately fixed. Next, the angle of point of aim was laid off against some well-known point, such as the top of Achi Baba. This angle was then set on a sextant. The gunnery officer then took the sextant to the armored director and sat in the trainer's seat. After this, the gunnery officer looked through the sextant to the top of Achi Baba and noted the correct point of aim reflected against it. He ordered the layer to train right or left until he had picked up this point. Finally, the gunnery officer corrected the range for the height of the point of aim and for the height of the target, if necessary, and fired. The first shot nearly always fell within the deflection scale for direction, and other rounds were then fired at the same point of aim, applying deflection, as the air observer called in the fall of shot. If there was too great an error in direction, another complicated series of steps altered the point of aim. One can sympathize with the navy's attempts to be accurate, which were of course even more difficult if the ship firing was also moving.[23]

The navy moved to solve these technical problems with the arrival of the monitors in late July, new planes in August, and an improved balloon ship. Balloons had been used early in the campaign, but suffered from one major defect: they could not rise higher than 3,000 feet (at first 1,500 feet), and so could not spot very well in thick bush, or tangled country, or at a distance. Yet fire from land tended to force the balloons out to sea, which obviously undercut efficiency. However, there were four phone lines to HMS *Manica,* the original balloon ship, so results of shoots with this ship, and its successors, came within twenty to thirty seconds, faster than via air observation, and as fast as FOOs on land for artillery. In addition, the monitors also used theodolites and flash spotting to triangulate enemy batteries, as well as FOOs, air/balloon observation, and photos, to plot targets. The range of monitors was also good, and one shoot on 17 August with 9.2-inch guns knocked out an enemy battery at 21,000 yards. Even greater range was produced by tilting the monitor to one side through the crew standing on one side of the deck, or even filling life boats with water to weigh down one side of the monitor! Disadvantages of the monitors were still the problem of observation, including the cutting of communication lines to shore; difficulty in getting good officers from the Fleet to train as observers; inefficiency of naval ratings at shore stations and as signallers; and lack of W/T between Cape Helles and ships. It was said that Allied troops feared naval fire near their trenches (indeed it was officially abolished soon after the 25 April landings), but by the end of the campaign troops welcomed monitor fire in front of their trenches.[24]

It seems that naval-infantry cooperation was hampered very much by technical obstacles to accuracy, and to a lesser extent by the different naval mind set. Another problem was inexperience in understanding just how many shells were needed to do serious damage to enemy defenses: it is instructive to see that on the key dates of 7 to 10 August 1915, when the navy was supporting the August offensives, total naval shell expenditure was around 2,000 shells per day. In contrast, during World War II, at Iwo Jima, the navy was firing more than 6,000 shells per day.[25] Even so, by late

1915, after the Gallipoli campaign was winding down, it appears that the naval monitors, plus more planes and improved balloons, had become, too late, capable of helping the infantry in a significant way.

Having reviewed the technical problems of artillery cooperation at Gallipoli, what then could the guns actually do? Basically, the guns could cut wire, and stop charging Turks with shrapnel, but found counterbattery, preliminary bombardment, and the barrage very difficult. Reviewing what could be done, it seems that wire entanglements were effectively cut before the Lone Pine attack on 6 August, when Hobbs noted "cutting away the wire myself—with the 1st New Zealand Field Artillery Battery, it was difficult but proved to be effective." On the question of defensive shrapnel fire against Turk attacks, Hobbs and other artillery officers are clear that field artillery shrapnel frequently did very great damage to the Turks, for example, throughout May and June, and at Lone Pine in early August.[26] However, the guns could not do counterbattery work sufficiently well to silence Turkish guns, especially when an attack went in. This was a crucial lack of technical ability, and emerges early on at Anzac. Time and again gunners report lack of success against enemy guns, due to inaccuracy, inability to observe, and well hidden Turkish guns. The artillery especially noted that it could get line, meaning direction, but not range, meaning exact distance, until more frequent air spotting allowed proper registration. When the Anzac 4.7-inch heavy gun was introduced in September, greater range was obtained, but it suffered from frequent prematures, and so whenever it fired a whistle blew so that all nearby could take cover! However, certain groups of Turkish guns were never silenced, such as the Olive Grove guns, and this had a very serious effect when Anzac attacks took place, for example, at Lone Pine. Here, Hobbs reported that on 6 August at 5:30 p.m. as the Australian surprise attack started "the enemy opened a terrific artillery fire as soon as the infantry attack was launched and this they have maintained all night and during the day and our batteries have had all their work cut out to cope with them." What this indicates is that Hobbs's guns and those of the navy were not capable of doing effective counterbattery work, and this was the case throughout the campaign at Helles as well as Anzac. Moreover, the value of counterbattery work was not always realized, thus at Suvla, in early September, the BGRA IX Corps stated that enemy guns had not yet been registered, "and so far the services of aeroplanes for this purpose have not yet been obtained or apparently asked for." And for long-range counterbattery, "there appears to be no definite scheme of registering ships on enemy's guns with a view to silencing them or driving them further afield." On the other hand, a counterbattery officer was first appointed at Helles on 16 October, which perhaps points up the differences between the two commands at Gallipoli.[27]

Two other artillery aspects were also beyond their capabilities: the artillery could not do a useful preliminary bombardment, nor a barrage. In order to get a sense of how very difficult these were, it will be useful to look briefly at the offensive of 4 June at Helles, then at the Anzac attacks in early August, and finally at the later offensives in August at Suvla and Hill 60.

On 4 June, six artillery problems stand out: first, the artillery was not accurate enough to target all of the Turkish front trenches, because they were too close to the Allied trenches (100 yards), yet the front enemy trench was precisely what the attackers needed to target most of all, and where most of the attackers' casualties would come from. This problem of accuracy applied particularly to naval guns, and the howitzers, but also to other guns, for one field battery shot short on 4 June. The artillery tried to overcome the accuracy problem with late registration to eliminate the error of the hour, not just the error of the day, but it was still not accurate enough. Second, several enemy machine guns escaped detection before and during the bombardment, and these caused heavy losses. Third, the bombardment on the Turkish lines of approach was reported as "good, but as there was not much movement of [Turkish] troops, it did not cause many casualties." In other words, only if the Turks actually came out of their reserve trenches, could artillery targets be found. Fourth, the naval supporting fire on 4 June was not useful. As the rear admiral wrote: "The difficulty of firing at targets which could not be seen from the guns, while under way, was very great, and I do not consider the actual support given was at all effective." Fifth, and a matter of inexperience, neither artillery nor infantry as yet realized the potential of the barrage, nor that the artillery should stay on the enemy trenches as long as possible as the attack went in. Thus, as an artillery officer pointed out in reference to the later Hill 60 attack on 21 August, but equally relevant on 4 June: "we had not learned the method of attack later followed in France which was to advance close up under the artillery barrage; by the time the troops had commenced to advance the artillery had lifted to their second target and the Turks immediately manned their trenches and could be seen standing waist high behind their parapet shooting at the attackers." Sixth, while it is true there were only 640 HE shells available, all the rest being shrapnel, yet the artillery failed to distinguish between the concept of destroying enemy trenches, impossible in World War I without a great deal of accurate HE, and the concept of suppressing enemy fire as the attack went in, which was feasible with shrapnel, even in the context of 1915. Finally, and not an artillery fault, there was poor staff work which permitted the 2nd Naval Brigade, for example, to attack at 12:15 p.m., when the artillery had already lifted to the rear targets at 12:00 noon.[28]

The Anzac diversionary attacks in early August illustrate the same artillery inabilities. At several points, enemy shrapnel caused serious loss, due really to lack of counterbattery success: at Lone Pine on 6 August in No Man's Land as the attack went in; at Sasse's Sap in disabling a key Australian machine gun; at Cook's and Youden's posts, due to not silencing the Wineglass Battery on 7 August, thus forcing retirements; and at Steele's Post on 6 August. Fear of inaccuracy also prevented artillery and naval fire from dealing with the Turkish front trenches at the Nek on 7 August, and according to the Australian official historian, Charles Bean, the bombardment at Lone Pine stopped three minutes before the attack due to concern by artillery officers that they could not fire accurately so close to their troops. Thus Bean recalled that artillery officers told the journalist P. F. E. Schuler just before or after the fight that "they were ordered to fire too close to their attacked troops on the Nek, & that they were not going to take any risk but would take good care that there was no chance of a mistake & hitting their own men. If so," wrote Bean, "there is a very important lesson in result." It is not entirely clear what attack Bean is referring to, but artillery inaccuracy is the point made. Then at the celebrated Australian Light Horse attack at the Nek on 7 August (poignantly described in the film *Gallipoli*), the preliminary barrage ceased seven minutes early, for reasons that are obscure. Bean considered the problem to be a mistake in the synchronization of watches, but it seems more likely to be naval and artillery fear of hitting their own men, plus faulty staff work.[29]

Thus a mixture of technical inabilities, and staff and artillery inexperience, created many difficulties for the Anzac diversionary attacks in August. Rather similar situations led to the misuse of technology in the preparations for the joint Anzac—British 21 August offensive at Suvla and Hill 60. According to the diary of Cunliffe-Owen, on 18 August "We had a sort of conference, and I was asked about artillery support from Anzac . . . de Lisle [IX Corps commander], who was to command, read out a plan of attack, there were no comments or discussions, no questions by Div. Generals. Then Hamilton said a few irrelevant words and at once the plan was accepted. Needless to say, it was sketchy in the extreme. . . . We went back to Anzac much depressed and not at all impressed with the Staff work of IX Corps." Two days later, on 20 August, Cunliffe-Owen remarked in his diary, "The IX Corps orders for their artillery was most vague, for the navy more so, and no allotment of objectives was settled. Most careful detail and allotment of all guns, with times and objectives by map is absolutely necessary," but was lacking.[30]

Cunliffe-Owen's apprehension was justified, since the adjoining commands of Anzac and IX Corps were still arguing late on the morning of 21 August about the allocation of artillery for the attack due to start that very same afternoon at 3:00 p.m. At the very last minute, the Anzac attack was postponed for half an hour, to 3:30 p.m., so that the Anzac artillery could support the IX Corps attack until 3:00 p.m., and then switch to support the Anzac attack from 3:00 to 3:30 p.m. The argument between the two commands was such that when the Anzac CRA received orders to switch his artillery to the support of IX Corps for the first half of the attack, and reported this fact to Major General Godley, the local Anzac commander, the CRA recorded that "Godley never forgave me," and in any case "Godley had no eye for country, and never could have been a good fighting soldier in mobile war." Under such circumstances, it is hard to imagine that the artillery could be accurate or provide useful fire, especially as fog also prevented useful support from the naval guns. The attack failed at Suvla, and partially succeeded at Hill 60, but counterbattery was thoroughly inadequate, and the initial bombardment did little damage to the entrenched Turks. This was not because of shortages, but through poor staff work, technical limitations, and inexperience in laying on a proper artillery bombardment. However, on other occasions, technical inability was more clearly at fault, as can be seen from the puzzled diary entry of the GOC Anzac, Birdwood, who wrote on 27 August that even after a heavy bombardment the enemy trenches still seemed full, so that a heavy fire broke out as the Allied troops advanced. A similar naivety occurs in late September, when the GOC 54 Division, Maj. Gen. Frederick Inglefield, penned draft orders for an attack on the crest of Hill 60, and wrote: "It would possibly assist our Artillery, if Aeroplanes were employed to locate hostile guns, and 'spot' for our fire." By this time, such spotting should have been obvious.[31]

Thus far, this study has stressed the technical inability of the artillery and naval gunnery at Gallipoli, together with staff problems and inexperience. One other factor seems important in considering the artillery, namely, the difficulty of the senior commanders and staff in deciding where technology fitted into warfare in 1915. In reality there were several different ways of understanding and conducting warfare at Gallipoli. There was, firstly, the idea that warfare was really quite simple, all that was needed was higher courage and better morale than the Turks, and success would follow. This solution was a favorite of Hamilton's, who recognized its limitations but kept on coming back to it. Therefore, while on 10 May, he told Clive Wigram, the King's ADC, that "The modern mechanism of machine-guns etc. organized by the Germans is too much for the old straightforward, dashing form of attack. We knew it was so in France but we had to try and see whether it was also so in Turkey." Yet by 7 June, Hamilton was telling Kitchener there was no need for strategy or tactics, just courage. Similarly, on 18 June, Hamilton told Churchill "But for this trench warfare no great technical knowledge is required. A high moral standard and a healthy stomach—these are the best . . . for he who aspires to fight his way to the front at the Dardanelles." At the end of June, Hamilton was chiding Churchill for overlooking the "morale of the Turk and his pride," which was just beginning to crack. In mid-September Hamilton still wanted the use of the bayonet in "a dashing assault"—one more try "before we subside definitely into this ghastly trench warfare." And at the end of September, Hamilton told Herbert Henry Asquith that "the Turks are morally on their last legs." It must be in this context that GHQ sent a message to the Anzacs on 1 July that in trench fighting, grenades must never replace the bayonet. Hamilton's Social Darwinism, and sense of war as a simple moral problem, did not provide a useful tool with which to defeat the Turks, since he mainly hoped that the "Turkish character" would break sooner or later. In fact this solution really led to attrition and a waiting game, and initially prevented the turn to technical options, such as the artillery.[32]

A second solution was to see France, that is, the Western Front, as the leading model to follow. Hamilton himself was respectful of Western Front experience, but did not like Western Front methods. Thus in May, Hamilton felt he could not force Hunter-Weston to attack at night because he had Western Front experience. Equally, he could not force Lt. Gen. Julian Byng to hurry up in September, since the latter wanted four days of bombardment: "All these fellows from France come here with this idea." Even Birdwood was uncharacteristically modest about being superseded by Gen. Sir Charles Monro: "his experience in France which I lack will be absolutely invaluable." However, what the France/Western Front model really meant was a considerable reliance upon artillery, and here the most conspicuous supporters were

Stopford and Maj. Gen. H. L. Reed (BGGS, IX Corps), who argued that they could not push on harder at Suvla, because they needed more artillery to defeat the Turks. At the Dardanelles Commission, Stopford reported that he had been studying methods in France, and "to attack the Turks in trenches strongly held unless you made preparations by artillery fire was absolutely useless waste of life." This slowly became a generally held opinion at Gallipoli: that to succeed, there needed to be set-piece, artillery-dominated attacks. In one case, on 6–7 August at Helles, this led to a refusal to obey orders when an officer accused the CO of the Royal Dublin Fusiliers of not having the guts of a louse to cancel the attack, since there was no proper artillery registration and bombardment.[33]

The France/Western Front model led easily to, thirdly, the idea of "more and more." If artillery was needed, and attacks in Gallipoli failed, then what was needed was simply more of everything. This was a very common reaction during and after Gallipoli. Hence, Lt. Col. H. Forman, CO "C" Battery, 29 Division, told Hamilton later that "with more men, more Howitzers, and with half as much ammunition as our French colleagues . . . we could have got through at the Helles end at any time." Hamilton himself, after the early May failure to take Achi Baba, called for "more and more munitions." Hunter-Weston blamed his failed attacks on not enough men and not enough HE shells. Later in France, toward the end of June 1916, and over-impressed with the Somme preliminary bombardment, Hunter-Weston told his wife that if he had the same number of guns, he would have been through at Gallipoli. Then a battery commander, referring to the 4 June and 28 June attacks at Helles, informed the British official historian Brig. Gen. Cecil Aspinall that the chief lesson of the artillery at Gallipoli was "that it was a mistake to set out to conquer the Turks with 8 Howitzers & little ammutn.[*sic*]" In his official Gallipoli history, Aspinall argued that more men in particular would have given success, but makes the surprising statement that the Turks had less artillery ammunition than the Allies, while Bean estimated that toward the end of the campaign the Allies were firing fifteen shells to every one the Turks sent over.[34]

However, if the three previous solutions of offensive warfare were not working, then it occurred to both Hamilton and Birdwood at the end of June that a fourth solution of "Bite and Hold" defensive warfare might be the key. Explaining this concept to Kitchener, Hamilton argued that "The old battle tactics have clean vanished. I have only quite lately realized the new conditions. Now, by skill, surprise, HE, etc., take a tactical position and the enemy will certainly counter-attack. Then you can begin to kill them pretty fast. To attack all along the line is perfect nonsense— madness!" It was a surprising reversal for Hamilton to give up the gallant bayonet charge, indeed he went so far as to tell Asquith that taking four lines of Turkish trenches was less important than resisting Turkish counterattacks, which caused the Turks huge losses. Hamilton seems to have been influenced by Birdwood, whose troops fought off a Turkish counterattack at the end of June, leaving 500 Turkish dead as opposed to 25 Anzacs. Birdwood wrote Hamilton "If only Enver would continue to get impatient like this and urge attacks, we should very soon be through here, for I do not think they could stand very many more of them . . . though they undoubtedly fight like tigers if you come up against their trenches." This was really another variant of the morale/attrition method, but clearly such a solution would take considerable time to achieve any final result.[35]

Subsequently, after the disappointments of Anzac and Suvla in early August, and the seeming drawbacks of the four solutions above, three of which were technology-based, a fifth solution seemed to emerge in late September. This can be termed the scientific answer to trench warfare, and was first mooted by Col. Maurice Hankey, secretary to the Committee of Imperial Defence. Hankey wrote a long letter to Birdwood in early October, outlining various technical methods of getting through the Turkish positions. These methods stemmed from Lloyd George's Trench Warfare Department at the Ministry of Munitions, and included water jets to wash away Turkish trenches, the new Stokes Trench Mortar (240 of which were allegedly being prepared for Anzac), smoke screens, and liquid fire. Hankey suggested using them all at one time to regain surprise and initiative. Birdwood replied saying that he had already considered water jets, though this would be a major undertaking. Birdwood was happy to take the smoke bombs and the mortars, but was reluctant to use liquid fire on the humanitarian grounds that "until they [the Turks] have shown some disposition to use gas, fire, etc., against us I think we should probably [be] indisposed to use these methods of warfare. The Germans however, are sure to make them [the Turks] adopt all their own frightfulness." Birdwood added that he was preparing large-scale mines, which might "have a very demoralizing effect." Used in combination, these methods could get the Turks on the run, "which I still maintain is the one way of attacking." Then in a letter to Kitchener a few days later, Birdwood asked for another 250,000 artillery shells, Stokes Mortars and 200,000 mortar shells, and 25,000 trench-howitzer 50-pound bombs. He planned to use all these plus extensive mining to move forward at Suvla. Kitchener promised artillery shells, but no mortars and no 50-pound bombs. Since mining was a long-term project, and artillery had not produced breakthrough results, this seemed to end the possibilities of scientific trench warfare. Indeed, the prospects for Gallipoli, apart from Kitchener's last ditch efforts, now came to an end.[36]

In conclusion, it would seem that artillery and naval gunnery during the Gallipoli campaign suffered to some extent from arms rivalry; ignorance of the needs of each other; and sheer inexperience. But a greater obstacle to the success of this major technology was simple technical inability to deliver what was required. At the same time, the evidence reveals that a learning curve was clearly under way at Gallipoli, which, however, occurred too late in the campaign to be useful. This is true of artillery organization, cooperation of the arms, heavier guns for counterbattery, improved naval accuracy, and more frequent air observation. Parallel to these events were the applications of the various solutions to warfare in 1915. At first, technology was not fully appreciated, and at least in Hamilton's mind, weapons were subordinated to the simple virtues of the bayonet. Subsequently, the Western Front model, and the idea of "More and More," relied to a much greater extent on artillery and naval gunnery.

When these solutions did not succeed in a breakthrough, the "Bite and Hold" and Scientific Trench Warfare conceptions tacitly acknowledged the inability of men and weapons to achieve what was expected, and sought more modest aims. In effect, while the first three solutions expected men and technology to adapt to plans and ideas, the last two solutions attempted to adapt plans and ideas to the available technology. Thus artillery and naval gunnery were initially expected to pursue aims which were technically incompatible with capabilities, and only in the later stages of Gallipoli was there a slightly better match of ideas and weapons.

However, the groping toward solutions in 1915 was not peculiar to Gallipoli, since these same options were also being considered and tried on the Western Front. Hence, although many explanations of Gallipoli have focussed on shortages of men, munitions, artillery and shells, it is important to note that the Western Front, at the same time, with much more men, artillery and munitions, produced an equal stalemate. Thus in 1915 the British offensives of Neuve Chapelle, Aubers Ridge, Festubert, Loos, all essentially failed to break through, as did the larger French offensives of Artois and Champagne. Why should Gallipoli be different? If on the Western Front, with all the artillery and men deemed necessary, offensives still failed, why would the same offensives not also fail in Gallipoli? Therefore, even if Gallipoli had received the artillery, munitions, and men requested, the chances would still be high that Gallipoli would also fail, despite geographical differences. It is probable, therefore, that failure both on the Western Front and in Gallipoli was actually systemic and structural, rather than just a question of more men and munitions. No doubt, massive amounts of men and munitions would eventually have worked, yet at the time, inexperience leading to poor tactics, the misuse of technology, and particularly the early limitations of artillery and naval gunnery, were really to blame for lack of success. As the potential centerpiece of technology at Gallipoli, artillery and naval gunnery simply could not do the job well enough to make success possible between April and August.

Nevertheless, the belief existed at Helles, Anzac, and Suvla, and among later historians, that if only there had been just another brigade or a little more artillery and ammunition at critical moments, Gallipoli would have succeeded. But in exactly the same way, the French army, after the basic failure of the Artois offensive of May and June 1915, believed that if only they had possessed just a little more artillery, and a few more men, the Artois offensive would have succeeded too. So perhaps the fairest conclusion to the campaign in Gallipoli is the belief of Brudenell White, BGGS at Anzac, who wrote later: "The wonder is, not that we failed, but that we did as much as was done."[37]

Notes

The author is grateful to the Australian War Memorial, Canberra, for a research grant that made much of this research possible, and to a University of Calgary Research Grant that greatly contributed to this research.

1. The literature on Gallipoli is vast and still growing. Hence this list will be highly selective. From the Australian side, the starting point is the two useful volumes by the Australian official historian, C. E. W. Bean, *The Story of Anzac*, vols. 1–2 (Sydney: Angus & Robertson, 1921–24; St. Lucia: University of Queensland Press, 1981). Of value also is idem, *Gallipoli Mission* (Canberra: Australian War Memorial, 1948). More recently, there is John Robertson, *Anzac and Empire* (Sydney: Hamlyn, 1990), and Denis Winter, *25 April 1915: The Inevitable Tragedy* (St. Lucia: University of Queensland Press, 1994). From the British side, the critical but censored official history is Brig. Gen. Aspinall-Oglander, *Military Operations Gallipoli: Inception of the campaign to May 1915*, vol. 1 (London: Heinemann, 1929), and *Military Operations Gallipoli: May 1915 to the Evacuation*, vol. 2 (London: Heinemann, 1932) (hereafter cited as Aspinall, *Gallipoli*). There is also the Dardanelles Commission, *First Report*, Command Paper 8490 (1917–18), and *The Final Report of the Dardanelles Commission*, Command Paper 371 (1919). However, these Reports are suspect because of the widespread collusion that took place among the participants. On the naval side is Sir Julian Corbett, *History of the Great War Based on Official Documents: Naval Operations*, vol. 2 (London: Longman's Green, 1921). Also useful is Capt. S. W. Roskill, *Documents Relating to the Naval Air Service*, vol. 1 (London: Navy Records Society, 1969). On the air side, H. A. Jones, *The War in the Air*, vol. 2 (Oxford: Clarendon, 1928; London: Hamish Hamilton, 1969), and Charles Samson, *Fights and Flights* (London: Ernest Benn, 1930). Of the modern interpretations, the best overall account is still Robert Rhodes James, *Gallipoli* (London: B. T. Batsford, 1965). James had access to most of the documents now available, the main drawback is his lack of footnotes. Further bibliography and interpretations of Gallipoli can be found in Timothy H. E. Travers, "Command and Leadership Styles in the British Army: the 1915 Gallipoli Model," *Journal of Contemporary History* 29 (1994). Most recently, Nigel Steel and Peter Hart, *Defeat at Gallipoli* (London: Macmillan, 1994), have argued that London never gave Hamilton the resources to succeed, given the strength of the Turks.

2. "Notes by Simpson Baikie on the Artillery at Cape Helles in the Dardanelles Expedition," Dardanelles Commission, CAB 19/28, Public Record Office, Kew (hereafter cited as PRO).

3. Ibid.

4. Ibid., 4; see Aspinall, *Gallipoli*, 2:45, for number of shells expended; Ian Hogg, *The Guns 1914–1918* (London: Pan/Ballantine ed., 1973), 50, says a gun for every 4 yards of enemy front, while Robin Prior and Trevor Wilson, *Command on the Western Front, The Military Career of Sir Henry Rawlinson, 1914–1918* (Oxford and Cambridge, Mass.: Blackwell, 1992), 33, say 6 yards. The latter is probably correct. For Turkish guns, see "Report of the Committee appointed to investigate the attack delivered on, and the enemy defenses of, the Dardanelles Straits," Naval Command Paper 1550, 1919/1921, (hereafter cited as *Mitchell Report*), 234, 51/39, Australian War Memorial (hereafter cited as AWM).

5. Major General Cunliffe-Owen, BGRA Anzac, Diary, 7 May, 13 July, and 8 Aug. 1915, CAB 45/246, PRO.

6. Cunliffe-Owen to GOC 1st Australian Division, 31 May 1915; Cunliffe-Owen to OC 3rd Bde. (through Aust. Divn.), 9 a.m., 10 June 1915; 25/367/49, AWM. Also Cunliffe-Owen, "Instructions for getting artillery fire on objectives quickly," 25 May 1915, Gallipoli—Phase I, 25/367/49, AWM.

7. Bean, *Story of Anzac*, 2:67–70; Colonel Hobbs (later Lt. Gen. Sir J. Talbot Hobbs), Diary, 19, 21, and 25 May 1915, PR 82/153/1, AWM. Discussion of the conflict and cooperation between artillery and infantry occurs in Bruce Gudmundsson, *On Artillery* (Wesport, Conn.: Praeger, 1993), 13, 69.

8. Brigade Training at Mena: "Exercise—Defense," n.d.; "Narrative Exercise," 20 Feb. 1915; "Decisions of GOC 1 Australian Division re; Conference on Signal exercise," 22 Feb. 1915; 25/941/3; Training Egypt 1915: "Notes by J. D. Irvine," 5 Mar. 1915, 25/941/6; AWM. On speed of shooting, see Rowe to Mother, 21 Jan. 1915, Rowe Letters, File #1, PR 90/090, AWM. On map-ranging and lectures from France, see Colonel Rosenthal, CO 3rd Australian Field Artillery Brigade, Diary, 23 Feb. and 2 Mar. 1915, Mitchell Library, Sydney. On the engineering side of

artillery at Mena, see Brigadier General Burgess, then CO 9th Battery, Burgess Papers, MSS 12, pt. 2, 25/224, AWM. On Western Front methods, see Coe to parents, 27 Mar. 1915, about huge use of ammunition in France, Letters, Coe Papers, 2 DRL 491; and Corporal Smith (later Major), Ammunition Column, 2nd Field Artillery, Diary, 9 Apr. 1915, and letter to "dear Mater," 2 Apr. 1915, about lecture from CO with notes from an officer from the Western Front, Smith Papers, PR 84/365; AWM. See also Brig. Gen. Cunliffe-Owen, "Artillery at Anzac in the Gallipoli Campaign," *Royal Artillery Journal* 46 (1919–20), 537, for infantry-artillery cooperation discussions, and lectures on experience in France at Mena.

9. On artillery problems at Anzac, see Bean, *Story of Anzac*, 2:73, 75. Parker to Cunliffe-Owen, 8 May 1915, app. 21 (B), 25/367/229, AWM. Hobbs, "Narrative of Operations (Artillery) 19-5-15," 1–8, War Diary, General Hobbs, 419/48/4, 3 DRL 2600, AWM.

10. Centralization and organization of the artillery can be followed in "Notes for Reference and Guidance of all Artillery Officers Australian Division Artillery," 30 May 1915; followed by Anderson, Brigade Major, "Notes for Artillery Officers 1 Australian Division," n.d., but the reference to the tactical centralization of all howitzers, mountain guns, mobile artillery, etc., under the Divisional CRA, makes it end of June or July 1915; 25/381/3, AWM. Colonel Rosenthal, Diary, Mitchell Library. Brigade Major, Corps. Arty., "Corps Artillery Signals. Memo No 1," 9 June 1915, 25/367/49, AWM. Cunliffe-Owen, Diary, 1 Nov. 1915, CAB 45/246, PRO. Memos of closer cooperation include: Brigade Major, Helles, "Artillery memorandum—20th August," 25/75/4; White to Hobbs, 5 Nov. 1915, 25/75/4; Brudenell White, BGGS, "General Staff Circular No. 2," including the order that although the CRA controls the artillery of the division, "a limited number of guns will be told off at the outset for immediate support of certain infantry units . . . ," 25/75/4; AWM.

11. Colonel Hobbs, Diary, 30 Apr., 13 and 17 May, 17 June, and 14 Oct. 1915, PR 82/153/1, AWM.

12. On planes available, see "Air Reconnaissance and Observation," 25/367/146, AWM. See also Charles Samson, *Fights and Flights* (London: Ernest Benn, 1930), 264–65. Technical difficulties and jamming in *Mitchell Report*, 506–9, 51/39, AWM. Dr. Alfred Price kindly informed the author that with the radio sets available at the time, jamming would be done by simply depressing the key, and so no such uncomplimentary words would be heard. Later jamming reported by air observer Captain Jopp on 31 July, "Report of month's work [*sic*] from 16th July to 19th Dec," 25/367/146, AWM. For clock code in France, see S. Bidwell and D. Graham, *Fire-Power: British Army Weapons and Theories of Warfare 1904–1945* (London: Allen & Unwin, 1982, 1985), 102.

13. Col. C. G. Miles, 2nd Australian Division, "Heavy Battery Reports," July–Nov. 1915, 25/367/131; Capt. A. H. K. Jopp, "Report of month's work," 25/367/146; AWM. Brigadier General Burgess, 9th Battery, MSS 12, pt. 2, 224, AWM.

14. Samson, *Fights and Flights*, 248. Jones, *War in the Air*, 57, 63. Cunliffe-Owen, Diary, end pages, "Reflections on Gallipoli Campaign," CAB 45/246, PRO. For provoking Turkish guns, and accuracy, see Report, 2nd Australian Field Artillery Battery, 1 July 1915; and 3rd Australian Field Artillery Battery, 2 July 1915; pt. 5, July 1915, 25/367/66, AWM. On range tables, see Stephen to HQ Anzac, 6 Aug. 1915, 25/75/4; Colonel Miles, Diary, Heavy Battery, 5 Aug. 1915, 25/367/131; AWM. On maps and survey, see Legge to 1st Australian Division, 18 Sept. 1915, 25/75/4; for survey, see *Mitchell Report*, 33, 36, 51/39, AWM.

15. 1 Australian Division Artillery Brigade Reports, 12 July 1915: five messages from 2nd FAB to CRA 1 Division, timed at 7:37, 8:55, 9:13, 9:35, and 9:47 a.m.; Johnston to CRA 1 Division, 13 July 1915; 8th Battery CO to Adjutant 2nd FAB, 13 July 1915; 1 Division Artillery Report, 13 and 20 July 1915; pt. 5, 25/367/66, AWM. Cunliffe-Owen, "Artillery at Anzac," 551.

16. Hobbs, Diary, 30 June 1915, PR 82/153/1, AWM; Maj. R. Casey, 1 Division Staff, Diary, 29 May 1915, MSS 6150, National Library, Canberra, cited in Winter, *25 April 1915*, 227–28. Brigadier General de Winton, CO 162 Inf. Bde., Suvla Bay Diary, CAB 45/241; Downing to Aspinall, 24 Feb. 1931, CAB 45/241; Smith to Aspinall, 26 Feb. 1931, CAB 45/244; PRO.

Turkish answers by Kiazim Pasha, Chief of the Turkish General Staff, 1919, to questions from the Australian Official Historian, cited in Bean, *Gallipoli Mission*, 352, app. 1.

17. Cunliffe-Owen, "Extracts from a lecture on Artillery at Anzac, given to the Royal Artillery Institution," 22 Jan. 1920, CAB 45/246; Cunliffe-Owen, Diary, 28 and 29 Apr., and 4 May 1915, CAB 45/246; PRO. Cunliffe-Owen, "Artillery at Anzac," 538. Samson, *Fights and Flights*, 234. Brig. Gen. Cunliffe-Owen, "Memorandum C.R.A. No. 1," to GOC New Zealand and Australian Division, 9 May 1915, App. 51 (a), 25/367/229, AWM. For vagueness of 971 spur, and air photo map, see *Mitchell Report*, 233, 242, 51/39, AWM.

18. Brig. Gen. Cunliffe-Owen, "For Artillery Observations Officers for Ships," 28 Apr. 1915, 25/367/49, AWM.

19. 3rd Brigade to *Queen*, 9:25 a.m., 26 Apr. 1915; 2nd Brigade to *Queen*, 10:18 a.m., 26 Apr. 1915; Anzac HQ Messages, 25/367/213, *AWM*; *Queen* to Base (W4), 2:30 p.m., 26 Apr. 1915; *Queen* to Base (W4), 3:37 p.m., 26 Apr. 1915; FOO (Anderson) to *Queen*, 26 Apr. 1915, Navy Signals, Anzac, April 1915, 25/367/26, AWM; and for the searchlight, see W5 to HZ, 3:15 a.m., 27 Apr. 1915; W5 to HZ, 3:25 a.m., 27 Apr. 1915; 1 Division Artillery HQ and Director Artillery Daily Reports, 25/367/189, AWM. For navy not shooting when requested, and inaccuracy, see Rosenthal, CO 3rd Battery to CO Division Artillery, for 7 to 8 June, 1915; Rosenthal, 3rd Battery, to CO Division Artillery, 23 July 1915; 1 Division Artillery Brigades, 25/367/186, AWM.

20. Hobbs, Diary, app., PR 82/153/1, AWM.

21. Vyvyan to "My Dear General White," 10 Nov. 1915, "Tables prepared by BGRA with Brigadier Generals to ensure close cooperation between Artillery and Infantry," 25/75/3, AWM; Grace to White, 1 Dec. 1915, "Navy Signals, November–December 1915," 25/367/26, AWM.

22. Report, BGGS VIII Corps, 15 Nov. 1915, *Mitchell Report*, 308, 51/39, AWM. HMS *Humber* to Anzac, 8 Dec. 1915; Boyle to White, 27 Nov. 1915; Captain of *Bacchante* to White, 10 Dec. 1915; "Navy Signals, November–December 1915," 25/367/26, AWM. In a postwar exploration of these Olive Grove gun positions, Bean found many emplacements, plus many dummies, but not much sign of battery destruction by Allied guns; Bean, *Gallipoli Mission*, 257–61.

23. Capt. C. Dix, "Anzac. Impressions of Landing," 33–34, 25/367/5, AWM; *Mitchell Report*, 243, 304, 391, 525, 51/39, AWM.

24. *Mitchell Report*, 210–11, 243–44, 265, 508–10, 515, 523–24, 528, 51/39, AWM.

25. Admiral de Robeck, "Expenditure of Ammunition 7th to 10th August," 13 Aug. 1915; and de Robeck to all ships, 13 Aug. 1915 (limiting expenditure of ammunition): Admiralty 137/2168, PRO. No doubt the fear of shell shortages also had an impact on these August 1915 figures. For Iwo Jima, see Rear Adm. Reed Carter, USN, *Beans, Bullets, and Black Oil* (Washington, D.C.: Department of the Navy, 1952), 289. The author is grateful to Jon Sumida for his generous help in providing this information. See also Jon Sumida, "British Naval Operational Logistics, 1914–1918," *Journal of Military History* 57 (July 1993), 454.

26. On wire cutting, opinion seems divided, but some batteries were capable of this, see Bean, *Story of Anzac*, 2:499–500; Hobbs, Diary, 6 Aug. 1915, PR 82/153/1; and Walker to Aspinall, 15 Apr. 1929, Cab 45/245, PRO. However, some letters argue against effective wire cutting, see Smyth to Bean, 8 Mar. 1931, Bean Papers, 3 DRL/7953/27, AWM; and Wynter to Aspinall, n.d., Cab 45/245, PRO. Many references to effective shrapnel defense are found in Hobbs, Diary, 24 and 31 May, 7 and 14 Aug. 1915, etc.; PR 82/153/1; AWM. Colonel Rosenthal also notes effectiveness of shrapnel, Diary, 31 May 1915, Mitchell Library.

27. For line but not range, see Col. Miles, Heavy Battery, Daily Reports, 20, 24, and 28 July, 3, 8, and 10 Aug. 1915. After this aeroplane shoots improved the range, but the Turks continually moved guns around and introduced new gun emplacements. The new 4.7-inch gun for longer ranges was introduced, but suffered many prematures, i.e, on 3, 18, and 20 Oct. 1915. On 21 Oct. 1915 the #83 fuse was condemned, the whistle warning is on 3 Nov. 1915; 25/367/131, AWM.

Bean, *Story of Anzac,* 2:81, 85, 504. BGRA IX Corps, 8 Sept. 1915, and counterbattery officer, cited in *Mitchell Report,* 306, 33, 51/39, AWM.

28. For not firing at front trenches, see Aspinall, *Gallipoli,* 2:47, and for Naval Brigade attack, see ibid., 49; for short shooting, undetected machine guns, and howitzer inaccuracy, see Grant to Aspinall, 2 Mar. 1929, CAB 245/241, PRO; for lines of approach shooting and navy problems, see *Mitchell Report,* 234, 51/39, AWM; for inexperience of barrage, see Dare to Bean, 16 July 1931, 3 DRL 7953/28, Bean Papers, AWM.

29. Bean, *Story of Anzac,* 2:504, 515, 545, 600–601. On the Nek attack, see Bean to Hobbs, 28 Feb. 1923, 3 DRL/7953/2, Bean Papers, AWM; Bean, *Story of Anzac,* 2:612–13. Note that (1) there was apparently a mix-up in what the three Australian batteries were supposed to do at the Nek, and they did not receive advance warning of the attack, possibly due to poor staff work by the CRA New Zealand Artillery; (2) barrage times for the Australian and New Zealand batteries were changed three times, and do not seem to mention the Nek; (3) Bean seems to have relied on Colonel Brazier's information, who gave Bean the idea that the Navy and the shore had not synchronized times, and Bean's account gives the impression that the land batteries did fire until 4:30; (4) yet the navy had a fire officer whose task was to send changes every clock hour to the ships; (5) Bean suggests only one destroyer was firing, while the orders show that four ships were firing; Bean, *Story of Anzac,* 2:611–13; "Naval arrangements" and "Anzac Third Phase," 25/367/75/5, AWM; Brazier to Bean, 13 Apr. 1931, 3 DRL 7953/27; Bean to Hobbs, 29 Jan. and 28 Feb. 1923; 3 DRL 7953/2, Bean Papers, AWM; Hobbs, Diary, 24 and 28 Aug. 1915, PR 82/153/1, AWM.

30. The ratio of Turkish and British shells is noted in Bean, *Story of Anzac,* 2:848n33. Cunliffe-Owen, Diary, 18 and 20 Aug. 1915, "Extracts from General Cunliffe Owen's Diary," CAB 45/246, PRO.

31. Winston to Aspinall, 11 Dec. 1930, CAB 45/245, PRO. Bean, *Story of Anzac,* 2:728ff. Bryan Cooper, *The Tenth (Irish) Division in Gallipoli* (London, 1918), reprinted *Black Rock* (Dublin: Irish Academic Press, 1993), 107–9, regarding 21 August. Lt. Gen. Sir W. Birdwood, Diary, 27 Aug. 1915, 29(a)/3 DRL 3376, Birdwood Papers, AWM. Major General Inglefield, "54th Division. Proposals for the assault and capture of the crest of Hill 60," 23 Sept. 1915, 3, 25/367/166, AWM.

32. Hamilton to Wigram, 10 May 1915, 5/1, Hamilton Papers, cited in Winter, *25 April 1915,* 219; Hamilton to Kitchener, 7 June 1915, Letters to Lord Kitchener, 5/1, Liddell Hart Centre for Military Archives, King's College, London University (hereafter cited as KCL); Hamilton to Churchill, 18 and 30 June 1915, "Important Letters," 5/6, Hamilton Papers, KCL; Hamilton to Kitchener, 5 May and 21 Sept. 1915, Letters to Lord Kitchener, 5/1, Hamilton Papers, KCL; Hamilton to Asquith, 30 Sept. 1915, 5/6, "Important Letters," Hamilton Papers, KCL. Hamilton's Social Darwinism is evident throughout his *Gallipoli Diary,* 2 vols. (London: Edward Arnold, 1920). GHQ to Anzac HQ, 1 July 1915, 25/367/73, AWM. In a variant of Hamilton's attitude, Birdwood spoke in August of Anzac high morale, and the need to get the Turks on the run: Birdwood, "Orders for Anzacs," 5 Aug. 1915, 3 DRL 3376/58, Birdwood Papers, AWM.

33. Hamilton, Diary, 5 May 1915, Mrs. Shields's 2d version, 6/9/4; Hamilton to Kitchener, 21 Sept. 1915, Letters to Lord Kitchener, 5/1; Hamilton Papers, KCL. Birdwood to Kitchener, 4 Nov. 1915, "Birdwood-Kitchener Secret Ciphers," 3DRL 3376/48, Birdwood Papers, AWM. For Stopford and Reed, see Stopford, "Report of Operations IX Army Corps," 26 Oct. 1915, 5, WO/106/707, PRO. Lt. Col. Aspinall, "Conversation with the BGGS IX Corps," Statement, n.d., 17/3/1/81, and "Questions to Stopford at the Dardanelles Commission," 23, 16/7/2/1, Hamilton Papers, KCL. Geddes to Aspinall, 8 Nov. 1930, CAB 45/244 and 45/245, PRO.

34. Forman to Hamilton, "Correspondence dealing with the Dardanelles Commission," Hamilton Papers, 17/3/1, KCL; Hamilton, *Gallipoli Diary,* 1:205; Hunter-Weston to wife, 29 Apr. and 5 June 1915, and 26 June 1916, Hunter-Weston Papers, British Library; Grant to

Aspinall, 2 Mar. 1929, CAB 45/241, PRO. Aspinall, *Gallipoli,* 2:335, 356, 485; Bean, *Story of Anzac*, 2:848n33. In contrast, Lt. Col. Stephenson (8th Manchesters) told Aspinall, in regard to the three battles of Kritihia, that there were enough men available, but that they were either not used or were too far off, Stephenson to Aspinall, 16 Jan. 1929, CAB 45/245, PRO.

35. Hamilton to Kitchener, 2 July 1915, Letters to Kitchener, 5/1; Hamilton to Asquith, 7 July 1915, "Important Letters," 5/6; Hamilton Papers, KCL. Birdwood to Hamilton, 1 July 1915, Birdwood Papers, 3 DRL 3376/62, AWM. Birdwood and Hamilton spent the day talking together on 2 July, no doubt this follow-up to Birdwood's letter firmed up Hamilton's "Bite and Hold" ideas; Birdwood, Diary, 2 July 1915, 3 DRL 3376/29(a), Birdwood Papers, AWM.

36. Hankey to Birdwood, 9 Oct. 1915, Birdwood to Hankey, 20 Oct. 1915, Birdwood-Hankey Correspondence, 3 DRL 3376/46; Birdwood to Kitchener, 21 Oct. 1915, Kitchener to Birdwood, 23 Oct. 1915, Birdwood-Kitchener Secret Ciphers, 3 DRL 3376/48; Birdwood Papers, AWM.

37. See Douglas Porch, "Artois, 1915," *MHQ: The Quarterly Journal of Military History* 5 (Spring 1993), 50; similar sentiments are expressed by Aspinall, among others, *Gallipoli,* 2:479–81. White to Bean, 27 Feb. 1931, 3 DRL 7953/27, Bean Papers, AWM.

A Clash of Military Cultures: German and French Approaches to Technology between the World Wars

James S. Corum

World War I was a period of extraordinarily rapid technological development. Military technology had changed dramatically in the decades before World War I, but even the most perceptive prewar military thinkers of 1914 were unable to predict the pace of wartime technical development. The militaries of World War I had to adapt to the emergence of the airplane, motor vehicle, tank, and poison gas as major new weapons. Commanders, general staffs, and war departments had to try to master the variety of new technology and adapt technology to operations.

The period between the world wars was a time of equally extraordinary technological development for the world's militaries. World War I had proven to be a technological catalyst. Such weapons as the tank, which was introduced in the war, enjoyed rapid development in the 1920s and 1930s. Accelerated by the war, motor vehicle development progressed, while every army engaged in motorization experiments. Radios and electronics passed out of their infancy as Britain and Germany developed radar. Most dramatic of all was the evolution of the airplane from a useful auxiliary weapon into a very powerful and decisive arm of the military.

The French military was extremely successful in adapting to technology during World War I. In many respects, the French effort in this regard equaled, and in several instances surpassed, the German. Deficient in heavy artillery at the start of the war, by 1917–18 the French Army had created a superb heavy artillery arm.[1] French vehicle production exceeded the German, and by 1918, the French Army was more advanced in motorization.[2] The most dramatic difference between the two armies was in the development of armored forces. The French General Staff initiated development of a French tank program in 1914.[3] During the war, the French produced 4,300 tanks, more than any other power.[4] In contrast, the Germans never deployed more than a few dozen tanks, and of these, only twenty were of German manufacture.[5]

Another French success story was their effectiveness in designing and producing high-quality aircraft. The military leadership and civilian industrialists demonstrated a high degree of innovation in effectively standardizing aircraft types and engines. France led the world in aircraft engine production and quality.[6] By 1918 the French had designed and built the first supercharged engine.[7] In design, the French equaled the Germans, with such rugged, effective, and swift aircraft as the Spad VII, Spad XIII fighters and the Breguet 14 bomber.[8] Indeed, the quality and quantity of French aircraft design and production enabled the Allies to gain air superiority in the latter part of the war. Due to the inefficiency of British aircraft design and production, the British had to rely on the French for many aircraft and engines. As late as 1918, British squadrons on the Western Front flew Spads and Nieuports into action.[9]

By the final year of the war, the French military showed considerable skill in utilizing new technology on the battlefield. The French Army demonstrated operational finesse in the successful counterattack at Soissons on 18 and 26 July 1918. French forces, supported by 571 tanks and masses of aircraft, drove a deep wedge into the German drive on Paris.[10] French infantry and tanks also took part in the Allied Offensive at Amiens on 8 August 1918 which decisively broke the German front. During the last two months of World War I, the American offensive at St. Mihiel and the Meuse Argonne relied on the French for tank, air, and artillery support.[11]

Twenty-two years later, the French had lost the technological edge they had enjoyed in 1918. The French ground forces were well equipped, but their commanders' inability to use the weapons effectively was apparent. In the field of aviation, the technological disparity was significant, with the Germans fielding a force which was both qualitatively and quantitatively superior. In numerous aspects of technology, the German military had shown itself far more innovative and effective in the development and employment of equipment.

This essay will explore the background of two contrasting military cultures, the French and the German, and compare their approaches to the development and adaptation of technology. It will first outline the state of comparative technological development of the French and German armies and their forces in 1939–40 and investigate some of the factors that brought those forces to their relative positions. Specific lines of investigation follow: first, the influence of doctrine on technology; second, the influence of military organizational systems on technology; and finally, the influence of the general staff cultures on technical development.

The State of French and German Military Technology in 1940

In the 1920s, the French had a clear lead in motorization and tank development. The French motor industry was one of the world's largest and was highly innovative in the military arena.[12] For example, in the 1920s Citroen produced the world's first practical half-track vehicles. The French half-track technology of the 1920s was so admired by the Germans that the German Maffei Company licensed the French half-track system in 1927 and produced a German model by 1930.[13] The military, following the success of French motor production and employment in the world war, had numerous commissions devoted to studying and implementing motorization.[14] In tank development, the French capitalized on their advantage gained in World War I, and several commissions were given a mandate to develop tanks. French tank design in the 1920s was clearly at the cutting edge with the design for the Char-B heavy tank. The Char-B would eventually be deployed in the 1930s and carried a 75-mm as well as a 47-mm gun. By 1940, 365 Char-B Tanks were built.[15]

In the air, the French also enjoyed a significant technological advantage. In the 1920s, the French Air Force was the largest in the world.[16] The French air motor industry led the world, and in the mid-1920s the French held many world records in aviation. The National Aeronautical Institute, founded in 1909, was nationalized and put under the direction of the Air Ministry in 1928. The Institute continued to produce highly qualified aeronautical engineers.[17]

Both in ground forces motorization and in aviation technology, the French lost their edge in the late 1920s–early 1930s. In Germany, the auto industry—which had been far smaller than the French auto industry of the 1920s—forged ahead. By the early 1930s, with subsidies and incentives from the new Nazi regime, the German auto industry became larger than the French.[18] The French were, however, able to keep a high standard in developing tank technology. For example, the French tanks of the mid-1930s were fitted with the world's first cast turrets, and the French tanks of the 1930s had the most sophisticated steering systems of the era.[19] The Germans, however, were able to match the level of French tank technology. For example, in 1936 the Panzer II and III were designed with the world's first torsion-bar suspension system.[20] In artillery, both countries demonstrated a rough parity in gun quality by 1940. In the post–World War I era, both the French and the German Armies had developed medium and heavy guns which were efficient and effective on the battlefield in 1940.[21] In other aspects of motorization, the French fell behind. Having once led the world in half-track production, the French virtually ended development of the half-track in 1933. Starting behind the French, the Germans licensed their technology and then forged ahead with their own innovative designs. By 1940, a wide variety of half-track vehicles were assigned to their armored and motorized divisions.[22] The Germans took the lead in other areas as well, including armored cars, armored command vehicles, and tracked assault guns, while French development in these areas remained relatively static.

The greatest disparity in ground-forces equipment between the French and the Germans was in communications. The French developed relatively few radio systems in the interwar period and devoted very little money to developing communications equipment.[23] They nonetheless assigned a high priority to the development of communications equipment which would be set in fixed installations along the Maginot Line.[24] The Germans also placed a high priority on developing communications equipment but, unlike the French, produced a wide variety of effective radios for ground forces, infantry, artillery, aviation, and tanks.[25] As of 1940, only French heavy tanks had radios, whereas all German tanks had radios, and numerous other armored cars and vehicles as well.

With the exception of communications, however, the French Army was not badly equipped in 1940. In fact, they possessed competent guns, good tanks for the era, and several armored and motorized divisions. The French Army of 1940 can be said to have had a modern level of motorization. In their approach to motorization, the German and French armies were actually very similar. Both armies studied motorization intensively, and were supportive of motorization; and both were influenced by national strategic considerations as net importers of oil, with concerns over supply in case of war.[26] This concern for oil, as well as the high cost of motorization, ensured that both armies would motorize gradually and would still use primarily horse-drawn transport for their infantry divisions in 1940.

One interesting aspect of the French and German concern about oil imports was that Germany and France were the only nations in the interwar period to develop diesel aircraft engines. Neither of these were effective, though they did have the virtue of using considerably less fuel than the high-performance engines of the era.[27] Both armies, employing common sense, chose the artillery branches as the first priority for army mobilization. Thus both France and Germany took a constrained and practical approach to motorization of ground forces.

While the disparity of technology between ground forces in 1940 may have been serious, in the air this disparity was, for the French, catastrophic. In 1940, the majority of the French aircraft in service were far inferior to their German counter-parts. For example, the Bloch 152, the Morane Saulnier 406, and the Curtis Hawk P-36 fighter aircraft were all inferior to the German Me-109 in 1940; and in the French bomber force, the Amiots and Farmans, did not come close in performance to the German Heinkels and Dorniers.[28] The only French bomber relatively equal to its German counterparts in 1940 was the Loire 45, roughly equivalent to the German Heinkel 111, Dornier 17, and Junkers 88 bombers.[29] The Dewoitane 520 fighter, which entered production in 1940, was the only aircraft that could hope to match the Me-109.[30] The French Portez 633 heavy fighter was inferior in speed and reliability to the Me-110 of the Germans.[31] In the field of dive bombers, the French belatedly manufac-tured and brought a mere handful of machines on line by 1940, in contrast to the Germans who employed over 300 Ju-87s for the campaign in France.[32] As an overall assessment, the French Air Force in 1940 was approximately three years behind the Germans in aircraft development and deployment. The Loire 45 entered serial produc-tion only in 1940, whereas the German Heinkel 111 and Dornier 17 had entered serial production in 1937. The Dewoitane 520, a project initiated at the same time as the Me-109, entered production only in 1940,[33] while the Me-109 entered serial production in 1937. In almost every case, it took the French two to four years longer to develop and deploy an aircraft model in the 1930s.

In the establishment of a basic aviation infrastructure, the French Air Force was as much as ten years behind that of the Luftwaffe. In 1939, France possessed only one paved runway in the entire country.[34] In 1933, France had only two radio bea-cons for aerial navigation.[35] The French belatedly launched an effort to modernize infrastructure in 1936, when they sought to improve aerial navigation for civil and military aviation; however, little was accomplished by 1940.[36] In contrast, even in the 1920s, the Germans had developed the most sophisticated aviation infrastructure in Europe. By 1927 Lufthansa was by far the largest passenger and cargo carrier in Europe.[37] In 1927 Lorenz all-weather landing systems were introduced for airport operations.[38] By 1931 seventeen airports had Lorenz systems.[39] German develop-ment of aviation instruments, including gyroscopic instruments, in the 1920s and early 1930s was equal to that of the United States,[40] and Germany in 1940 possessed numerous paved runways.[41]

The only aspect of aviation where a rough equality existed was in aircraft engine development. The French had always had a strong engine industry, and in 1940 had some effective 2,000-hp engines in development.[42] In other areas, however, the French declined even to compete with the Germans. The most dramatic examples of this were in the development of radar and the jet aircraft. At the outbreak of the war, the French Navy had developed radar technology, but the French Air Force had no radar program.[43] In contrast, the Germans in 1939 were already producing advanced radar sets, and were deploying radar for air defense.[44]

Jet development followed a similar pattern. During the interwar period, perceiving that the piston engine had specific speed limitations which were rapidly being approached by the major powers, Heinkel initiated production of both jet engines and jet aircraft. With little government financing and only a handful of engineers, Heinkel developed the first jet engine program in Germany in the mid-1930s.[45] The culmination of their efforts came in August 1939, when the Heinkel 178 became the first jet aircraft to fly.[46] The French had no comparable development program.

Finally, in the matter of operational forces, there are other notable disparities between the French and the German interwar forces. In 1935, the Germans—following the lead set by the Soviet Union—began to develop airborne forces.[47] By the outbreak of the war in 1939, a full airborne division had been formed by the Luftwaffe, and other large units were in the process of formation.[48] In contrast, by 1939 the French Air Force had formed only a small detachment of 175 airborne soldiers, and employed them on one maneuver. Otherwise, the French Army and Air Staff exhibited little interest in airborne or air-landing troops.[49]

In antiaircraft technology, France and Germany, which had been approximately equal at the end of World War I, developed similar families of light antiaircraft guns of 20–37 mm. In production, however, the French placed little emphasis on the antiaircraft arm, and by the outbreak of the war, the French, relative to the Germans, possessed a mere handful of light antiaircraft guns. In the field of heavy antiaircraft guns, the French in 1922 had developed an excellent high-powered 90-mm gun, but then halted further development of heavy antiaircraft guns and stayed with the low-velocity, obsolete 75-mm.[50] As for the Germans, by 1932, Krupp and Rheinmetall had developed the famous 88-mm antiaircraft gun. By the beginning of World War II, the French Army had only 4,000 antiaircraft guns of all types, versus over 3,000 heavy and 10,000 light antiaircraft guns for the German forces.[51]

The Influence of Doctrine on French and German Interwar Technology

Both the French and the German armies of the interwar period provide useful illustrations of the dominance of doctrine in the process of developing military technology. Both the French and the Germans developed a clear and explicit doctrine, and in both cases, the armies adopted technologies that fit their doctrines.

The interwar French operational doctrine, as expressed in the *Provisional Instructions On the Tactical Employment of Large Units* (1921),[52] is described as the *bataille conduite*, or the methodical battle. In the 1921 regulations, the French General Staff expressed the view that technology had so changed the battlefield that firepower was now the primary element in warfare.[53] Firepower made the defense extremely powerful. The French Army, however, also determined that only the offense could bring victory and a successful conclusion to the campaign. Therefore, a great part of the French doctrinal thought was tied up in the methodical battle, which was, in essence, an offensive doctrine. The French offensive doctrine of the interwar period had the following characteristics: (1) strict, centralized control by the corps and the army, with little room for initiative of junior commanders; (2) massive, centralized, and concentrated artillery support since firepower dominated the battle; and (3) forward movement by the infantry in short bounds of five kilometers or so, under massive artillery support, and at that point, the advance would halt in accordance with specific phase lines, so that the artillery could deploy forward, and the battle could be rejoined, on successive days. Under the terms of the methodical battle, commanders such as Maurice Gamelin, Henri Petain, and Maxime Weygand believed that the correct employment of doctrine could ensure victory.

The methodical battle had its origins in the campaigns and methodology of 1918. After the disasters of 1916 and 1917, it seemed that the French Army had finally discovered the secret of success on the battlefield: carefully planned offensives with massive firepower. Forerunners of the methodical battle had proved effective in the summer and fall of 1918. In essence, the tactics of late 1918 were geared to minimize casualties in the French Army. Studies on the armored force developed the corollary to the French dogma: namely, that infantry would not be able to advance without strong tank support.[54]

The French Army doctrine was couched in the terminology of science, or more accurately, pseudo-science. Articles and discussions within the army concerning the methodical battle contained numerous tables and formulae, which were published as appendices to the doctrine.[55] For example, various attacks required specific gun frontages per square kilometer before an attack could be initiated.[56] The effectiveness of the Maginot Line defenses, in another case, was illustrated by tables demonstrating the number of rounds from German heavy guns that were necessary to knock out each armored casement. In this example, the number of rounds was so large one could conclude that the Maginot Line was effectively unbreakable.[57] In yet another case, mathematical formulae were used to prove the effectiveness and superiority of antitank guns defending against a German armored attack.[58]

The French Army spent considerable time, effort, and money during the interwar period to develop the necessary artillery and tank arms which would support the methodical battle. Given the high priority of antitank weapons in French doctrine, the Army in 1940 was equipped with good antitank guns.[59] Since rapid movement and maneuver were not part of the French interwar doctrine, however, very little effort was devoted to developing radio communications as a natural corollary to mobility.

The development of French motor vehicles provides another useful illustration of the ascendancy of doctrine over technology. The Army used early-model half-tracks in experiments of the 1920s, and these greatly impressed the German observers.[60] Half-tracks were most suitable for rapid operational maneuver and motorized units, which, however, at the time were not emphasized in French doctrine. Thus, deployment of half-tracks was dropped for lack of interest. Armored cars played a relatively minor role in French doctrine as well, because reconnaissance had less importance. As a result, Panhard armored cars of the 1930s were given a low priority, although the quality of the product was technically equal to that of the Germans.[61]

Yet another case for the primacy of doctrine over technology is seen in antiaircraft guns. French antiaircraft was the responsibility of the army's Artillery Directorate. The army placed little confidence in airpower having a decisive effect on the battlefield. Gamelin himself believed that the losses of aircraft in the first few weeks of the war would be so heavy, that airpower would cease to be an important factor in the battle.[62] Therefore, the French lagged behind in the development of antiaircraft guns.[63]

French Air Force Doctrine and Technical Development

The technological development of the French ground forces was limited by the rigid doctrine of the army; the French Air Force suffered from the opposite situation: a lack of clear doctrine providing consistent paradigms for the development of technology.

In the decade after World War I, French air doctrine developed little from the operations of 1918. Until 1933, the French Air Force was part of the army, and the army was primarily interested in reconnaissance, observation, close interdiction, and air defense. The use of airpower was not a priority in army doctrine of 1921; indeed, aviation was scarcely mentioned in army regulations. In the 1930s the commanders in chief, Generals Weygand and Gamelin, demonstrated little knowledge or informed interest in military aviation.

French air forces were largely left to their own devices in developing doctrine, but even after it became an independent service, the air force's generals lacked the status both within the military and within the French political system to further specific doctrines or approaches to technology. The army commander, as the military's supreme commander, had the status to establish doctrinal and strategic guidelines for all the services but showed a lack of interest in aviation doctrine. This meant that the French air doctrine of the 1920s and 1930s became, by default, the responsibility of civilian air ministers; and with changes in air ministers, doctrine changed dramatically.[64]

Though the French air arm had mainly a ground support function in the late 1920s, the strategic bombing theories of the Italian general, Giulio Douhet, gained a wide acceptance among the air force officer corps. In the late 1920s, a program to produce a "battle plane" in accordance with Douhet's doctrine was initiated. Known as the "BCR" (Battle, Combat, Reconnaissance) aircraft, this multiseat, two-engine craft was to carry out army support functions and also be able to reinforce the heavy bomber force in long-range, strategic operations. This attempt to apply Douhet's doctrine to technology resulted in a series of thoroughly inferior multipurpose aircraft, which proved to be mediocre in each mission.

In January 1936, Pierre Cot became French aviation minister and inaugurated a series of major rearmament programs for the air force. Due to the poor performance of the aviation industry in developing and manufacturing aircraft, Cot initiated a program to nationalize and rationalize French aviation production. By infusing the aviation industry with large amounts of new capital, he hoped to create the large air force France needed. From 1936 to 1938, under Cot's ministry, the primary focus of the French Air Force was in building a strategic bomber force. Cot firmly believed in the offensive mission of the air force and was an enthusiast for Douhet's doctrine.[65] In 1938, however, when the government changed and Cot was removed, the new air minister, Guy LeChambre, began a new armaments plan for the Air Force, known as "Plan 5." Plan 5 rejected the emphasis on bomber production and instead placed the development and production emphasis on fighter planes.[66] Guy LeChambre's vision of airpower was essentially the same as General Gamelin's, in that the priority of the French Air Force was to form a defensive line to protect Army operations.[67] Bombing became a secondary mission. Thus, by the outbreak of the war, the French had in one decade undergone three major changes in operational doctrine, all instituted by the Air Ministry. The nationalization program by the war's outbreak had produced results in increased aircraft production. Nevertheless, French aircraft production continued to lag behind that of the Germans. The German policy tended toward the standardization of a few kinds of aircraft for specific missions. For example, the Germans built only one, single-engine fighter in quantity before World War II: the Me-109. The French, however, distributed aircraft production among the various aircraft companies, and ordered small quantities of many different aircraft models. This practice is in sharp contrast to the French policy of producing few types in large quantity that proved so successful in World War I. The French were unable to achieve anything resembling economies of scale in the 1930s, so that by the war's outbreak, the French were flying a half-dozen different single-engine fighters to Germany's one. The same situation existed for bombers and reconnaissance aircraft.[68] Naturally, this resulted in far more complicated logistics, supply, and procurement problems for the French Air Force, and consequently low operational rates.

On the military side, the French air force chiefs of staff in the 1930s showed little interest in operational innovation. The two air force chiefs of staff prior to World War II: General Philippe Frequant (October 1936–February 1938) and General Joseph Vuillemin (February 1938–July 1940) initiated no programs for the air force beyond traditional technologies of standard bombers, fighters, and reconnaissance models. Thus, in 1940, the French Air Force had no radar and little in the way of radio navigation equipment. In particular, after little serious study or experimentation, the French Air Force rejected the concept of the dive bomber.

What is especially remarkable about the French Air Force is that, even in fields of aviation where doctrine demanded specific technical solutions, very little effort was taken to link the aircraft with doctrine, with the possible exception of the ill-fated BCR Program. In a nation with a strategic bombing doctrine, as France had until 1938, one might have expected an emphasis on long-range navigation and instrument flying, yet the French in the 1930s were far behind the Germans, Americans, and British in developing basic navigational instruments.

The Effect of Doctrine on German Technology

The Germans, like the French, firmly believed that doctrine should ideally help guide technological development. In the German case, both ground-force and airborne technology demonstrated the impact of operational doctrine on development and procurement of equipment.

Immediately after World War I, the German Army, led by Hans von Seeckt as chief of the general staff, instituted a massive study of the lessons of World War I with the intention of developing new operational doctrine for ground and the air forces. In the period of 1919–21, 500 German officers—commanders, general staff officers, and technical experts—were put to work on committees analyzing every aspect of the operations of the war, from mountain operations to bombing; from tanks to fighter defense. By 1921, the 500 officers, including 130 airmen, had completed their thorough study of the war's lessons, and out of this was distilled the German operational doctrine, Army Regulation 487, *Leadership and Battle with Combined Arms.*[69]

Von Seeckt, who coordinated the postwar study of doctrine, took an entirely different position from the French military. Von Seeckt believed that maneuver, not firepower, was the dominant element in warfare, and that to execute maneuver, mobility was essential. Unlike the French, the Germans believed much more strongly in movement to effect offense. The Germans sought the means to get the armies out of the trenches and static battles, which were seen ultimately to Germany's strategic disadvantage, and to win campaigns and decisions quickly, by maneuver and mobility. In contrast to the French, who affirmed the essential unity of the army,[70] von Seeckt and the Reichswehr maintained a doctrine of warfare which essentially relied on the establishment of two different armies. One army would be an elite force, heavily armed with the latest weaponry and highly mobile. This army would be the offensive force, which would seek the decisive battle and out-maneuver its opponents. The second army would be primarily infantry, not as well armed, and would for the most part consist of reservists. This second army would act mainly in a defensive capacity.[71]

Again in contrast to the French, Army Regulation 487 emphasized in its preface that war was an art—albeit, a rational art—and not a science. There was little of the pseudo-scientific emphasis on tables and formulae to be found in the French operational regulations of 1921 and 1936. Army Regulation 487 outlined a number of general principles to be followed, but no formulae. For example, the German operational doctrine decentralized the operational leadership and not only allowed, but insisted, that junior officers would possess considerable initiative in command. Artillery, which in France was highly centralized, was decentralized in the German Army. The emphasis was not on deploying large numbers of guns, as with the French Army, but rather on rapidly deploying smaller numbers of guns. Army Regulation 487 emphasized combined-arms operations, and airpower played an important part in the German Army operational regulations. In addition, due to the emphasis on offensive maneuver, tanks were given an important role in German doctrine of the 1920s.[72]

This operational doctrine led the German Army in a number of technological directions. For instance, the mobile battlefield required mobile communications. Accordingly, the Germans set about in the 1920s developing an entire family of army and air force radios for rapid communication. German doctrine also implicitly emphasized coordination of the air and ground forces at the operational level, as well as the use of tanks and armored vehicles. Finally, the doctrine of maneuver warfare pushed the German Army toward an intensive study of motorization. By 1926, the first table of organization and equipment for a motorized division and tables of organization and equipment for mechanized brigades had been set out in detail by the general staff.[73]

German Air Doctrine and Technological Development

During the 1920s, in the expectation of eventual rearmament and the creation of an independent air force within the German Army, a secret air staff was set up to perform the functions of an air force general staff. Between 1919 and 1921, the air staff carried out a comprehensive study of World War I, and established a number of principles of air war, which would form the basis for German air doctrine in the interwar period.

The German air doctrine of the 1920s was set out in the *Directives for the Operational Air War,* written in 1926.[74] The conclusion of the air staff from their study of World War I was that airpower was intrinsically most effective in the offense, not the defense. Even though German airmen had fought a defensive air war during World War I, and had enjoyed an extremely high kill ratio versus the Allies, they discovered that a strong aerial defense did not lend itself to decision in war.[75] Therefore, the Germans concluded that bombers were the primary weapon of the air arm. In an air campaign, the first priority of the air force would be to gain air superiority. The air superiority battle would be won by taking the war to the enemy and destroying his air force, preferably on the ground. Once air superiority had been gained, the air force would move to the interdiction and strategic bombing missions.

As with the army, the air force doctrine essentially outlined the creation of two air forces: one air force would be an army support force, flying short-range reconnaissance missions, providing fighter defense for the army, and carrying out ground attacks. The second air force would be an independent air force serving under the strategic but not operational direction of the High Command. The independent air force had a strategic mission. The operational air force's strategic mission was not solely to bomb cities or industries deep in the enemy heartland—although this was a possible mission. The German concept of strategic air war was based on strategic effect. The independent air force, essentially a bomber force and a long-range force, would be directed toward those targets that would produce the most decisive effect. This could sometimes entail bombing the enemy armaments industry, but in other cases, it could entail direct bombing of the enemy army or his transportation.[76]

German air doctrine of the 1920s applied some of the Prussian Army's most traditional principles to the new aerial weapon. First of all, airpower would be used in mass and not be distributed in small packets. Second, the strategic air force would maintain a large operational reserve, ready to exploit opportunities. Third, airpower would be used at the decisive point.

In 1935, the reestablished Luftwaffe published a new operational regulation, Luftwaffe Regulation 16: *Conduct of the Air War.* The essential principles of air war that had been developed in the 1926 were outlined in greater detail. However, the basic lines of doctrinal evolution remained unchanged. This doctrinal stability was of great benefit in developing weapons and equipment. The Luftwaffe that went to war in 1939–40 was the bomber-heavy force that was called for by the doctrine of the interwar years.[77] The dive bomber, under study and development since the mid-1920s, was available for the close air support (CAS) mission and also to hit strategic targets. The German doctrine called for fighter escort of the bombers, and the Me-110 long-range fighter was developed for this purpose. Army/air force cooperation was an important part of doctrine, so a considerable organization equipped with mobile communications systems was created.[78]

Consistent with the French, the deficiencies in the German doctrine also resulted in deficiencies in equipment development and procurement. The best illustration of this principle in the Luftwaffe is found in the lack of interest in doctrine for naval aviation before World War II. At the outbreak of the war, modern combat aircraft designed for long-range antishipping strikes and torpedo attacks were not available. The naval air arm had to make do with relatively obsolete, low-performance seaplanes. It was a deficiency that would hurt the Germans considerably when they went to war against Britain.[79]

The Influence of Organization on Technical Development

The French Army and Air Force were poorly organized to develop and oversee the production of modern equipment. From the end of World War I until the early 1930s, the French Army possessed no centralized office specifically charged with the responsibility for developing and evaluating new technology. In the postwar era, each of the major branches of the army (infantry, cavalry, engineers, artillery, etc.) had its own technical office and was responsible for developing the equipment that pertained to that branch. The infantry branch, for example, had responsibility for tank development, as tanks were considered to be an auxiliary of the infantry. Development of other armored vehicles such as armored cars were, however, the responsibility of the cavalry branch. No section of the general staff carried the authority to coordinate branch weapons programs or to ensure an objective program of testing.[80] While each branch technical office contained some officers who were qualified and even highly talented in technical matters, there existed no comprehensive or systematic program of cooperation between the branches. Development in the French Army was carried out in a fragmented, compartmentalized manner.

In the early 1930s under the initiative of General Weygand, then vice-president of the *Conseil Supérieur de Guerre,* some attempt was made to bring order to the process. A Consultative Council on Armament was created consisting of the senior branch inspectors, general staff department heads and the chief of the general staff. A Technical Cabinet was created to act as a central office for research, testing and manufacture. Yet, the reform had only a partial effect. The actual development of weapons and the establishment of the specifications for the equipment still resided within the branches of the army. In 1933, the energetic and capable War Minister Eduard Daladier created a new department for the manufacture of armaments which would execute the armaments plans formulated by the branches, but supervision over the departments themselves was still lacking.[81] Finally, in 1935, the Technical Cabinet was replaced by a new section for armaments as a permanent office of the general staff. This new section, however, carried relatively little formal authority to coordinate the development and procurement process.

The French Air Force's development of equipment suffered from organizational problems within the French aviation industry and especially within the Aviation Ministry. The first problem of the air force was one of command authority. In wartime, the air force was subordinate to the army. In peacetime, however, the air force operated under the Ministry of Aviation. In 1928, when the Aviation Ministry was created, the air force was still a branch of the army, and the French aviation industry was in a state of decline. In the 1920s the aviation industry lived primarily off small orders from the military. Although from the 1920s into the 1930s, the French commercial airlines received the highest subsidies in Europe,[82] the French aviation industry made little progress in developing modern and competitive transport planes or an infrastructure of modern airfields.[83] Indeed, mismanagement, waste, and even criminal fraud seem to have soaked up funds provided to French civilian aviation. The Aéropostale scandal of 1933, in which airline officials were found guilty of graft and theft, was one of the embarrassments that triggered the nationalization of the aircraft industry in the mid-1930s.

While there were good arguments for nationalizing the aircraft industry, there are many examples of the negative effect that the nationalization had on production and development of aircraft. Ministry politics seem to have played as large a role in the development and production of aircraft as the requirements of national defense. Marcel Bloch-Dassault, owner of Bloch Aircraft Company and one of the leading aircraft designers in France (Bloch-Dassault would later design the Mirage Jet) was removed as director of his company when nationalization came. A year later, Bloch was asked to return, but he was dismissed again in 1939.[84] The Air Ministry, a bloated and poorly organized agency (when it was established in 1928 there were over 1,000 ministry employees in the Paris headquarters alone), had notoriously poor relationships with manufacturers and commercial organizations.

While the Air Ministry performed poorly in its duty to develop aviation technology, the French Air Force leadership deserves much of the blame for France's position in 1940. While the air force carefully followed the development of aircraft, there was little planning in the air staff for the material and personnel necessary maintain the aircraft and equipment required for a modern air force. At the outbreak of the war, the French Air Force had only 40 percent of their required radiomen and 23 percent of the required mechanics. A special commission was set up by the air force to determine personnel needs on 26 September 1939—almost a month after the outbreak of war.[85] Other examples of poor industrial planning by the staffs abound. At the outbreak of the war, French production plummeted due to the call-up of skilled aircraft factory technicians to serve as reserve infantrymen at the front. Later in 1939, many soldiers were released from duty in the army to return to war production.[86] Numerous similar occurrences happened in the manufacture of armaments for the army. Renault's largest tank plant was almost closed in September, 1939 due to the call-up of skilled workers.[87] With decades to plan for a major war, neither the French Air Force nor the French Army had developed a staff or adequate plans for coordinating the economic side of warfare.

The failure of the French Air Staff to plan or organize for the broader requirements of technology was directly translated into extremely low readiness rates for French aircraft in May 1940. Exact figures for aircraft operational rates are not available (another sign of French disorganization) for May, 1940; but a fair estimate from the numbers of aircraft that flew on missions is an average operational rate of about 50–60 percent for fighter units and no more than 40 percent for bomber units.[88] Even today, no one is sure of how many aircraft were grounded for lack of bombsights, radios, machine guns or other basic equipment—but even the official histories imply that the numbers were in the hundreds.[89] The Germans began the campaign in 1940 with significant aerial superiority. Although in May 1940, the French Air Force had over 2,200 aircraft available,[90] the lack of coordination and planning by the French Air Force ensured the Germans a decisive margin of superiority.

German Organization and Technology

The Reichswehr's approach to developing equipment in the 1920s and 1930s was almost opposite to the French methodology. General von Seeckt, as army commander in chief from 1920 to 1926, reorganized the German Army Headquarters and General Staff to provide clear lines of responsibility for technical development as well as a centralized agency for technology within the army. As with the French Army, the German General Staff (*Truppenamt*) consisted of the normal departments:

Army Organization, Training, Intelligence, Operations and Logistics. Under the general staff came the inspectorates for the various branches of the army. A parallel organization to the Truppenamt, the Waffenamt or "Weapons Office" was created. The Waffenamt, which had as its chief a general of equal rank to the chief of the Truppenamt, had approximately as many personnel as the Truppenamt.[91] Under the Waffenamt stood inspectorates for weapons development which corresponded to the Truppenamt's inspectorates. It was the duty of the general staff inspectorates to develop ideas, doctrine, and training programs for new equipment as well as requests for research and specifications for new weapons. The inspectorate worked together with its corresponding Waffenamt inspectorate which would develop prototypes and conduct testing. The responsibility for the development of weapons and equipment lay directly in the hands of the chief of the Waffenamt who reported directly to the army commander in chief and the defense minister.[92]

Unlike the compartmentalized .French system, German development of technology was coordinated as a whole, with several inspectorates cooperating on the development of some items of equipment. For example, the chief responsibility for the development of tanks lay with the motor vehicle section of the Waffenamt, which worked together with the Inspectorate of Motor Vehicles in the Truppenamt. However, representatives from the Inspectorate of Communications Troops were also assigned to the armor projects in order to ensure radios were developed for the tanks. The artillery inspectorates were assigned the responsibility for developing tank guns and also had members assigned to tank development projects.

Realizing the importance for production planning in modern warfare, the Germans created a war economics office in 1926; it reported to the army commander. The assignment of the office was to maintain contact with armaments industries, collect information, and carry out planning for industrial mobilization.[93]

This system, which the Nazis inherited in 1933, worked fairly rationally to create prototypes of equipment which matched doctrinal requirements with considerably less duplication of effort than the programs of the French General Staff. The French, for example, had two sets of tank programs in the 1930s—one in the Infantry Inspectorate and the other in the Cavalry Inspectorate. Although the Economics Planning Office could never effectively meet the enormous task given to it, in contrast to the French, at least a rudimentary personnel plan had been set which exempted skilled civilian workers in the armaments industries from military service at the outbreak of war. Consequently, the Germans in the early months of the war experienced only minor reductions in some areas of production due to the call-up of reservists.[94]

Due to the ban on military aviation required by the Versailles Treaty, the Germans had to rely on a more awkward organizational system for the development of military aviation technology. The shadow air staff had representatives spread throughout the general staff. For example, the Intelligence Section contained one or two airmen who specialized in air intelligence. Sections for aviation were distributed within the Weapons Office. Also involved in development of aviation technology was an aviation branch of the Ministry of Transportation. The Aviation Department was responsible for regulating all aspects of German civil aviation.

Despite this awkward system, which did not provide for any single agency for development of aviation, the Germans managed to forge ahead in development and, by 1927, develop aircraft technology superior to that of France.[95] Behind the successful rebirth of German aviation in the 1920s was the extraordinary level of cooperation between the armed forces, the Aviation Department of the Transportation Ministry, and the civilian manufacturers. The German undersecretary for aviation from 1923–34 was Capt. Ernst Brandenburg, wartime commander of the 1st Bomber Wing which had carried out the strategic campaign against London with Gotha bombers in 1917–18. Brandenburg, who was appointed at the insistence of the Army Commander von Seeckt, worked to develop German civil aviation as a basis for later aerial rearmament.[96]

Civilian aviation was staffed throughout by former pilot-officers who retained a reserve status within the shadow Luftwaffe. Information on all the latest developments in aviation, navigation equipment, and foreign aircraft types was collected by the air staff. Ehrhard Milch, a wartime captain of the Air Service and a director of Lufthansa, carried on an extensive correspondence with the air staff in the 1920s, providing Lufthansa experience in long-range navigation, aircraft engines, new instruments, and so forth.[97] Manufacturers such as Ernst Heinkel worked closely with the air staff to develop new aircraft at a time when the Inter-Allied Control Commission had shut down German aviation.[98] In 1926, senior Lufthansa manager (and wartime air officer) Dr. Robert Knauss made a pioneering long-distance flight from Berlin to Peking. Upon his return to Berlin, he delivered his flight logs to Lt. Col. Helmuth Wilberg, chief army staff officer for aviation.[99]

The Civil Aviation Department used its money wisely to develop the necessary basic infrastructure for modern aviation.[100] The level of research on aviation carried out in German institutes of the 1920s and early 1930s was equivalent to the level of French research, despite a smaller budget. By the late 1920s the Germans had planned for a limited rearmament, and the secret air staff accelerated work in developing prototypes for a reborn German Air Force. With a realization that the air staff component of the Reichswehr General Staff was diluted among too many offices, General Blomberg in 1932 prepared and won approval for a plan to consolidate all the Reichswehr air activities into one department which would be called the "Air Defense Office."[101] Col. Wilhelm Wimmer, active in the development of military prototypes since the 1920s, became head of the Technical Office under the new organization.[102]

Under the new Air Ministry, established in 1933, Wimmer and the army's aviation experts argued for building up the very small German aviation industry (battered by the depression, German aircraft manufacturers had only 3,200 workers in 1932[103]) by contracting for moderate numbers of relatively mediocre aircraft already developed such as the He-51 and Arado 68 fighters and Ju-52 and Do-23 bombers. While the aircraft industry expanded to fill the initial orders, a second generation of high-performance aircraft was developed. This second generation developed under the tenure of Wimmer as chief of the Technical Office included the Me-109, Me-110, Do-17, He-111, and Ju-87—all aircraft on the cutting edge of technology for the era.

There were numerous inefficiencies in the Luftwaffe's prewar rearmament program, not least the appointment of Ernst Udet to serve as chief of the Technical Office. Whereas the second generation of German aircraft was developed with great rapidity, the third generation was plagued by delays caused by continual redesigns. In one case, a very effective bomber, the Ju-88, was delayed by Udet's insistence that it be capable of dive-bombing. With the exception of the FW-190 fighter, the third generation of Luftwaffe aircraft (Hs-129, Me-210, He-177) was disappointing. The problem with overcentralized control is that when the person at the center is incompetent, the damage can be great. However, with the exception of Udet's appointment, Reichsmarshal Hermann Goering, who had little understanding of modern aviation, interfered little with the technical decisions made by his generals in the prewar period. Even Udet could make some technical decisions more astute than those of the French leaders. For example, in 1936 Udet decided that the Me-109 would be Germany's only single-engine fighter. This ensured that mass production and economies of scale would give the German fighter force superiority over the French in 1940.

Military Culture and Technology

The tradition and culture of the German General Staff enabled it to adapt at a rapid pace to technical development and experimentation in the interwar period. The tradition of the general staff, which was maintained and strengthened under the tenure of Hans von Seeckt, emphasized a logical and critical approach to questions of operational doctrine, military organization, and equipment. The general staff corps had great prestige within the army, and the officers selected for the general staff were allowed a considerable degree of freedom to question, criticize, and propose new ideas. While the senior commanders of the army were expected to have strategic and operational vision, even the junior officers of the general staff were allowed and encouraged to make modifications and contributions to the ideas proposed by their seniors. In short, the general staff corps saw itself as a collective body which had the responsibility to develop ideas.

One of the most characteristic expressions of the German military culture was the *Denkschrift*, literally "Thought Paper." Officers would propose ideas or critique ideas in essays circulated throughout the general staff. The tradition of the general staff was such that these essays were read by the senior commanders, and often acted upon. At the very least, the Denkschrift would provide a framework for debate.

In the immediate aftermath of World War I, two important thought papers concerning war and technology were circulated among the High Command. The first was von Seeckt's proposal for a small, elite and highly mobile professional army. Maneuver warfare, von Seeckt argued, required a professional force, because only a highly trained elite force could hope to use the complex modern weapons effectively and move with rapidity to gain the decision.[104] The second influential Denkschrift was

"The Technical and Tactical Lessons of the World War," written by Col. Kurt Thorbeck in 1920. In it, Thorbeck ruthlessly criticized the general staff for not having officers conversant with the technical and material demands of war. He called the general staff's lack of technological familiarity "the basic mistake of the war."[105] Thorbeck's critique would lead to a major reform of military education and culture under von Seeckt.

The tradition of the Denkschrift, combined with a tradition of critical examination, meant that the role of technology was emphasized in the comprehensive study of the lessons of the war carried out in 1919–20.[106] One of the most important, and also the most overlooked, reforms that von Seeckt made in the German General Staff was a new program to provide technological expertise. Before World War I, entry into the general staff had been by competitive examination and the three-year course at the Kriegsakademie (War College). Starting in the early 1920s, officers could enter the general staff by attending a civilian university or technical college, and earning a technical degree. Approximately 10 officers per year (out of a total officer corps of 4,000) were selected by the Reichswehr for attendance in civilian engineering programs, their fees paid by the army. When the course was completed, the officer would return to regular duty, usually serving as a specialist with the Waffenamt.[107] Many of the officers, however, were also given the opportunity for troop commands. Some of the most senior Wehrmacht leaders of World War II took engineering degrees in lieu of the Kriegsakademie—notably, Field Marshal Wolfram von Richthofen, who earned a doctorate in engineering at Reichswehr expense.[108]

Along with the program in technical education, von Seeckt also insisted on an informal system of technical education within the officer corps. Dissatisfied with the level of knowledge of technology and foreign weapons demonstrated by the general staff officers, von Seeckt instituted a program of bimonthly seminars for the officers in the Truppenamt, who would spend the morning or afternoon being briefed on the latest technical developments by experts from the Waffenamt.[109] Through his insistence on technological literacy, and by encouraging the serious study of engineering by the officer corps, von Seeckt helped ensure that the officer corps as a whole would be infused with an interest in, and appreciation for, technology.

One example of technology developed within the general staff tradition is the effort of Erich von Manstein toward promoting the concept of the assault gun, or *Sturmgeschütze* for the German infantry divisions. In 1935, von Manstein wrote a Denkschrift outlining his idea for creating an assault gun: a heavy gun mounted without a turret on a tracked chassis, that could provide limited armored support for infantry divisions which were faced with fixed enemy defenses. Such a gun would be cheaper and simpler to build than a tank, and would be of enormous value to the infantry in defense or offense. Manstein argued that a detachment of such guns should be assigned to each infantry division, while tanks should remain concentrated in the panzer divisions. Heinz Guderian and the staff of the Inspectorate of Panzer Troops strongly opposed the idea of putting assault guns in the infantry divisions, but the infantry and artillery inspectorates saw merit in the idea.[110] By 1937 development and testing of the assault guns was underway, and trials proved their feasibility. By the outbreak of the war, production of the assault guns was underway, and a detachment of the first models, with a 75-mm gun mounted on the chassis of a Panzer III tank, proved their worth in battle.[111] The assault gun, developed in many versions, would become one of the most valuable battlefield weapons of the German Army in World War II. Examples such as the assault gun demonstrate the importance of open and honest debate within the general staff, and the possibilities for nontechnical specialists to develop innovative solutions to operational problems. The interest of von Manstein in the details of new weaponry also illustrates the wide dissemination of technical knowledge and interests throughout the army in the interwar period.

The Luftwaffe: Leadership and Technology

The Luftwaffe was the most technologically oriented of the German services in the interwar period. The nature of the ban on military aviation during this period meant that the Luftwaffe had to be created in the early 1930s from a small corps of airmen retained by the Reichswehr after World War I. By necessity, the Luftwaffe had to rely on the former officers and pilots of the wartime air service who had served in civil aviation from 1919–34. When rearmament came, there were several hundred wartime officers with combat experience working for Lufthansa, for the aircraft companies, for the Air Department of the Transportation Ministry, and in the civilian flight schools. These officers were eager to reenter the military and join the new Luftwaffe. Just how important these reserve officers were to aerial rearmament is demonstrated by the proportion of senior leaders of the Luftwaffe who came from interwar civil aviation: of the 600-plus generals of the Luftwaffe serving between 1935 and 1945, approximately 150 had been involved with civil aviation between 1920 and 1934.[112] Erhard Milch, state secretary for aviation and field marshal, served as a director of Lufthansa before rejoining the military. General der Flieger (of fliers) Robert Knauss, later to be commander of the Luftwaffe General Staff College, also came from the Lufthansa board of directors.[113] Col. Gen. Alfred Keller, commander of the First Air Fleet from 1940–43, worked for Junkers and ran a flight school before 1934.[114] Lt. Gen. Werner Junck, wartime commander of Jagdkorps II, worked for Heinkel before joining the Luftwaffe.[115] Lt. Gen. Theodor Osterkamp, a World War I *Pour le Mérite* holder, managed a seaplane station prior to returning to the Luftwaffe. Osterkamp would become the air commander for North Africa in 1941–42.[116] Officers with a specialized technical background in civil aviation were enlisted into the technical and special staffs of the new Luftwaffe, and many rose to high rank.

Several historians, including Richard Overy, have suggested that the Luftwaffe suffered from serious leadership problems, since such a large proportion of the Luftwaffe officers were brought in from civil aviation.[117] Other writers, noting the several hundred officers transferred from the Army in 1934–35—Walther Wever,

Albert Kesselring, and Hans-Jürgen Stumpff, to name a few—refer to the senior leadership of the Luftwaffe as "amateur aviators."[118] Overy argues that there was a clash between the "Prussians," the regular officers who had remained with the Reichswehr, and the "Outsiders," who had reentered the military in 1934–35. This clash of cultures and viewpoints seriously damaged the Luftwaffe.[119]

Of course, there were serious personality clashes between senior officers of the Luftwaffe—as with any military service—but there is no evidence of animosity on the basis of "Prussian" or "Outsider" status. That Milch was disliked by many was more a function of his own personality than his service with Lufthansa. Even those who disliked him regarded him as highly competent.[120] Walther Wever came from the army but was nevertheless highly respected by the professional airmen of the Reichswehr.

I would argue that the influx of hundreds of reserve officers from civil aviation in the first stages of the German aerial rearmament was one of the great advantages that the Germans enjoyed in the interwar period. The officers from Lufthansa or Junkers were probably better informed about the nature of modern aircraft technology and the conditions of long-distance flying than regular French officers, who had led an air force garrison life of staff and flying jobs during this same period. The Luftwaffe's superior use of human resources is one factor which enabled the Germans to gain the technical advantage over the French in 1940. The director of airfield construction for the Luftwaffe was brought in from civilian life as an airfield construction engineer and professor of architecture at a civilian university.[121] No professional military airmen knew more about the management side of aviation, nor had they achieved the same degree of success, as Erhard Milch. As to the amateur status of officers brought from the army—many, like Kesselring, learned to fly and later proved themselves to be excellent operational air commanders.

The French Military Culture

The French General Staff tradition was significantly different from that of the Germans. In the French tradition, the staff was no more than an organization to assist the commander. The direction, ideas, and vision all flowed from the commander. Open debate of operational concepts was not part of the French military culture. In contrast to the comprehensive and critical effort of the 500 officers who worked to develop German operational doctrine after World War I, the French operational regulation was drawn up by a committee of 13 officers.[122] When Charles DeGaulle initiated a debate about the organization of a large armored force in 1934, he was punished by having his promotion to colonel delayed.

Jenny Kiesling has argued that the French interwar army discouraged debate because, in an army dependent on large numbers of reserve officers, bringing the doctrine of the army into question would indicate a lack of confidence and thus undermine morale.[123] Other factors may also have ensured a less critical approach by the French officer corps. If the French Army had initiated a comprehensive examination of the lessons of World War I, they would have had to discuss and attempt to come to grips with the scandalous mass mutinies of 1917, when half the divisions of the army were incapacitated by their refusal to attack. Although these mutinies were quelled by hundreds of secret executions, the French Army has, to this day, refused to examine those events.[124] An honest evaluation of the performance of the army and its senior officers would have probably crippled the army in its relations with the government.

Thus, in the interwar period, the French Army was dependent on the understanding of technology possessed by its senior officers. The two most important commanders of the interwar period; Marshal Petain, vice president of the War Council from 1920 to 1931, and General Gamelin, army commander from 1935 to 1940 and chief of the Defense Staff from 1938 to 1940, were knowledgeable about most aspects of ground-force equipment. Neither, however, had any understanding of aviation technology, and accordingly the French air weapon suffered from the senior commanders' neglect. Such German Army commanders as von Seeckt, Beck, von Blomberg, and von Fritsch, on the other hand, demonstrated a strong interest in, and support of, military aviation.

The French also suffered from the lack of vision displayed by its commanders regarding motorization. To be sure, from the 1930s on, Gamelin placed a high priority on the creation of motorized divisions, but his concept of motorization was essentially flawed. Gamelin was interested in motorization as a purely strategic concept: the creation of a motorized reserve force which could move quickly to Belgium in order to deploy against the Germans. Gamelin, moreover, had no concept of operational mobility: motorization helped infantry divisions move quickly by truck; once they arrived, they would dismount and fight like any other line infantry division.[125] Due to this lack of an operational concept, the French did not develop communications, armored carriers, self-propelled guns, and so on, like the Germans—though the French did have the required expertise and the industrial base.

The culture of the French Air Force command played a central role in that force's poor position in 1940. During the interwar period, the French Air Force in many ways resembled a pilots' club. Commanders were interested in developing aircraft types, but little thought was given to creating the infrastructure of an operational force or in planning for industrial mobilization. Like the Germans, the French possessed a large reserve of experienced airmen from World War I who had entered civil aviation. As rearmament accelerated in the 1930s, however, the only interest that the regular air force officers had for their reserve officers involved their flying proficiency.[126] There was no search for skills outside the narrow field of piloting. France produced as many skilled aircraft engineers in the interwar period as Germany, but the French Air Staff had little interest in recruiting such men. The French interwar air force is one of the best examples in history of the misuse of excellent human resources. If Marcel Bloch-Dassault had been a German, instead of being fired he would have likely ended up a general in the Luftwaffe.

Conclusion

Many elements contribute to the technology of an armed force. The economic base of a nation, its educational system, and its financial position all are decisive factors, and their importance should not be underrated. A comprehensive approach to the study of interwar technology would require a hefty volume. I have therefore confined myself to a few factors concerning the effect of military doctrine, organization, and culture on technology. My conclusion is that these factors have as much bearing on the development, procurement, and employment of weapons as the objective economic and scientific factors. Furthermore, the comparison of the French and Germans in this period illustrates the importance of the individual military commanders on the development of technology. One can plausibly assign some of the blame for the poor state of the French Army and Air Force in 1940 to the politicians and to the economy, but this does not absolve the commanders from failing in their duty to oversee the development of an effective doctrine and effective weapons to match that doctrine. Despite the complaints of the official French histories concerning the lack of funding for the military in the interwar period, this was not the major cause for technological deficiency.[127] In those areas where the French provided higher funding than the Germans—notably, civil aviation—they still went to war with inferior technology.

It is certainly not my intention to claim that the German approach was foolproof, or even particularly efficient. The Germans built their share of bad weapons and aircraft. As Richard Overy points out in *War and Economy in the Third Reich,* the German war economy of 1939–40 was extremely inefficient.[128] Volumes have already been written on the technological mistakes the Germans made before and during the war.[129]

The purpose of this essay has been to compare the two approaches to technology, and the German approach still comes across by far as the most successful. It was primarily successful due to a military culture that encouraged innovation, discussion, debate, and a comprehensive approach to the study of war.

Notes

1. In 1917, the French Army deployed the Canon de 155-mm GPF as the standard heavy gun. With a range of 19,500 meters and a 43-kg shell, it was highly respected by the Germans and the Allies. See Peter Chamberlain and Terry Gander, *Heavy Artillery* (New York: Arco, 1975), 17. On other French guns, see pp. 14-19.

2. Werner Oswald, *Kraftfahrzeuge und Panzer der Reichswehr, Wehrmacht und Bundeswehr* (Stuttgart: Motorbuch Verlag, 1975), 10-11. In 1918, the French Army employed 100,000 motor vehicles, not including tanks, on the Western Front. The German army motor vehicle total during the war did not exceed 40,000 in use.

3. See Kenneth Macksey, ed., *The Guinness Book of Tank Facts* (Enfield, England: Guinness, 1980), 28.

4. Ibid., 45.

5. Oswald, *Kraftfahrzeuge und Panzer,* 36-39.

6. John Morrow, *The Great War in the Air* (Washington, D.C.: Smithsonian Institution Press, 1993), 369-71. In World War I, the French built 52,000 planes and 88,000 engines. The Germans built approximately 48,000 planes and 43,000 engines.

7. Charles Christienne and Pierre Lissarrague, *A History of French Military Aviation* (Washington, D.C.: Smithsonian Institution Press, 1986), 117.

8. On French aircraft at the end of the war, see Morrow, *The Great War in the Air,* 363-71, and Christienne and Lissarrague, *History of French Military Aviation,* 117-22, 155-57.

9. In 1918, Squadrons 1, 19, 23, 29, and 60 of the Royal Flying Corps and RAF flew Nieuport 17s, Spad VIIs, and Spad XIIIs on the Western Front. See Christopher Shore and Norman Franks, *Above the Trenches* (London: Grub Street, 1990), 30-36.

10. Macksey, *Guinness Book of Tank Facts,* 37.

11. At the St. Mihiel Offensive of the U.S. Army in September 1918, the French provided 185 tanks and the United States 174 tanks. The French supported the U.S. offensive in the Meuse-Argonne in September/October 1918 with 750 tanks. See Macksey, *Guinness Book of Tank Facts,* 231.

12. See Patrick Fridenson, "Les relations entre les industries automobiles française et allemande des anneés 1880 aux anneés 1960," in Yves Cohen and Klaus Manfrass, eds., *Frankreich und Deutschland: Forschung, Technologie und industrielle Entwicklung im 19. und 20. Jahrhundert* (Munich: C. H. Beck'schen Verlag, 1990), 334-42, esp., 335-36.

13. See Walter Spielberger, *Die Motorisierung der deutschen Reichswehr 1920-1935* (Stuttgart: Motorbuch Verlag, 1979), 145-51.

14. For a detailed review of French motorization efforts 1919-1939, see Ministère de la Défense, *Les Programmes d'Armament de 1919–1939* (Château de Vincennes: French Defense Ministry, 1982), 260-329.

15. See Christopher Foss, ed., *An Illustrated Guide to World War II Tanks and Fighting Vehicles* (New York: Arco, 1981), 16-19.

16. In 1923, France was rated as the strongest aeronautical power. It had an air force of 123 squadrons, with 1,050 modern aircraft. See Aeronautical Chamber of Commerce of America, *Aircraft Year Book 1924* (New York: n.p., 1924), 185-93.

17. The National Aeronautics Institute produced about 100–150 aeronautical engineers a year in the 1920s and 1930s, although many only had a two-year course in engineering. See *L'Ecole Nationale Supérieure de l'Aéronautique: Cinquante anneés d'existence (1909-1959)* (Paris: National Aeronautics Institute, 1959), 59-60, 65-66.

18. Maurice Larkin, *France Since the Popular Front* (Oxford: Clarendon Press, 1988), 389. Between 1925 and 1929, France produced an average of 207,000 motor vehicles per year to Germany's 90,000; between 1930 and 1934, France produced 193,000 per year to Germany's 101,000; and between 1935 to 1939, France produced 200,000 per year to Germany's 304,000.

19. R. M. Ogorkiewicz, *Armoured Forces* (New York: Arco, 1970), 177-78, 336.

20. Ibid., 343.

21. On German and French artillery of the interwar period, see Peter Chamberlain and Terry Gander, *Light and Medium Field Artillery* (New York: Arco, 1975), 13-30, also Chamberlain and Gander, *Heavy Artillery,* 14-26.

22. See note 21.

23. Between 1923 and 1939, the French military devoted only 0.15 percent of their military budget to communications equipment. See Robert Doughty, "The French Armed Forces, 1918-40," in *Military Effectiveness* (Boston: Unwin Hyman, 1988), 2:39-69, esp. 58. The French had commercial technology which could have been exploited for the military. In 1931, a commercial UHF link was opened between Britain and France. The UHF frequencies, however, though used extensively by the Germans, were not developed by the French. See Tony Devereux, *Messenger Gods of Battle* (London: Brasseys, 1991), 84.

24. In the interwar period, one of the largest of the French radio procurement programs was

for the OCTF, and for the R and F type radios and receivers (650 radios) for the Maginot Line fortifications. Only a handful of radios were planned for the Char-B and R-35 tanks. See Ministère de la Défense, *Les Programmes d'Armament,* 416.

25. During the 1920s and early 1930s, the Weapons Office and Communications Inspectorate developed a broad family of effective radios for the tactical use of the Army. See Adolf Reinicke, *Das Reichsheer 1921-1934* (Osnabrück: Biblio Verlag, 1986), 196.

26. On oil and French strategic planning in the interwar period, see R. Nayberg, "La Problématique du revitaillement de la France en carburant dans l'Entre-deux-guerres: naissance d'une perspective géostratégique," *Revue Historique des Armées,* no. 4 (1979). For a good overview of the strategic effect of the oil supply on German motorization, see Richard DiNardo, *Mechanized Juggernaut or Military Anachronism?* (Westport, Conn.: Greenwood, 1991), 7-9.

27. By the 1930s, the Germans had developed the Mercedes-Benz DB-602, a 16-cylinder diesel aircraft engine rated at 1,320 hp. The French developed the Clerget 16H, a 16-cylinder diesel aircraft engine rated at 2,000 hp. See Paul Wilkinson, *Aircraft Engines of the World: 1941* (New York: Paul Wilkinson, 1941), 104-5, 168-69. The Germans took the world lead in diesel aircraft engines. The Ju-86 bomber, which first flew in 1934, was powered by the Jumo 205 diesel engine. It had a low power-to-weight ratio, but low fuel consumption. See William Green, *Warplanes of the Third Reich* (New York: Galahad, 1970), 414.

28. The Bloch 152 fighter, a mainstay of the Armeé de l'Air in 1940, had a maximum speed of 316 mph, and an armament of two 20-mm cannon and two machine guns. It was slower than even the German Me-110 heavy fighter. See Kenneth Munson, *Fighters 1939-45* (London: Blandford, 1969), 39. The Amiot 143 bomber, used by the French in 1940, was designed in the late 1920s, and had a maximum speed of 193 mph and a bomb load of 1,300 kg. The Bloch 210 was designed in 1932, had a maximum speed of 200 mph and a bomb load of 1,600 kg. The primary German bombers of 1940 were the Heinkel He-111 and the Dornier Do-17. The Heinkel He-111 had a maximum speed of 252 mph and a bomb load of 2,500 kg. The Do-17 was faster, at 255 mph. Both clearly outclassed most of the French bomber force. See Enzo Angelucci and Paolo Matricardi, *Combat Aircraft of World War II 1933-1937* (New York: Military Press, 1987), 22, 30.

29. The LO-45 (also known as the Leo 451) was a good medium bomber, with a maximum speed of 250 mph and a bomb load of 1,500 kg. Although it compared well with German aircraft in 1940, only 5 were operational as of September 1939, and perhaps only 110 were operational by June 1940. See Enzo Angelucci, *Rand McNally Encyclopedia of Military Aircraft* (New York: Gallery Books, 1990), 281-82.

30. The Dewoitane D-520 had a maximum speed of 326 mph and carried a 20-mm gun and four machine guns. The Me-109E aircraft it faced was faster, at a maximum speed of 357 mph, and had two 20-mm cannon and two machine guns. See Munson, *Fighters 1939-45,* 56, 64.

31. Ibid., 82, 88. The Portez 63, a twin-engine fighter/attack aircraft, had a top speed of 264 mph, six machine guns, and a 180-kg bomb load. The Me-110 fighter was better powered than the Portez 63, with a top speed of 336 mph, and carried heavier armament: two 20-mm cannon and five machine guns. The Me-110 also had a longer range and a much heavier bomb load of 1,000 kg. See also Hans Redemann, *Innovations in Aircraft Construction* (West Chester, Pa.: Schiffer Military History, 1991), 58-65.

32. Peter Smith, *Dive Bomber! An Illustrated History* (Annapolis, Md.: Naval Institute Press, 1982), 92, 101. In May 1940, the French Navy possessed five squadrons, sixty aircraft each, of Loire LN-410 and Vought Vindicator dive bombers. Both the Vindicators and Loires were far inferior in bomb load and performance to the Ju-87; for example, the LN-410 had only a 500-lb. bomb load to the Ju-87B's 1,100 lbs.

33. Angelucci, *Rand McNally Encyclopedia of Military Aircraft,* 222.

34. See Christienne and Lissarrague, *History of French Military Aviation.*

35. Ibid.

36. Ibid.

37. Ernst Kredel, "Der deutsche Luftverkehr," in Ernst Jünger, ed., *Luftfahrt ist Not!* (Leipzig: Deutschen Luftverbandes, 1931), 264-77.

38. Helmuth Schmidt-Reps, "Das Funkwesen in der Luftfahrt," in Jünger, ed., *Luftfahrt ist Not!*, 278-89, esp. 282.

39. Ibid., 283.

40. See Martin Mäder, "Technische Hilfsmittel für die Navigation und Steuerung an Bord neuzeitlicher Verkehrsflugzeuge," in Jünger, ed., *Luftfahrt ist Not!*, 305-22. By 1930, it was common for German civil aviation to use artificial horizons, gyrocompasses, ADF, and Lorenz beam navigation systems.

41. By 1937, German civil aviation had a network of twenty-nine ADF beacons and twenty-nine radio control stations, as well as eighteen illuminated civilian airports. Heinz Orlovius, ed., *Die Deutsche Luftfahrt Jahrbuch 1937* (Frankfurt aM: Verlag Fritz Knapp, 1938), 94. By 1938, the network of directional beacons had grown to thirty-four, while sixteen airports had ultra short-wave instrument landing systems—and these figures do not include the navigation systems of the Luftwaffe. See idem, ed., *Die Deutsche Luftfahrt Jahrbuch 1938* (Frankfurt aM: Verlag Fritz Knapp, 1939), 108-124.

42. Wilkinson, *Aircraft Engines of the World*, 120-21. By 1940, the French Air Ministry had developed the Hispano-Suiza 24Y, rated at 2,200 hp.

43. In 1939, the French had developed seaborne radar for the liner *Normandie*. See Kenneth Macksey, *Technology in War* (New York: Prentice-Hall, 1986), 120.

44. Werner Niehaus, *Die Radarschlacht 1939-1945* (Stuttgart: Motorbuch Verlag, 1977), 29-34, 73-75. The German radar program started under a naval contract in 1929. By 1934, primitive radar sets had been tested and in 1936, the Luftwaffe initiated the radar that would become the "Freya." An early Freya was tested in the 1937 Wehrmacht maneuvers. By 1939, Freya radars had been deployed to detect British bomber raids against Wilhelmshafen.

45. Walter Boyne, *Messerschmidt Me 262* (Washington, D.C.: Smithsonian Institution Press, 1980). A small research team led by von Ohain began developing a jet engine, the He S-3B, in April 1936, and completed and ran the engine in March 1937. The total cost was approximately $20,000.

46. Hans Redemann, *Innovations in Aircraft Construction*, esp. 106-9.

47. Volkmar Kuhn, *German Paratroops in World War II* (London: Ian Allan, 1978), 8-14.

48. In 1938, the German paratroop units were organized as the 7th Air Division. In 1939, the 22nd Infantry Division started retraining as an air landing division. In 1940, these two airborne divisions would lead the German assault on the Netherlands. See ibid., 17-19.

49. P. Buffotot, "La Perception du Réarmement Allemand par les Organismes de Renseignements Français de 1936-1939," *Revue Historique des Armeés*, no. 3 (1979).

50. Christienne and Lissarrague, *History of French Military Aviation*, 310.

51. Michel Forget, "Die Zusammenarbeit zwischen Luftwaffe und Heer bei den französichen und deutschen Luftstreitkräften im Zweiten Weltkrieg," in Horst Boog, ed., *Luftkriegführung im Zweiten Weltkrieg* (Herford: E. S. Mittler Verlag, 1993), esp. 510-11.

52. Ministère de la Guerre, *Instruction Provisoire sûr l'Emploi Tactique des Grande Unités*, 6 Oct. 1921.

53. Ibid., chap. 3, par. 115: "Fire is the most important factor in battle. It destroys or cripples the enemy. Attack means carrying the fire forward. Defense is fire that stops."

54. E. C. Kiesling, "Reform?—Why?: Military Doctrine in Interwar France" (paper presented to the Society of Military History, 8 Apr. 1994), 11.

55. A typical product of the French scientific approach to war is found in the 1930s writings of Gen. Narcisse Chauvineau. An attack on a continuous front required a 3:1 superiority in infantry, a 6:1 superiority in artillery and 15:1 superiority in shells. See Alvin Coox, "General Narcisse Chauvineau: False Apostle of Prewar French Military Doctrine," *Military Affairs*, 37 (February 1973), esp. 16.

56. Robert Doughty, *The Seeds of Disaster* (Hamden, Conn.: Archon, 1985), 102-3.

57. The French calculated the destruction of each point of a fortress' outer works would require 100-150 rounds of 280–400-mm artillery shells. An armored strongpoint in a fortress required 400 rounds of 320-, 370-, or 400-mm mortar shells to be destroyed. See "Französiche Anschauungen Über Angriff und Verteidigung an Festungsfronten," *Militärwissenschaftliche Rundschau*, no. 5 (December 1939), 702

58. Kiesling cites a 1937 French study that overestimated the range and stopping power of a 25-mm gun. The French posited a 1,000-meter effective range, and a rate of fire of 15 rounds per minute, with a 25 percent hit rate. Thus, the French estimated, 19 of 30 German attacking tanks would be destroyed by a single French antitank gun in a model battle. See Kiesling, "Reform?— Why?," 14-15.

59. On the French antitank gun program, see Ministère de la Défense, *Les Programmes d'Armament*, 342-51. General Gamelin made production of the 25-mm antitank gun a top priority in the 1938 army armaments budget. See Henry Dutailly, "La Puissance Militaire de la France en 1938," *Revue Historique des Armées*, no. 3 (1983), 5-9. See also Franz Kosar, *Panzerabwehrkanonen 1916-1917* (Stuttgart: Motorbuch Verlag, 1980), 55-60.

60. German officers observing the 1922 and 1924 French maneuvers were impressed by the new French vehicles and equipment, but held a low opinion of the French tactics for the equipment. See T-3 Truppenamt, "Die französischen Herbstmanöver 1922," 11 Sept. 1923, in Bundesarchiv/Militärarchiv, Freiburg, Germany (hereafter cited as BA/MA), RH 2/1547, also T-3 Truppenamt, "Die französischen Herbstmanöver, 1924," 10 Dec. 1924, ibid.

61. In 1940, the Germans had about twice as many armored cars as the French: 350 French to 600-plus German. See Ogorkiewicz, *Armoured Forces*, 432-34.

62. In 1938, General Gamelin commented, "The role of aviation is apt to be exaggerated, and after the early days of war the wastage will be such that it will more and more be confined to acting as an accessory to the army," as cited in Anthony Adamthwaite, *France and the Coming of the Second World War* (London: Frank Cass, 1977), 162

63. Robert Frankenstein, *Le Prix du Réarmament Français 1935-1939* (Paris: Publications de la Sorbonne, 1982). In the 14-billion-franc rearmament program of September 1936, only 4.3 percent of the equipment funds were devoted to anti-aircraft defense. Up to 1940, the mainstay of the French antiaircraft force was a slightly improved 75-mm gun from World War I. The armament programs of 1937–39 funded only 356 new 75-mm antiaircraft guns. See Ministère de la Défense, *Les Programmes d'Armament*, 182-183.

64. General Gamelin, appointed as supreme commander of the armed forces in 1938, was unsure of the parameters of his authority to command the air force. In 1938–39, he requested, but did not receive, clarification from the government. See Martin Alexander, *The Republic in Danger: General Maurice Gamelin and the Politics of French Defense, 1933-1940* (New York: Cambridge University Press, 1992), 168.

65. On the influence of Douhet on Pierre Cot and the French Air Force, see Thierry Vivier, "Pierre Cot et la Naissance de l'Armeé de l'Air," *Revue Historique des Armées*, no. 181 (December 1990), esp. 109–10. Pierre Cot's views on strategic bombing and nationalization are explained in his book, *L'Armeé de l'Air* (Paris: Editions Bernard Grasset, 1939).

66. This was a regression to the French combat doctrine of the 1920s. The 1928 Air Service Operational Doctrine stressed the need to gain air superiority. Air Superiority would not, however, be gained by bombing enemy airfields or infrastructure, as in German doctrine. Air superiority in French doctrine would be gained by masses of fighter planes over the front. See Ministère de la Guerre, *Règlement Provisoire de Manoeuvre de l'Aéronautique* 2 (1928), pars. 1-4, 6-7.

67. General Gamelin referred to the air force in the 1930s as "The Shield of the Army." See Alexander, *Republic in Danger,* 150. Air Minister Guy LeChambre reported to the Aeronautical Commission in February 1938, "In the initial phase of the war, however, what we'll need above all is to put our airspace under lock and key, as we've done for our frontiers" (p. 163).

68. According to Emmanuel Chadeau, *De Blériot à Dassault: Histoire de l'Industrie Aéronautique en France 1900-1950* (Paris: Fayard, 1987), 343, "In May 1940 . . . the French forces employed 23 aircraft types, 38 models in 42 versions."

69. James Corum, *The Roots of Blitzkrieg* (Lawrence: University Press of Kansas, 1992), 37-43, 144-55.

70. In response to DeGaulle's 1934 book advocating a separate, elite mechanized army, General Weygand replied, "Two armies, not at any price. . . . We already have a mechanized, motorized, organized reserve. Nothing need be created." Cited in Paul-Marie De la Gorce, *The French Army: A Military-Political History* (New York: George Braziller, 1963), 273.

71. Corum, *Roots of Blitzkrieg,* 28-34, 200-201.

72. Ibid., chap. 6.

73. On the organization of motorized division and motorized brigades, see *Truppenamt T-4 Winterkriegsspiel 1926-1927,* BA/MA RH 2/2822. On TOEs for armor regiments, see *Heeresdienstvorschrift 487 Part II* (1923), pars. 524-25.

74. *Richtlinien für die Führung des Operativen Luftkrieges,* May 1926.

75. Ibid., pars. 1-7. Ibid., par. 40: "A delaying action in the air or a purely defensive approach does not describe the true character of the air force."

76. Ibid., pars. 83-85 and 91-95.

77. In April 1940, the combat forces of the Luftwaffe included a total of 1,620 fighters, 1,726 bombers, 419 dive bombers, and 46 ground attack planes, for a ratio of 1.4:1 bomber/attack aircraft to fighters. See Williamson Murray, *Strategy for Defeat: The Luftwaffe 1933-1945* (Maxwell AFB: Air University Press, 1983), 32-33.

78. Forget, "Die Zusammenarbeit," esp. 511-12.

79. In October 1939, Navy Commander Admiral Raeder wrote to Reichsmarschall Göring to complain of the poor performance and capabilities of the Do-18 and He-115 seaplanes, which were the backbone of the Naval Air Arm. Raeder argued that the Navy urgently needed aircraft with effective range, plus torpedo and bombing capability. See Raeder to Göring, 31 Oct. 1939, in BA/MA RM 7/168.

80. Eduard Daladier complained in May 1937 of the extreme delays in equipment development and production caused by the army branch inspectorates. The Armaments Council lacked both a proper staff and audit powers. See Alexander, *Republic in Danger,* 118-19.

81. See Robert Doughty, *Seeds of Disaster.*

82. Ronald Miller and David Sawers, *The Technical Development of Modern Aviation* (London: Routledge, 1968), 15. The French airlines, while the most heavily subsidized, were also the most inefficient. In 1928, the French airlines received only 10.6 percent of their income from purely commercial activities. Lufthansa, which carried far more traffic than French airlines, earned 30 percent of its income from purely commercial activities, not including mail service, in 1929.

83. French commercial aircraft design of the interwar period lagged far behind that of the other major powers. The Germans made great advances in the use of metal and in the construction of wings, in such aircraft as the Junkers F-13, W-33, and W-34. The Junkers trimotors, G-24 and Ju-52, were very popular outside of Germany. By the 1930s, the Ju-86 and Focke Wulf Condor were in use in German commercial aviation, and were sought by other countries. The only French commercial transport of note was the Bloch 220—and only 16 of these were built before World War II. See Peter Brooks, *The Modern Airline: Its Origins and Development* (London: Putnam, 1961), 52-57, 88-93, and Enzo Angelucci and Paolo Matricardi, *World Aircraft: Commercial 1935-1960* (Chicago: Rand McNally, 1979), 188-89, on the Bloch 220.

84. Jack Gee, *Mirage* (London: MacDonald, 1971), 8.

85. Christienne and Lissarrague, *History of French Military Aviation,* 335.

86. Alistair Horne, *To Lose a Battle: France 1940* (London: Penguin Books, 1969), 127-28.

87. Alexander, *Republic in Danger,* 371.

88. Murray, *Strategy for Defeat*, 36, estimates a French operational rate of no more than 40 percent for many squadrons before May 1940. French historians indicate a very low operational rate during May 1940. Group attacks were consistently made with 40–50 percent of the group's official strength. See Pierre Paquier, *l'Aviation de Bombardement Française en 1939-1940* (Paris: Berger-Levrault, 1948), 8-9, 208-35, for a log of French air activity during 10–20 May 1940.

89. Christienne and Lissarrague, *History of French Military Aviation*, 336-70.

90. See Lucien Robineau, "Die französiche Luftpolitik zwischen den beiden Weltkriegen und die Führung des Luftkrieges gegen Deutschland (September 1939 bis Juni 1940)," in Boog, ed., *Luftkriegführung im Zweiten Weltkrieg*, esp. 739. In May 1940, the RAF in France had about 400 aircraft. See Derek Wood and Derek Dempster, *The Narrow Margin* (Washington, D.C.: Smithsonian Institution Press, 1990), 126. The German Air Force had approximately 3,600 aircraft to oppose the Allies in May 1940.

91. In the mid-1920s, the Waffenamt employed sixty-four officers, including two major generals, two colonels, and twelve lieutenant colonels. An additional twenty-one officers worked at test sites for the Waffenamt. See Wehrministerium, *Rangliste des Deutschen Reichsheeres* (Berlin, 1925).

92. A good overview of the Waffenamt operations can be found in Erich Schneider's "Waffenentwicklung: Ehrfahrungen im deutschen Heereswaffenamt," *Wehrwissenschaftliche Rundschau* 3 (1953).

93. On the economic mobilization plans of the German Army, see George Thomas, *Geschichte der deutschen Wehr und Rüstungswirtschaft (1919-1943/45)* (Boppard am Rhein: Harold Boldt Verlag, 1966), 53-57.

94. See Richard Overy, *The Air War: 1939-1945* (New York: Scarborough House, 1980).

95. By 1928, Germany had taken from France numerous official aviation world records for altitude, distance, and duration. A German-crewed Junkers W-33 aircraft crossed the Atlantic before a French team in 1928. See L. Hirschaier, ed., *l'Anneé Aéronautique 1928-1929* (Paris: Dunoud, 1929), 98–100, 167, 245. In 1929–30, the Germans developed the four-motor Junkers G-38 aircraft, the largest commercial aircraft of the time, with 3,200-hp and 38-passenger capacity. See Peter Supf, *Das Buch der deutschen Fluggeschichte* (Berlin: H. Klemm, 1935), 2:613.

96. Friedrich von Rabenau, *Seeckt: Aus seinem Leben* (Leipzig: Koehler Verlag, 1940), 529.

97. The German military archives contain numerous letters and reports written by Milch, as Lufthansa director, to the chief air staff officer. For example, in 1928 Milch wrote Air Staff Officer Major Sperrle a friendly report on Lufthansa's experience with the reliability of various radio models. See BA/MA RH 2/2222, *Correspondence File of 1928.*

98. Ernst Heinkel, *He 1000* (London: Hutchinson, 1956), 78-79.

99. Interview by author with Hansgeorg Wilberg, son of Gen. der Flieger Helmuth Wilberg, Swisttal-Buschhoven, Germany, 18 June 1992.

100. Between 1926 and 1932, the German aviation industry received 321 million Reichsmarks in government subsidies, investment, research funds, etc. According to the exchange rate of the time, this amounted to U.S.$12.42 million per year. See Ralf Schabel, *Die Illusion der Wunderwaffen* (Munich: Oldenbourg Verlag), 103.

101. Mathew Cooper, *The German Air Force 1933-1945* (London: Jane's, 1981), 2-3.

102. Edward Homze, *Arming the Luftwaffe* (University of Nebraska Press, 1976), 59.

103. Ibid., 73.

104. Hans von Seeckt, *Denkschrift an der Heeresleitung, February 18, 1919*. U.S. National Archives German Records. Von Seeckt Papers, File M-132, Roll 21, Item 110.

105. Col. Kurt Thorbeck, "Die Technische und Taktische Lehre des Krieges," 12 Apr. 1920, BA/MA RH 12-2/94.

106. Erich von Manstein, who worked in the General Staff Operations Section after World

War I, recorded that the top priority of the Operations Section in the 1920s was developing new weaponry, especially armored vehicles and the motorized transport for the army. See Erich von Manstein, *Aus Einem Soldatenleben* (Bonn: Athenäum Verlag, 1958), 110.

107. Reinicke, *Das Reichsheer,* 312.

108. Some of the notable German senior officers who were sent to receive engineering degrees by the army in the interwar period include: Field Marshal von Richthofen (Engineering Ph.D.); Maj. Gen. Robert Fuchs, Commander, 1st Air Division; Gen. der Flieger Johannes Fink, Commander, II Air Corps; Maj. Gen. Friedrich Deutsch, Commander, 16 Flak Division; Lt. Gen. Gerhard Conrad, Air Commander, XI Air Corps; and Lt. Gen. Richard Schimpf, Commander, 3rd Paratroop Division.

109. Hans von Seeckt to Waffenamt, and *Inspektionen,* 21 Jan. 1924, BA/MA RH 12-2-21.

110. Von Manstein, *Aus Einem Soldatenleben,* 246-49. Manstein's account of the Sturmgeschütze also provides a good picture of debate within the general staff, and how the Army Weapons Office could move efficiently and quickly to develop an effective new weapon.

111. Walter Spielberger and Uwe Feist, *Sturmartillerie* (Fallbrook CA: Arco, 1967), pt. 1.

112. Information is from Karl Friedrich Hildebrand, *Die Generale der deutschen Luftwaffe 1935-1945,* vols. 1–3 (Osnabrück: Biblio Verlag, 1990). Hildebrand has published the official service records of all 688 men who reached the rank of general in the Luftwaffe.

113. Ibid.

114. Ibid.

115. Ibid.

116. Ibid.

117. Overy, *The Air War,* 137, asserts that the inclusion of so many reactivated officers and officers from the army "had the unfortunate consequence of dividing the air officer corps into those who regarded themselves as heirs of the Prussian tradition, and those who came from an unorthodox, particularly technical background. Part of the hostility felt between regular soldiers and the parvenus arose from the fact that the newcomers were given high military office without having followed the normal army channels."

118. "The result of all this was that the Luftwaffe was shaped by aviators who were amateur soldiers, and soldiers who were amateur aviators." See Telford Taylor, *The March of Conquest* (Baltimore Md.: Nautical & Aviation Press, 1991 Reprint), 25-26.

119. Overy, *The Air War,* 136.

120. Field Marshal von Richthofen disliked and distrusted Milch for his "love of intrigue," but respected him as highly competent in aviation. From the author's interview with Götz Freiherr von Richthofen, son of Field Marshal Wolfram von Richthofen, Aumühle, Germany, 21 June 1992.

121. Chief of construction operations for the Luftwaffe was Ministerialdirigent Professor Doktor Ingineur Heinrich Steinmann, a World War I pilot who taught construction engineering at the Technische Hochschule in Braunschweig from 1926 to 1934. See Steinmann's record in Hildebrand, *Die Generale der deutschen Luftwaffe.*

122. *Report to Ministère de la Guerre, October 6, 1921,* from Preface to Ministère de la Guerre, *Instruction Provisoire sur l'Emploi Tactique des Grandes Unités,* report of 6 Oct. 1921. The committee that wrote the report consisted of General Georges (chairman); ten army generals, three colonels, one lieutenant colonel, and one major. Only one officer of the board, General Pujo, was an airman.

123. Kiesling, "Reform?—Why?," 7.

124. Between April and September 1917, serious mutinies broke out in fifty-four French divisions. The French Army was temporarily crippled. Over 20,000 men were found guilty of crimes by military courts. Hundreds of French soldiers were secretly executed—even today, the French will not release details. For an account of the mass mutinies in the French Army, see Richard Watt, *Dare Call It Treason* (New York: Simon & Schuster, 1963), esp. 299-303.

125. On French concepts of motorized armor forces, see Jean DeLaunay, "Chars de Combat et Cavalerie (1917-1942): la naissance de l'arme blindeé," *Revue Historique des Armeés,* no. 155 (June 1984), esp. 10-11. The concepts of mobility of Gamelin, Weygand, and other senior generals is well described in Paul Reynaud's memoirs, *In the Thick of the Fight 1930-1945* (New York: Simon & Schuster, 1955), 158-61.

126. See Thierry Vivier, "Les Réservistes de l'Air (1919-1939)," *Revue Historique des Armeés,* no. 174 (March 1989).

127. General Wegand, French commander in chief after Gamelin's relief in June 1940, argued that the ineffective armament of the French Army in 1940 was the responsibility of the military commanders—not just the politicians. See Gen. Maxim Wegand, *Histoire de l'Armeé Française* (Paris: Ernest Flammarion, 1953), 418-20.

128. Richard Overy, *War and Economy in the Third Reich* (Oxford: Clarendon Press, 1994), esp. chaps. 8–9. Edward Homze, *Arming the Luftwaffe,* 262-65, concludes that the German aviation industry was poorly managed in the 1930s.

129. A useful recent work outlining many of the inefficiencies of German aircraft production is Willi Boelcke, "Stimulation und Verhalten von Unternehmen der deutschen Luftrüstungsindustrie während der Aufrüstungs und Kriegsphase," in Boog, ed., *Luftkriegführung im Zweiten Weltkrieg,* 81-112.

Rocket Science: Engineering Better Artillery

John F. Guilmartin, Jr.

The use of rockets for military purposes antedates the industrial age by many centuries. Indeed, though the evidence is thin and equivocal, rockets were probably used in war not long after the first primitive guns. But the earliest guns and rockets were dependent on the same source of propulsive energy, black powder; and, wonderful stuff that it is, black powder—the simple mixture of potassium nitrate, charcoal, and sulfur—has inherent limitations which constrained the performance of rockets more than that of guns. At the most basic level, black powder contains insufficient propulsive energy per unit mass to lift payloads of any great weight when used in rockets, the rub being that the propellant must lift its own mass plus that of the casing and nozzle, as well as the payload. By contrast, substantial masses of black powder could be confined in the chamber of a gun, where the mass of propellant and launcher were of concern only for reasons of logistics and mobility, and used to drive inert projectiles with militarily meaningful velocities and impact energies. Simply put, guns were hauled by human and animal muscle, but rockets had to haul themselves and black powder had limited haulage potential.

Beyond considerations of energy per unit volume, the efficiency of rocket motors varies as an inverse function of the molecular weight of the decomposition products, the relevant parameter being specific impulse, abbreviated I_s, measured in pounds thrust per pound mass of fuel per second. Although expressed in seconds, I_s is actually a dimensionless measure of efficiency, the larger the value the more efficient the fuel.[1] As the equation indicates, massive decomposition products result in less efficiency, and the decomposition products of black powder are massive indeed, consisting of only 43 percent gas and 57 percent solid particles by weight.[2] In contrast, the decomposition products of modern liquid rocket propellants and nitrocellulose-based gun propellants are almost entirely gaseous except for traces of water vapor.[3] Putting numbers on it, black powder used as a rocket fuel has an I_s of about 150, while nitrocellulose-based solid propellants similar to those used in the first modern battlefield rockets in World War II have an I_s of about 260.[4] By way of comparison, liquid oxygen and kerosene, the fuel combination used by the Atlas ICBM, has an I_s of about 300, and the liquid oxygen/liquid hydrogen combination used in the Space Shuttle's main engines, has an I_s of about 390.[5] With regard to guns, black powder's limitations imposed an upper limit on muzzle velocity of about 2,000 fps regardless of size or design[6] whereas modern rifles and cannon typically have muzzle velocities in the 2,500–3,000 fps range.[7] I cannot put numbers on black powder rocket velocities, but the performance disparity between black powder and modern rocket propellants was even greater than was the case with guns.

Black powder's final limitation lay in the fact that very large charges of black powder are unstable, particularly in rockets where the charge must burn gradually from the face. Black powder worked well in small rockets, partly because small casings are relatively stronger due to scale effect; but the charges of large black powder rockets were prone to crack from stress and vibration, causing the flame front suddenly to spread along internal fissures causing catastrophic explosion. Very large black powder guns could be made; very large black powder rockets could not. Unlike liquid-fueled rockets, black powder rockets could not be scaled up to carry larger payloads, a practical reality defied in the late 1920s and 1930s by a handful of German daredevils with more intestinal fortitude than engineering perspicacity who decorated the landscape with their remains and those of a variety of rocket-propelled vehicles.

Before leaving black powder for modern propellants, I would emphasize that efficiency and effectiveness are not the same thing—a point which engineers seem to grasp more readily than systems analysts. Black powder ordnance was extraordinarily effective in its day and had a profound impact on the course of history. Black powder cannon laid the technical groundwork for the steam engine by forcing the development of heavy-duty, precision machine tools for cannon-boring in the mid-eighteenth century.[8] Black powder ordnance forged the link between science and design engineering by prompting the development of experimental internal ballistics, among them Benjamin Robins's invention of the ballistic pendulum in the 1740s, Thomas Jefferson Rodman's development of gauges capable of measuring pressures within a gun in the 1850s, and Andrew Noble's perfection of the electro-mechanical ballistic chronograph in the 1870s.[9] The same massive decomposition products which limited efficiencies could also produce remarkably large internal pressure peaks by means of refracted shock waves, something which Rodman, the father of scientific internal ballistics, discovered inadvertently in the striking irregularities in his data, and which numberless medieval and early modern gunners discovered in the final instant before their departure for Valhalla, Fiddlers's Green, or wherever, but I digress.[10]

In the final analysis, black powder rockets were useful mainly for incendiary and signaling purposes. To be sure, a well-designed black powder rocket could carry a tactically useful explosive warhead to ranges which matched those of a field artillery piece firing a projectile of comparable mass. Moreover, the rocket had the advantage of being more mobile in that the launching trough or tube was far lighter than the carriage of a field piece, and spin-stabilized rockets suffered from no great disparity in accuracy in comparison with smooth bore artillery. But black powder rocket artillery had a crippling liability: the lack of close-in firepower. The dominant characteristic of field artillery from the age of Vauban and Frederick the Great to the end of the black powder era was the ability to decimate troops in the open at short to medium ranges with a hail of case shot or canister, and this was a capability which rocket artillery did not possess. Even short-barreled howitzers designed to fire exploding shells, which in many respects had a tactical profile similar to that of rocket artillery, could fire case or canister in a pinch.

So, aside from the odd city burnt and a stirring line in the American national anthem, rockets had little impact on war in the black powder era. Not until the Second Industrial Revolution of the mid-nineteenth century, when science was linked to the industrial process, did rockets with important military potential make their debut. The pivotal event was the development of nitrocellulose-based propellants[11] and high explosives in the final decades of the nineteenth century. Beginning with nitrated cotton—guncotton—developed by an Austrian army officer, chemists developed an array of gun propellants which offered important performance advantages over black powder. Over the long term, the most important of these was a dramatic increase in muzzle velocities, a product of higher energy densities, less massive decomposition products, and specific impulses nearly double that of black powder. The armies and ordnance establishments of Europe and the industrialized world were quick to perceive the advantages of the new propellants, and within a very short time had standardized on remarkably effective, and remarkably similar, bolt-actioned infantry rifles firing jacketed lead projectiles of about 0.3 inches diameter from fixed, necked-down brass cartridges at muzzle velocities on the order of 2,500–2,800 fps.[12] Artillery applications were not far behind. Over the short term, the biggest perceived advantage was the absence of smoke, another product of black powder's preponderance of solid decomposition products, from the battlefield, hence the term smokeless propellants. At the same time, high explosives, of which nitro-glycerin was the first, offered far greater destructive effect per unit mass upon impact than black powder and were quickly applied to artillery shells.

As with black powder, the new propellants were easier to use in guns than rockets, and a considerable period of time passed before they were rendered sufficiently docile to be used in rockets. Nor was there much incentive to do so. The effectiveness of quick-firing artillery with precision hydro-pneumatic recoil mechanisms which appeared in the waning years of the nineteenth century, beginning with the justly famous French 75 of 1898, was enormously greater than anything which had gone before. Military and naval establishments needed a generation to become accustomed to the enormous destructive power at their disposal, and in the meantime there seemed little need for novel solutions to the tactical problems which artillery was deemed capable of solving. Through the end of World War I, rockets were used for signaling and for throwing life-lines in maritime rescue operations. Aside from a handful of Le Prieur black powder rockets used by the French in aerial attacks on German observation balloons, that was about that.

The Versailles Treaty and the insatiable curiosity of German scientists and engineers in the interwar years changed things. There was a certain inertia and inexorable logic in the accelerating accumulation of scientific and technological knowledge unleashed by the Second Industrial Revolution, and bombardment rockets with high-performance propellants and warheads would no doubt have appeared somewhere sooner or later. The Versailles prohibitions on German aviation and ordnance development combined with the intellectual and social ferment of the Weimar Republic to dictate that it would be in Germany. A key factor, which I have yet to see even hypothetically explained, was the appearance in Germany, between the last decades of the nineteenth century and the ascendancy of the Nazis in the 1930s, of a remarkable profusion of strikingly competent and innovative design engineers. Denied powered aircraft and heavy artillery by treaty, the Germans turned to rockets, intellectually in surprisingly realistic dreams of manned space flight and militarily in the form of bombardment rockets.

First off the mark in terms of operational hardware was the development of a family of solid-fuel battlefield bombardment rockets of 280-mm, 210-mm, and 150-mm diameter which first entered service in 1941.[13] Spin-stabilized and produced with a variety of high-explosive, smoke, and chemical warheads, they were launched from their packing crates or from light five or six-tubed launchers. The engineering design compensated for the instability of the diethylene glycol dinitrate fuel by careful dividing of the charge into multiple sticks.[14] The most widely used of these weapons was the 150-mm *Nebelwerfer* (literally smoke projector, a code name used to disguise the program's real purpose), later infamous among GIs as the "Screaming Meemie." Though substantially less accurate than equivalent tube artillery systems, *Nebelwerfer* batteries could produce extremely high volumes of fire in a short period of time. The concomitant liability was one common to all multiple-tube bombardment rocket systems, the huge cloud of dust and smoke raised by a battery in action which acted as a magnet to counterbattery fire and Allied airpower. The 210-mm rocket was adapted to aerial use in 1942–43 and was launched from underwing tubes on Luftwaffe fighters to break up USAAF heavy-bomber formations.

The German solid-fueled bombardment rockets were nearing operational service when World War II began and were the most advanced extant, but solid fuels represented only part of the story. Almost from the beginning, serious students of rocket propulsion recognized the limitations of solid-fueled rockets, and from the 1920s explored the possibilities of liquid fuels as a means of achieving the specific impulses and total thrusts required for high-thrust/high-performance applications, notably manned space flight. American scientist Robert Goddard was actually the first to fly a liquid-fueled rocket in 1926,[15] but a combination of government disinterest, public ridicule of his visions of space flight, and his own secretive nature pushed his impressive engineering achievements to the sidelines. In the meantime, German space flight societies had begun to experiment actively with liquid-fueled rockets.

The critical turning point for German rocketry came in 1932 when then-captain Walter Dornberger offered army support for further experimentation, coopting rocket engineer Wernher von Braun, and several of his associates in the process. The organizational story of the development of the A-4/V-2 rocket by Dornberger, von Braun, and their organization, working first at an abandoned artillery range near Berlin and then at the Peenemünde proving ground, has been told and retold and needs no repetition here.[16] A few words on the technical content and strategic potential of the program, however, are in order.

Inspired by the so-called "Paris Gun" of World War I, the *Kaiser Wilhelm Geschütz* which shelled Paris from a distance of some 75 nautical miles in the spring of 1918, the

V-2 was a technological marvel. In military terms, however, like the Paris Gun, the V-2 program represented a monumental waste of strategic resources. The real importance of the V-2 lay in the advance of high-performance booster technology by at least two orders of magnitude and in the development of practical solutions to some of the most vexing problems of space operations, among them guidance and control. The V-2 was the first operational space vehicle ever and, in its operational versions, was guided by the first reprogrammable, fully electronic analog computer.[17] Though they never used it in operational vehicles, V-2 engineers discovered ablative cooling, later an essential component of the development of the reentry body for the Atlas ICBM (intercontinental ballistic missile) nuclear warhead.[18]

Nor were von Braun and his group the only German engineers advancing rocket technology in the 1930s and 1940s. Helmuth Walter, originally working on a Navy contract to build a wakeless torpedo, developed the HWK (Helmuth-Walter-Kiel) hydrogen peroxide-fueled rocket engines. Used as drop-off boosters for the Ar-234 jet bomber and to power the Me-163 rocket fighter, they were the first operational liquid-fueled rockets ever. Hydrogen peroxide had limited potential as a primary rocket fuel, and the engines themselves were a dead end, but Walter developed efficient turbopumps to force fuel and oxidizer into the combustion chambers, the first positive-feed rocket fuel system. Walter turbopumps were an essential ingredient of the V-2's technical success.[19]

Austrian engineer Eugen Sänger worked for several aerodynamic research organizations of the Third Reich on a variety of rocket and ramjet propulsion projects, and, with his mathematician and later wife Irene Bredt, freelanced the design of an aerodynamic spacecraft of radical design, the so-called skip-bomber which was to be the inspiration and lineal technical ancestor of all subsequent lifting reentry vehicles, both notional and actual, from the X-20 Dyna Soar to the Space Shuttle. All of these projects were to have far-reaching consequences.

But before addressing the long-range consequences of prewar developments in rocketry, a brief discussion of the relative technical and tactical merits of military rocket systems and tube artillery is in order. The discussion belongs at this point in my analysis because such an assessment would have been impossible prior to World War II: the technology on which to base it simply did not exist. It also belongs here because a significant number of tactical rocket systems with important implications for the future were developed during the war, and their development and adoption, or nonadoption, was driven by the basic factors considered here. Finally, in terms of launch and impact energies, rocket and tube artillery systems came of age in World War II. Although the advent of precision guidance systems has acted to redress rocketry's inherent disadvantage in accuracy, the balance sheet of advantages and disadvantages dictated by Newton's laws was established and at least implicitly understood in World War II. It has not changed materially since.

The basic advantage of rocket systems was–and is—that the mass of launcher and projectile, ready to fire, was considerably less than that of a tube artillery system capable of firing a projectile of the same mass the same distance. This generalization did not apply to small arms; because of scale effect,[20] the barrels and chambers of pistols, rifles, and machine guns are proportionately much stronger than those of artillery pieces, and relatively light weapons can safely impart considerable kinetic energy to a small projectile. Moreover, they do so from the muzzle whereas rocket systems accelerate more gradually and need distance to achieve their full velocities. In essence, very small rockets were not worth the effort. Very large rockets were, or at least could be, for they produce their thrust gradually with far lower chamber pressures than guns, thus circumventing scale effect. This final point was appreciated in principle from the 1920s, but was demonstrated as a practical reality only with the A-4/V-2 program, and in that reality lies the V-2's importance as a technological benchmark.

For pieces firing warheads larger than hand-grenade size, where great kinetic energy was not required, rockets offered compelling advantages in overall system weight, unless there was a need for continuous rapid fire. The down side was that rockets were less accurate than rifled artillery pieces and that individual rounds were considerably heavier; in addition, rocket systems gave away their positions on firing with the flash of motor ignition and the smoke, dust, and debris raised by the exhaust. No great disadvantage in aerial or naval combat, this could be a serious matter in ground combat.

Though it took time to translate the above relationships into engineering and production decisions, they dictated that rocket systems would have important advantages when the tactical requirement was for the relatively infrequent delivery of a relatively massive projectile at modest velocities. The tactical implication in the above technological prescription was that rocket weapons would be most useful against well protected, high-value targets at short ranges. The logic in that prescription was underwritten by the discovery, or rather rediscovery, early in World War II of the Monroe effect: the penetration pattern produced by the directional, inward focusing of a high-explosive charge formed in the shape of a hollow cone. Terminal ballisticians learned that the effect could be magnified by lining the inside of the cone with a thin coating of heavy metal, typically copper, to create a thin, hyper-velocity stream of hot gasses and molten metal which could cut through armor, concrete or whatever, like an instant welding torch. The antitank potential of the Monroe effect was first noted by American engineers, fortunately so in light of the ineffectiveness of prewar American antitank guns, and led to the development of the 2.36-inch rocket launcher, the famed bazooka. The bazooka saw combat in 1942, proved effective, and was quickly imitated by the Germans as the *Panzerschreck*, a rocket launcher of similar design which entered service the following year. The definitive system in this category was the German *Panzerfaust*, a rocket-propelled antitank grenade which dispensed with the niceties of a reusable launcher in favor of larger warheads and penetrating capability.

The Germans had the early lead in rocket technology, but the Allies closed fast. American engineers at the Guggenheim Aeronautical Laboratory of the California Institute of Technology (GALCIT), precursor to the NASA Jet Propulsion Laboratories, made particularly swift progress, moving up from experiments with crude black powder JATO (jet-assisted take off) rockets in 1941 to the development of a liquid-asphalt/potassium perchlorate-based solid propellant with good performance and extended storage life in 1942.[21] It was this discovery which made large solid rocket motors safe and feasible, paving the way for silo and submarine-based strategic missiles with long shelf lives, a matter beyond the scope of the present analysis.

By 1944, Britain and the United States fielded a variety of bombardment rockets, used mainly from shipboard to cover amphibious assaults, but also fired from multiple-tube launchers mounted on tanks and other vehicles. The British developed efficient air-to-ground rockets, used with considerable effect against German armor in the Normandy campaign. The Soviets, meanwhile, fielded *Katyusha* bombardment rockets fired from truck-mounted multiple tube launchers in considerable numbers, and rocket-firing Il-2 *Shturmovik* ground-attack aircraft became a staple of Soviet airpower. Finally, the Germans developed rocket-boosted, radio-guided antiship missiles and used them in small numbers with some success before war's end.

The neat calculus of costs and benefits sketched out above, of course, was understood only empirically and imperfectly, and as the war progressed the contending military and ordnance establishments responded to the tactical challenges of the moment on the basis of incomplete knowledge. The results were a tribute to human ingenuity. Competing with conventional tube artillery, in addition to rocket systems, was a bewildering array of rifled and smooth-bore mortars; squeeze-bore guns; spigot mortars; rifle grenades; recoilless rifles and; in the waning days of the war, the high/low pressure gun. In the event, most of these systems proved to be curiosities with little enduring value. The perfection and expanded use of mortars as infantry-support weapons was the only exception of note, all the more remarkable since some of the most successful designs went back to World War I.[22] The spigot mortar principle, most prominently used in the British shoulder-fired PIAT (projector, infantry, anti-tank), is a case in point.[23] Squeeze-bore antitank guns produced marvelously high velocities, but were more complex and expensive than competing systems and were rendered obsolete by HEAT (high-explosive antitank), discarding sabot, and "squash head" ammunition for conventional tube ordnance.[24] The recoilless rifle was a technical success, but has been relegated to a relatively minor role in land warfare, analogous to that of the howitzer in the wars of the French Revolution and Napoleon.

Battlefield rocket systems had marginal impact on the conduct of World War II with the major exception of infantry antitank munitions. Beyond doubt, the bazooka, *Panzerschreck, Panzerfaust,* and their various derivatives robbed armored forces of the ability to penetrate infantry defenses more or less with impunity which they had enjoyed in the early stages of the war. This was as true on the Russian steppes as in the Normandy bocage, and the net result was a decided decline in the strategic fluidity of the ground war, save in conditions where one force or the other was essentially broken in attritional struggle before the other achieved an armored breakthrough. With this one, major, exception, then, battlefield rocket systems demonstrated more potential for the future than actual performance on the battlefield in World War II.

A final category of rocket system deserves mention, if only by way of a caution-ary note. The rocket plane exercised a powerful grip on imaginations, not only among the general public—witness the popularity of *The Rocketeer*, an American comic strip character of the 1930s—but among engineers who should have known better. The basic problem was one of endurance: while Russian and German engineers, at least, were capable of designing rockets capable of imparting more velocity to an aircraft than the aerodynamic knowledge of the day could handle, fuel consumption was extremely high, and ranges were correspondingly short. The Soviets, with a perfectly functional prototype rocket interceptor in 1941, had the good common sense to recognize that point defense interceptors were of no strategic value to them and dropped the project.[25] The Germans were not so astute and proceeded with the development of no less than three competing designs. The Messerschmitt Me-163, the only one to see operational service, was powered by a Walter HWK-109 hydro-gen peroxide rocket, a technically superb engine, but one which used highly volatile fuels prone to spontaneous explosion.[26] Ludicrously, the Me-163 was armed with a pair of low-velocity 30-mm "pavement buster" Mk-108 cannon, so called because the weapon sounded like a jack hammer. Between the weapon's short effective range and the Me-163's high speed, a pilot making a stern intercept on an American bomber had only a few seconds between entering effective range and having to break off the intercept to avoid ramming his target.[27] Unsurprisingly, Me-163s shot down few bombers; their principal impact on air operations was to raise the pulse rates of Allied aviators who saw them for the first time.

A more serious proposition was the Bachem Ba-349 "Natter," a cleverly de-signed vertically launched interceptor which was, in essence, a surface-to-air missile with a human guidance system. Launched from vertical rails, the Natter was powered by the same Walter engine as the Me-163, but parallels between the two designs stop there. The Natter was accelerated to flight velocity by two jettisonable, solid-fuel boosters and was armed with a battery of twenty-four nose-mounted 73-mm folding fin rockets.[28] The expendable airframe was made of wood. Natters were to be launched into American bomber formations, where the pilots would salvo their rockets at a single bomber, break off the attack, and bail out after activating a mechanism which extracted the rocket engine from the aircraft by parachute. The expensive compo-nents were thus preserved for re-use; best of all, the pilots did not even have to learn how to land. Granted, the Ba-349 had a significantly shorter radius of action than the Me-163, but at least it was a credible bomber destroyer.[29] With its lethal offensive ordnance, the Natter could have posed serious problems for the USAAF if deployed in quantity. Fortunately for the Allies, Willy Messerschmitt's influence with the Nazi authorities outstripped Erich Bachem's, and the Ba-349 program was starved for resources: a single Natter battery was emplaced before war's end, and there is no evidence that it fired.

More portentous were a number of German surface-to-air missile designs which were in a state of advanced development by 1945. Notable among these was the *Wasserfall,* a radio-guided derivative of the A-4/V-2 capable of reaching altitudes in excess of 50,000 feet and ranges of sixteen miles with a 154-lb. explosive warhead.[30] Whether or not there was any direct linkage in design lineage, the *Wasserfall* was an accurate predictor of the Soviet SA-2 of the Cold War and Vietnam.

War is a terribly unpredictable, messy, and inexact business and, particularly with respect to war, history is the worst of all predictive tools . . . except for the others. If nothing else, I am convinced that the careful study of history can provide an analytical framework for developing questions which can usefully guide our future conduct. I will therefore conclude by looking ahead.

There can be little doubt that "rockets," as we ordinarily use the term, are near the technical limits of their development. Barring the development of structural materials and propellants which violate the laws of physics as we now understand them, something which I do not totally discount, we are unlikely to see any dramatic improvements in rocket performance. This is particularly true of the high end of the performance and power spectrum, where the Space Shuttle's main engines and jettisonable external fuel tankage press technology very close to absolute physical limits. The Shuttle's main engines and tankage are remarkable engineering achievements; to cite a relevant statistic, less than 4 percent of the external tank, the largest portable, free-standing cryogenic container ever built, is load-bearing structural weight. But they represent only marginal advances in performance over earlier systems. The Titan III, a historical dinosaur powered by an ungodly mixture of unsymmetrical dimethyl hydrazine and nitrogen tetroxide, is still our first-line heavy booster for unmanned payloads. The even more ancient liquid-oxygen and kerosene-fueled Atlas remains a perfectly serviceable booster, and the venerable Agena upper stage, still shielded by security restrictions in its operational military variants, remains the most successful spacecraft ever built. Granted, marginal improvements in propulsive efficiency can be obtained with exotic fuels and oxidizers, but they entail disproportionate penalties in cost, complexity, and flexibility. Substituting liquid fluorine for liquid oxygen in liquid hydrogen engines would entail immense corrosion, containment, and handling problems for a gain of less than 5 percent in specific impulse.[31] There are no doubt exotic corners of the strategic and tactical world where such gains are worth the price—strategic reconnaissance is a case in point—but they are likely to remain highly specialized and involve small numbers of operational vehicles.

More to the point of my analysis, air-breathing systems promise to offer more for less over the short and medium term, and probably the long term as well. That is clearly true at the high end of the size and performance spectrum where hybrid rocket/ramjet propulsion systems will surely power the SR-71 replacement if they do not already. Unconstrained by the laws of ballistic trajectories, such vehicles not only come close to matching rocket-boosted systems in terms of raw speed; they more than make up for any speed deficiency with greater maneuverability and lower susceptibility to detection by virtue of operating at lower altitudes. The advantages of air-breathing systems are even more apparent at the small end of the spectrum, where highly efficient turbine and turbo-compound engines can drive very small aerodynamic vehicles for very long distances. The Tomahawk and ALCM cruise missiles of the Persian Gulf War, effective as they were, will look very large and crude when compared to their replacements.

At the strategic end of the spectrum, rockets possess capabilities which other systems can match only with difficulty, and in some cases not at all. Intercontinental ballistic missiles and intermediate range, or theater-level, ballistic missiles can deliver destruction over great distances more rapidly than piloted aircraft or cruise missiles. Hypersonic cruise missiles will cut into the ballistic systems' advantage in swift delivery over long ranges, but only at considerable disadvantages in cost and complexity. By contrast, ballistic missiles, particularly mobile ballistic missiles, offer considerable capability for the price and will be with us at least until antimissile systems capable of reliably intercepting ballistic warheads in reentry come into widespread use, and probably even afterwards.

Air-breathing lower stages have demonstrated a significant capability in orbiting payloads of modest size, and the importance of such systems will increase, particularly with regard to reconnaissance systems, so long as the miniaturization of electric and mechanical components continues to reduce payload mass and bulk. Indeed, it is easy to envision a runway-based operational booster system with rocket/ hypersonic ramjet propulsion reaching service within the next decade.[32] That having been said, rocket boosters not all that different from those developed in support of the Soviet and US manned space programs will be needed to boost very large orbital and extraorbital payloads for the indefinite future: at the very large end of the spectrum, scale effect works to the advantage of rockets and to the detriment of other systems, or at least that is how I read it.

Insofar as battlefield systems are concerned, rocket-propelled systems occupy several vital niches and will continue to do so for the foreseeable future. Precision-guided air-to-air, surface-to-air, air-to-ground, and antiarmor missiles are a case in point. Particularly with respect to homing air-to-air ordnance, only rockets can provide the required modest initial acceleration, payload size, range, quick response, and long shelf life.[33] It is not inconceivable, though the prediction has been made in error several times before, that improved motors, aerodynamics, and sensors may enable air-to-air missiles to replace guns entirely in aerial combat. Similar considerations apply to surface-to-air missiles, with the added proviso that rockets have a significant advantage over antiaircraft artillery in warhead size. Although gun systems have important advantages for use against low-altitude targets in speed of response, rate of fire, initial velocities, and minimum effective ranges, their maximum range limitations, an unavoidable consequence of the effect of aerodynamic drag on a ballistic projectile, relegate them to the close-in defense role.

For reasons addressed earlier, unguided bombardment rockets can deliver a higher volume of fire for a short period of time for a lower total system weight and cost than conventional tube artillery. In addition, rocket systems are particularly well suited for distributing "smart" submunitions. But conventional tube artillery can project equivalent amounts of firepower quickly and accurately at considerably less cost per round and will not become obsolescent any time soon. Rocket-propelled grenades for antitank use and bunker busting are perhaps the most important of rocket battlefield munitions, and the only one with no real substitute. Cheap, simple, and man-portable, they offer lots of tactical bang for their weight, bulk, and cost; like the brass small arms round with a jacketed lead bullet, they press the laws of nature, tactics, and economics to the limit and will be with us as long as the laws of physics as we now understand them apply. There will no doubt be significant enhancements in capability in all of these categories of rocket weaponry in the years and decades to come, but the bulk of the improvements, and the most important ones, will be in sensor and guidance systems rather than propulsion.

Notes

1. Dieter K. Huzel and David H. Huang, *Design of Liquid Propellant Rocket Engines*, 2d ed., NASA Special Pamphlet SP-125 (Washington, D.C.: NASA, 1971), 10–11.
2. H.M. War Office, *Treatise on Ammunition*, 8th ed. (London: H.M.S.O., 1905), 29.
3. Modern solid rocket fuels are a special case in that they are often tailored to have massive exhaust products to maximize initial thrust at the expense of ultimate velocity in a manner analogous to first gear on an automobile. The Space Shuttle's strap-on, jettisonable boosters are a case in point: Aluminum is added to the fuel to obtain greater thrust and energy density at the expense of I_s.
4. The values were extrapolated from exhaust velocity data for black powder, smokeless powder, and a representative selection of liquid oxygen/fuel combinations given by Willy Ley, "Evaluating the Vaunted V-2," *Aviation* 44 (February 1945): 213, and exhaust velocity and I_s values for standard liquid oxidizer/fuel combinations from Huzel and Huang, *Design*, 25–27, tables 1-6–1-9. I make no claims for precision, but am confident that my figures are accurate indicators of relative performance. For comparison, the I_s of hydrogen peroxide (H_2O_2), the least efficient rocket fuel considered by Huzel and Huang, is 140.
5. Huzel and Huang, *Design*, 27, tables 1-9, 1-10.
6. The highest black powder muzzle velocity of which I am aware was 2,497 fps obtained with an experimental British 5.87-inch gun, John F. Guilmartin, Jr., "Ballistics in the Black Powder Era," in Robert D. Smith, ed., *British Naval Armaments*, Royal Armouries Conference Proceedings 1 (London: H.M. Royal Armouries,1989), 277n5, citing Andrew Noble, *Fifty Years of Explosives, Royal Institution of Great Britain Weekly Evening Meeting, Friday, January 18, 1907* (London: Royal Instituion of Great Britain, 1907). Muzzle velocities of about 1,800 fps were the limit for operational heavy ordnance firing spherical-grained powder.
7. The generalization, based on a survey of standard reference works, applies to infantry rifles and cannon proper (that is, artillery pieces designed for long-range fire) of World War II vintage when the technology had matured. Howitzers and mortars, designed to fire more massive projectiles for shorter ranges, are deliberately designed with lower muzzle velocities. Conversely, aircraft, antiaircraft, tank, and antitank guns have considerably higher muzzle velocities.
8. For the development of cannon-boring, see William H. McNeill, *The Pursuit of Power: Technology, Armed Force, and Society since A.D. 1000* (Chicago: University of Chicago Press, 1982), 167–70. For a definitive treatment of the process, see Carel de Beers, ed., *The Art of Gunfounding: The Casting of Bronze Cannon in the Late 18th Century* (Rotherfield, East Sussex, England: Jean Boudriot Publications, 1991).
9. The seminal development was Robins's invention of the ballistic pendulum, published in *The Principles of Gunnery* (London: J. Hourse, 1742). By making it possible to calculate the impact velocities of projectiles with simple calculations based on Newtonian physics, the ballistic pendulum permitted direct experimentation with the efficiency and effectiveness of

propellants. This led directly to the optimization of powder charges (they had been far larger than necessary) and the development of superior powder made with charcoal burned in enclosed cylinders rather than pits or open ricks. Rodman's pressure gauges permitted the direct measurement of peak pressures at various points along the bore and made it possible to match stress with strength in cannon design, an area in which Rodman excelled. Nobles's ballistic chronograph made possible the direct measurement of the acceleration of the projectile within the bore, permitting calculation of average, as opposed to peak, pressures within the bore as a function of distance and time. Ironically, Robins's hopes of making a science of external ballistics were dashed by the fiendish complexities of transonic drag rise. For a summation, see Guilmartin, "Ballistics," 73–98.

10. There is a serious point in my digression: one of the most sophisticated testing methodologies is run-to-failure, in which the machine is run until something fails and the failed component redesigned until something else fails first. By purely empirical means, the methodology is capable of detecting and correcting for problems which are not theoretically explainable.

11. I have used the term "nitrocellulose-based propellants" as shorthand for all propellants based on nitrated organic compounds, including glycerin, glycol, etc.

12. See Ian Hogg and John Weeks, *Military Small Arms of the Twentieth Century* (Chicago: Follett, 1973), 3.01–3.02, for a perceptive summation. Interestingly, small arms designers were apparently at first reluctant to take full advantage of the new propellants, and the first such weapons offered only marginal improvements over their black powder predecessors. By about 1900, the period of gestation was over and the technology had matured to reflect the values indicated.

13. U.S. War Department, *Handbook on German Military Forces,* TM-E 30-451 (Washington, D.C.: U.S. Government Printing Office, 15 March 1945), vii–88.

14. T. J. Gander, *Field Rocket Equipment of the German Army, 1939–1945* (London: Almary Publications, 1972), 14, 27. Solid rocket fuels commonly used in World War II had scaling limitations not all that different from their black powder predecessors; they did, however, have substantially greater energy densities and could thus loft considerably larger payloads per unit of fuel mass.

15. G. Edward Pendray, "Pioneer Rocket Development in the United States," in Eugene M. Emme, ed., *The History of Rocket Technology: Essays on Research, Development, and Utility* (Detroit: Wayne State University Press, 1964), 22.

16. The standard account, though bordering on hagiography, is Frederick I. Ordway III and Michael R. Sharpe, *The Rocket Team* (New York: Thomas Y. Crowell, 1979). For a technically informed overview, see Gregory P. Kennedy, *Vengeance Weapon 2, the V-2 Guided Missile* (Washington, D.C.: Smithsonian Institution Press, 1983).

17. James E. Tomayko, "Helmut Hoelzer's Fully Electronic Analog Computer," *Annals of the History of Computing* 7 (July 1985): 227–40.

18. Plywood used to insulate radio antennae in developmental models was found to have survived the heat of reentry by charring. This serendipitous discovery led to the development of advanced ablatives in the United States during the 1950s; author's interview of Wilhelm Raithel, former A-4/V-2 engineer, Bethesda, Md., 25 July 1984.

19. Roy Healey, "How Nazi's Walter Engine Pioneered Manned Rocket-Craft," *Aviation* 45 (January 1946); "V-2's Powerplant Provides Key to Future Rocketry," ibid. 45 (May 1946); and author's interview of Eberhard Rees, Marshall Space Flight Center, Huntsville, Ala., 17 July 1985. Rees explicitly confirmed the link between Walter technology and the V-2 turbopumps, ignored by published sources.

20. Scale effect is based on the fact that the ultimate strength of any material is based on the strength of the bonds which hold its constituent molecules together; the greater the number of molecules and bonds, the weaker the structural member in terms of its mass. Scale effect explains why an ant can lift a far greater proportion of its mass than an elephant and why a toothpick can support a proportionately greater load than a two-by-four.

21. Frank K. Malina, "Origins and First Decade of the Jet Propulsion Laboratory," in Eugene M. Emme, ed., *The History of Rocket Technology: Essays on Research, Development and Utility* (Detroit: Wayne State University Press, 1964), 11–12.

22. The smoothbore mortars with bore diameters in the 3-inch to 82-mm range which served as standard battalion-level infantry support weapons in the British, American, Soviet, German, Italian, and Japanese establishments were straightforward developments of the British 1915 Stokes infantry mortar; Ian V. Hogg, *The Encyclopedia of Infantry Weapons of World War II* (New York: Thomas Y. Crowell, 1977), 102–15.

23. The spigot mortar reverses the function of barrel and projectile. The projectile, normally finned, consists of a warhead mounted atop a tube containing primer and propellant. The "barrel" of the launcher is a solid steel rod, and firing is accomplished by driving the rod into the projectile or the projectile onto the rod. The part of the apparatus subjected to the greatest strain, and therefore the most heavily built, is the "barrel," that is, the tail of the projectile; this increases the inert mass of the projectile relative to that of the launcher, reducing muzzle velocity and increasing the weight of the individual round. These disadvantages notwithstanding, the PIAT, though inferior to the bazooka, *Panzerschreck* and *Panzerfaust* in terms of mass of warhead delivered per unit of total system mass (all used shaped charge warheads) was an effective weapon.

24. Squeeze-bore ordnance works on the Venturi principle which dictates that reducing the diameter of a duct will increase the velocity of gas flowing within it proportionately. This principle was harnessed by firing a flanged shell down a tapered bore, the barrel swaging the flanges against the sides of the projectile. The 28/20-mm antitank gun designed around this principle by the German designer Gerlich attained a muzzle velocity of 4,600 fps for a total system weight of only 260 pounds in the stripped-down, airborne version, *Handbook on German Military Forces*, viii, 31–32.

25. The aircraft in question was the Berezniak-Isaev BI-1, powered by a 1,100-lb. thrust bi-fuel liquid-fuel rocket with sufficient fuel for eight to fifteen minutes of powered flight; William Green, *War Planes of the Second World War, Fighters,* vol. 3 (London: Hanover House, 1961), 125–26.

26. The fuel was 80 percent hydrogen peroxide/20 percent water (C-stoff) oxidized by a mixture of hydrazine hydrate, methyl alcohol, and water), J. R. Smith and Antony Kay, *German Aircraft of the Second World War* (London: Putnam, 1972), 508–20, 517.

27. Ibid., 517.

28. Ibid., 53–59.

29. Information to the author from Alfred Price, USAF Academy Military History Symposium, October 1994.

30. Smith and Kay, *German Aircraft,* 656–60.

31. Huzel and Huang, *Design,* 27, fig. 1-10; the respective figures for I_s are 391 and 410 seconds.

32. See Bill Sweetman, *Aurora* (Osceola, Wis.: Motorbooks International, 1994), for a technically informed discussion of the engineering and aerodynamic aspects of air-breathing hypersonic vehicles.

33. By modest acceleration, I mean modest compared to that sustained by projectiles fired from conventional guns, the difficulty being the vulnerability to shock of avionics and sensor components. The resistance of avionics and sensor systems to acceleration forces and vibration can be reduced and has been; the replacement of vacuum tubes with transistors made a difference in reliability of at least an order of magnitude in this respect. Transistor systems have been "toughened" to the point that precision-guided artillery shells are a feasible proposition. That having been said, such toughening is achieved at a price in performance and cost, and the relatively gentle launch acceleration profiles of rockets will continue to be advantageous.

A New Look at the "Wizard War"

Alfred Price

The purpose of this essay is to cast new light on the action that Winston Churchill termed the "Wizard War": the jamming of the radio beams intended to guide German bombers to targets in Great Britain during the dark days of 1940 and 1941. In the historical context this action is important because it saw the first protracted use of the techniques that are now termed Electronic Combat.

The story began in the early 1930s, when the German Lorenz Company pioneered a VHF radio-beam system to assist airliners to reach airports in bad weather. The principle of operation was similar to that of the radio-range system widely used in the United States. The German beam comprised two fine overlapping beams. Morse dots were radiated in one beam and Morse dashes in the other; and where the beams overlapped, the dots and the dashes interlocked to produce a steady note.[1]

In the 1930s it was believed that VHF waves traveled in straight lines. That meant they would not follow the curvature of the earth, so a VHF beam system would have only a limited usable range. However, when German engineers conducted tests with the VHF beams they discovered that the usable range was about 30 percent greater than expected. In fact, the beams transmitted on these frequencies did bend slightly to follow the curvature of the earth; an aircraft flying at 20,000 feet could pick up the beam signals at a distance of up to 250 miles from the transmitter.[2]

To exploit this discovery, the Luftwaffe contracted the Telefunken Company to design and build a high-powered long-range beam transmitter code-named Knickebein. To produce a steady note lane that was only .3° wide, the transmitter fed a huge directional antenna 315 feet across and 100 feet high, which ran on a circular railway track so it could be aligned on the target. Knickebein transmitted on one of three frequencies: 30, 31.5, or 33.3 Mhz.[3]

To guide bombers to their target at night or in poor weather, two such beams were employed. The aircraft flew along the steady note lane from one transmitter, and released their bombs as they passed through the steady note lane from a second transmitter which crossed the first at the bomb release point. One of the operational advantages of Knickebein was that the bomber did not need special equipment to pick up the beam signals. The airfield beam approach receiver, a regular item of equipment fitted to all German multi-engined bombers, could also pick up the Knickebein signals. Knickebein was easy to use because, as a matter of course, bomber pilots were already practiced in the use of their airfield approach radio-beam system. Knickebein was readily accepted by bomber crews and it appears that there was no serious interface problem between the scientists and the operators.

In August 1940, the Luftwaffe began its large-scale day-and-night bombing campaign against targets in Great Britain. As well as the Knickebein beam transmitters in Germany, the Luftwaffe erected others in France, Holland, and Norway. Thus any target in the British Isles could be marked with Knickebein beams. The Battle of Britain was of course fought out primarily by day. The night blitz on Britain, however, began at about the same time and is less well known, though it ran for far longer and caused considerably more damage and casualties.

In June 1940, shortly before the start of the Battle of Britain, a RAF Elint (electronic intelligence) aircraft picked up Knickebein signals for the first time. The German beam-attack system seemed to pose a grave threat to Britain's cities—the RAF had dealt harshly with attacking forces coming in by day, but at night the British air defense system was weak to the point of impotence. Briefed by his scientific advisors of the seriousness of the menace, Churchill ordered the formation of a radio-jamming organization at the very highest priority.

Exactly two months later the jamming organization, designated No. 80 Wing (equivalent to a U.S. group), was ready to begin operations. That shows how rapidly a nation can react, if people are frightened enough. Commanded by Wing Commander Edward Addison, a signals specialist, the unit rapidly built up to a strength of 180 men and women.[4]

Dr. Robert Cockburn, at the Telecommunications Research Establishment at Swanage, headed the small team that hand-built many of the jamming transmitters. Initially, makeshift jammers were produced by modifying hospital diathermy sets into spark transmitters giving a power output of about 150 watts. But soon a purpose-built jammer (code-named Aspirin) was built to counter Knickebein.[5]

Aspirin radiated Morse dashes on the German beam frequencies at a power of 500 watts. The Morse dashes were *not* synchronized with the German signals, rather they were superimposed on top of them. The idea was that when a German pilot entered the dash zone, he would turn in the required direction. But when he arrived in what should have been the steady note lane he continued to hear dashes. When the pilot reached the dot zone, he heard simultaneous dots and dashes which failed to resolve themselves into a clear note.[6]

Both during and since the war there have been stories that the RAF deliberately bent the German beams. These stores are not true. The normal method of jamming Knickebein was to radiate *unsynchronized* Morse dashes on the beam frequencies. It is possible that on occasions the British dashes and German dots came together to produce some sort of bent beam. But there was never any deliberate "beam aiming."[7]

Knickebein never achieved anything like its full potential. Having spoken to several German bomber crewmen who flew over Britain at that time, this author believes that the low-powered jammers were only partially effective in concealing the Knickebein signals. Aspirin did render Knickebein unusable for many of the German bomber crews, but in a quite different way. Whether or not it concealed the beam signals, by its very presence the jamming showed that the defenders knew of the existence and the location of the beams. German bomber crews feared that the RAF would send night fighters to fly along the beams to pick them off. In fact the RAF tried this a few times and found that it did not really work. The German bombers flew in a very loose stream. The night fighters cruised at about the same speed, which meant that in practice they simply maintained position with their prey in front of them but beyond radar range. The chances of making an interception in this way were slim, but the German bomber crews were worried about the possibility, and many of them refused to fly in the beams over enemy territory. Knickebein fell into disrepute, as fewer and fewer crews used the system.[8]

In the fall of 1940, the British Intelligence service discovered that German bombers were using a new type of radio-beam system over Britain, code-named the X-Gerät. Originally conceived by Dr. Hans Plendl in the late 1930s, this had been developed into an operational system by Dr. Rudolf Künold of the Lorenz Company. The Luftwaffe Signals Service formed a special unit to conduct the flight tests of the system commanded by Lt. Col. Heinrich Aschenbrenner, a World War I bomber navigator who had later become an air signals specialist. Aschenbrenner was uniquely qualified to bridge the gap between the scientists who developed the system and the aircrew who were later to use it in combat, and he played a key role in the operational introduction of X-Gerät.[9]

The X-Gerät was essentially similar in operation to Knickebein, but it worked in a much higher part of the frequency spectrum, between 66 and 75 Mhz. Because of this, X-Gerät was considerably more accurate than Knickebein, and it had a steady-note lane only .05 degree wide. The X-Gerät transmitters radiated complex fans of beams, but only four of them were necessary for the operation of the system. One beam marked the approach path to the target, and three more crossed the approach beam to provide accurate way points short of the target. As the bomber passed through the last two cross beams, a special clock was run to measure the ground speed very accurately, and when the plane reached the computed bomb-release point, the bombs were released *automatically*.[10]

X-Gerät was much more difficult to use than Knickebein and only one German bomber unit was trained and equipped to use it. The unit, Kampfgruppe 100 (equivalent to a U.S. group), was the lineal successor to Heinrich Aschenbrenner's operational test unit and it was the world's first night precision-attack unit.

As the British Intelligence service learned about X-Gerät, work began on building a jammer. Code-named Bromide, this radiated omnidirectional Morse dashes with a power of about 100 watts. By mid-November 1940 the first few jammers had been rushed into service.

That was the situation on the night of 14–15 November 1940, when Kampfgruppe 100 led the large-scale bombing attack on the city of Coventry. Thirteen of the unit's Heinkel bombers led the attack, dropping canisters of incendiary bombs to start fires to guide in the rest of the German bomber force. A Bromide jammer was located on the bombers' run in, but it proved ineffective. When the jammer was designed, insufficient information had been known about X-Gerät. Although the jammer probably emitted on the correct frequency, the note it transmitted was modulated not at the 2,000 Hz used by the new system, but at the 1,500 Hz used by Knickebein. The filter circuits in the X-Gerät receivers were sensitive enough to separate the beam signals from the jamming. In the ensuing attack Coventry suffered heavy damage. More than five hundred people were killed, four hundred injured, and large sections of the city were burned to the ground.[11]

Following this catastrophe, RAF intelligence officers examined a captured X-Gerät receiver and discovered the deficiency in Bromide. At top priority the jammers were modified to radiate on the correct modulation. During the months that followed, Kampfgruppe 100 led several attacks. But never again would a British city suffer the concentrated devastation inflicted on Coventry. It was reasonable to conclude—as British Intelligence did conclude at the time—that the modification to Bromide caused the sharp decrease in effectiveness of the German Pathfinder unit.[12]

The above account might be termed the "conventional wisdom" regarding the "Wizard War." This author also believed it, until he attended a reunion of Kampfgruppe 100 aircrew in Germany a few years ago. One of the great things about studying contemporary history is that one can talk to the very people who took part in it. And it goes without saying that one does not have the full story of a military engagement until one possesses detailed accounts from all of the sides taking part. At the reunion one of the German radio operators was asked what he thought of the effectiveness of British jamming of X-Gerät—how much trouble had he had with it? He said he did not remember any jamming. The morse-dash jamming put out by Bromide was described, but he said he had never heard such jamming. He called over other one-time radio operators and soon there were half a dozen of them discussing the matter. And none of them remembered hearing the Bromide jamming.

The radio operators were not stonewalling. They were genuinely intrigued and they really wanted to help. They racked their brains. They described buzzing noises, musical breakthrough, unscreened engine ignition noises, and various other sorts of interference. But none of it sounded anything like the British jamming, and almost certainly all of it had been unintentional.

Later this author was introduced to Hubert Langerfeld, who had been the unit's expert on the X-Gerät system and was the officer responsible for setting up the beams. He was told about the British attempts to jam X-Gerät with Bromide. But his radio operators had not heard any such jamming. What had gone wrong?

Tall and austere, Langerfeld is one of those fellows who pauses for about twenty seconds after hearing the question, before he delivers his considered reply. When his answer came it was worth the wait. He said that when X-Gerät had originally been conceived, the possibility of enemy jamming had been considered. To counter this threat, in addition to the main beams, spoof beams were to be transmitted to give the enemy something to jam. During the attacks on Britain the spoof beams were turned on early and ostentatiously tuned in; those were the ones that the RAF monitoring stations had picked up. The "real" beams usually came on a few minutes before the leading bombers reached their target.[13] Eventually No. 80 Wing discovered what was happening, but the jamming of X-Gerät was far less effective than thought.

Langerfeld's disclosure answered one question, but it raised others. If the X-beams that marked the targets were not being jammed even after the modulation of Bromide had been corrected, why were the later attacks led by Kampfgruppe 100 less effective than the one against Coventry? If the Bromide jamming was generally ineffective, one should expect to have seen concentrated damage inflicted on other British cities.

To find the answer to that question, we need to examine the weather conditions and the state of the moon on the night of 14–15 November 1940, when Coventry was attacked. The attack took place on a very clear night with a full moon. In fact the main part of the German bomber force was able to find the target relatively easily, even without assistance of Pathfinder aircraft. Also, Coventry possessed many old timbered buildings. It was exceptionally vulnerable to attack by fire and its flak defenses were weak (those factors would also be present when the German city of Dresden was burned out, later in the war). In fact it was the rare combination of perfect weather, a full moon, a highly combustible target, and weak defenses that sealed the fate of Coventry, rather than the use of radio beams and Pathfinder marking. Germans who flew with follow-up bomber units that night have expressed doubt that the Kampfgruppe 100 marking made much difference; the moon was so bright that they would have found the city anyway.[14]

It should also be pointed out that the early German Pathfinder methods were, by later standards, very naive. Using a dozen or so planes to drop regular incendiary bombs, like the 2-pound stick-shaped weapon with its notoriously poor ballistics, was not a good way to mark a target. Since the follow-up aircraft dropped these same weapons, even if the Pathfinder marking was accurate, its effect was quickly diluted.

Compared with the Pathfinder methods used by the RAF later in the war, Kampfgruppe 100's efforts were puny. During the large-scale RAF night attacks, Pathfinder aircraft comprised up to 10 percent of the attacking force. The RAF Pathfinder aircraft dropped specially developed marker bombs that had excellent ballistics to produce distinctively colored spot fires at the target. Relays of Pathfinders renewed the marking at regular intervals throughout the period of the attack.[15]

So when Kampfgruppe 100 led attacks on nights when there was poor weather and no moon, and the target was not so combustible as Coventry, it is not surprising that more often than not the results were poor. During 1940 and 1941 the technique for delivering accurate night attacks was in its infancy. Even when there was no effective jamming of the X-Gerät, the chances of achieving a concentrated attack were poor.

In conclusion, there are five points to be made. First, it should be noted that Kampfgruppe 100 of the Luftwaffe and No. 80 Wing of the Royal Air Force, the main opponents in this skirmish in the radio waves, had many points in common. Both were very small and highly specialized units and, because of this, they were able to employ top-rate signals experts in their key positions. These people spoke "the same language" as the scientists and engineers who built their specialized electronic equipment, so communication between the two was not a major problem.

Second, both Kampfgruppe 100 and No. 80 Wing operated close to the state of the art in their respective spheres, and they had to develop their operational techniques from first principles as the battle progressed. Often they had to respond quickly to the tactical situation, using whatever equipment was available, on the basis of incomplete and often misleading information on what their opponent was doing. In such circumstances it is hardly surprising that in many cases the methods adopted were not the most effective.

The third point stems from the second. We now know that the hastily improvised British jamming effort was far less successful in concealing the Knickebein and X-Gerät beam signals than was thought at the time. However, even the ineffective jamming of Knickebein did have the valuable effect of persuading German bomber crews to abandon using their beams for fear that they would serve as a focus for patrols by enemy night fighters.

Fourth, as the RAF and the USAAF discovered later in the war, for the effective marking of targets it is necessary to have dedicated units with specialized radio or radar equipment and specially developed marker bombs. During a night attack lasting an hour or more, the RAF employed as many as 70 Pathfinder planes to initiate the marking and renew it at regular intervals. And even then things sometimes went spectacularly wrong. Seen against this background it is clear that the odds of achieving success were stacked against Kampfgruppe 100, which never sent more than 20 planes to mark a target and usually operated with about half that number.[16]

The fifth and final point is that this action shows how very difficult it is to quantify the operational effectiveness of a particular electronic combat technique or tactic. It is a case of trying to deduce reasons why things fail to happen. There can be any number of reasons why things fail to happen in wartime, of which jamming may be only one and perhaps not the most important.

Notes

1. Alfred Price, *Instruments of Darkness* (London: William Kimber, 1967), 20.

2. Kenneth Wakefield, *The First Pathfinders* (London: William Kimber, 1981), 19.

3. Ibid., 20.

4. Price, *Instruments*, 31.

5. Ibid., 35.

6. Ibid., 36.

7. Ibid.

8. Conversations with ex-Luftwaffe bomber pilots Günther Unger of KG 76, Heiligenhaus, Germany, 1977; and Otto von Ballasko of KG 1, Gants Hill, London, 1970.

9. Wakefield, *First Pathfinders*, 20 et seq.

10. Price, *Instruments*, 38 et seq.

11. Ibid., 44.

12. Ibid., 45 et seq.

13. Conversation Hubert Langerfeld, ex-officer with KGr 100, Würzburg, Germany, 1979.

14. Conversations with Günther Unger and Franz Bergmann of KG 76, Heiligenhaus, Germany, 1977.

15. Gordon Musgrove, *Pathfinder Force* (London: Macdonald & Janes, 1976). This source gives a very detailed description of the evolution of RAF night Pathfinder tactics during World War II.

16. Wakefield, *First Pathfinders*, 204 et seq.

Fermis as the Measure of War: Neutrons, Electrons, Photons, and the Sources of Military Power*

Alex Roland

Is there a defining technology of modern war? By modern war I mean conflict since Hiroshima. By "defining technology" I take the meaning that J. David Bolter employed in *Turing's Man*, his cultural history of the computer. Bolter believes that certain technologies in certain ages have had the power not only to transform society physically but also to shape the way in which people understand their relationship with the physical world. "A defining technology," says Bolter, "resembles a magnifying glass, which collects and focuses seemingly disparate ideas in a culture into one bright, sometimes piercing ray."[1] Can any technology be said to have done this for modern war?

Defining technology must first be distinguished from deterministic technology. Karl Marx introduced the notion of technological determinism in his oft-quoted observation that "the hand-mill gives you society with the feudal lord; the steam-mill, society with the industrial capitalist."[2] Marx's defenders insist that he was not a technological determinist,[3] but this is a tough case to argue. Fundamentally, Marx insisted that history was shaped by control of the "forces of production," that is, by material forces. His vantage point in the cradle of the industrial revolution provided evidence at every turn.[4]

In the twentieth century, other authors have become still more alarmed by technological determinism.[5] Technology, it seems, is gaining ever more power over human agency. Worse still, technology seems to develop a momentum that feeds upon itself, reducing still further the ability of humans to intervene.[6] Our future seems headed in the direction of Hal, the runaway computer in Stanley Kubrick's classic film *2001: A Space Odyssey*.

Bolter's vision of a defining technology, however, is less sinister than this. Defining technologies, for him, are simply those technologies that characterize an age. His examples are the potter's wheel for the ancient Greeks, the clock in the late Middle Ages, the steam engine in the nineteenth century, and the computer today. These were certainly powerful cultural artifacts, and they surely shaped the course of history. They did not necessarily, however, determine how that history will proceed.

---

*1 Fermi=1 Femtometer=.000000000000001 meters.

173

It should be noted, furthermore, that the term "technology" is an anachronism before the seventeenth century. When it first appeared, it referred only to systematic study of arts and crafts. Marx, the greatest of technological determinists, did not use the term. Nor did most thinkers before the twentieth century conceive of technology as a generic descriptor of what I call the "systematic, purposeful manipulation of the material world."[7] Rather, most technologies were crafts, each considered a separate sphere, handed down from generation to generation by oral tradition through a process of apprenticeship. The smith and the stonemason, the weaver and the potter, never thought in terms of technology.

Still, what we now understand as technology did exist. And technology is a useful category for exploring human history. At times, as Bolter suggests, it even defines some of that history—whether or not it determines it. The examples are not limited to Bolter's.

There have been defining military technologies as well. The chariot was the defining technology of warfare in the West in the second millennium B.C.[8] Fortification defined what we would now call the national security of the great empires of the ancient and classical world.[9] The castle and the accoutrements of the mounted knight defined feudal warfare; the stirrup, which Lynn White elevated to near-deterministic status, was a catalyst of feudal warfare, not a determinant.[10] World War I has been called the chemists' war and World War II the physicists' war.[11] In one of the great insights of military history, Walter Millis said that the defining technology of World War II was the internal combustion engine.[12] By this he meant not that the internal combustion engine determined the course of the war but that it was an indispensable component of the major instruments of the war, the least common denominator of the most important war machines. Since motorized envelopment in three dimensions characterized success in operational art, concerns for fuel dominated strategy for the Axis powers during World War II.

The technology that seems to define the postwar era does not have a name. It is akin to nanotechnology, which is technology at the molecular or atomic level.[13] Atoms are about a third of a nanometer in diameter, that is one-third of a billionth of a meter.[14] Coating the hard disk in a normal desktop computer today is a monolayer that is manufactured to a thickness of one nanometer, give or take a few angstroms, that is, tenths of a nanometer. We already function commercially at this level.[15] But the technology I envision operates at a still smaller scale. Its dimensions are measured in femtometers, that is, millionths of a nanometer or 10^{-15} meters. Neutrons are this size; electrons and photons are smaller still, less than 10^{-18} meters, the current limits of our ability to measure. Modern war is defined by the ability to manipulate nature on this scale.

The argument operates at several levels. At times it appears to contradict itself. Before exploring the specifics, then, some large perspectives are in order.

The world wars, the wars of the first half of the twentieth century, may be seen as wars of industrial production. Battles and weapons, strategy and tactics, men and machines drove the combatants to victory or defeat. But in the end, the United States won both wars by outproducing its enemies. The losers lost not because they ran out of room or people or will, but because they ran out of stuff. The defining technology of these wars may be seen as the large technological systems that Thomas P. Hughes associates with the golden age of America, from 1870 to 1970.[16] The systems builders—Thomas Edison, Henry Ford, Alfred Sloan, and other giants of American industry—taught the country how to produce in massive quantities the electricity and the trucks and the planes and the ships that swamped our enemies. These wars were won in our factories and shipyards and power plants just as surely as they were won on the plains of Europe or the islands of the Western Pacific. Walter Millis's internal combustion engine was simply the universal icon of the American mill.

The desideratum of World War II was quantity, not quality. The Germans had better tanks and faster planes—but not enough. The Japanese had better torpedoes and better planes—but not enough. Surely a qualitative race took place during the war, but it did not determine the outcome. Rather, it bred the perception that the *next* war would be won by superior quality. The Germans fielded ballistic missiles and jet aircraft by the end of the war. The United States and Great Britain fielded radar and the atomic bomb. Research and development appeared to be displacing industrial production as the measure of military strength. This perception assured that the defining technology of warfare in the second half of the twentieth century would be different from that of warfare in the first half.

But has it been? Viewed from one perspective, warfare in the last fifty years seems to have been dominated by vast quantities of large-scale arms and equipment. The superpowers produced tens of thousands of nuclear weapons. To deliver them they built and deployed thousands of ballistic missiles, thousands of aircraft, and hundreds of nuclear-powered submarines. Arrayed behind these strategic weapons were the tanks and ships and planes that poured into the international arms bazaar over the last four decades. While the superpowers and a handful of other nations monopolized nuclear weapons and their delivery systems, countries around the globe scrambled to arm themselves with the arsenals of World War II. The paradigmatic wars of the second half of the twentieth century, those between the Arabs and Israelis, looked indistinguishable from the Battle of the Bulge. Indeed, the 1973 Arab-Israeli War seemed to turn on mass consumption of the tools of war just as surely as the collapse of the Third Reich had done thirty years earlier.[17] What had really changed?

What changed, I would maintain, is that quality had replaced quantity as the deciding factor. Quantity was surely still important, but the qualitative arms race proved more important than the quantitative. That qualitative race took place not in the factories and shipyards, but in the laboratories and universities. And it aimed not at large-scale systems—though these were surely important—but at the mastery of femtotechnology. The new large-scale system was not the industrial base so much as the research infrastructure that discovered how to manipulate neutrons, electrons, and photons.

To test this hypothesis, it is necessary to examine the nature of warfare in the second half of the twentieth century. In this period, there are no defining wars such as the world wars. Rather, warfare has become diversified and heterogeneous.[18] For purposes of analysis, I propose to discuss it in three categories: nuclear war, conventional war, and what has come to be called low-intensity conflict. Femtotechnology has worked differently in each of these realms.

The most important type of war in the second half of the twentieth century has been nuclear war. Ironically, this war was never fought; it simply loomed. Its weapons hung over the decades like a Mahanian fleet in being. Save for the two bombs dropped on Japan at the end of World War II, these weapons have never been fired in anger. Yet they shaped all warfare in the second half of the twentieth century just as surely as if the superpowers had unleashed their arsenals. The shared conviction that these weapons could not be used to advantage prevented World War III. History offers no example of two major powers like the United States and the Soviet Union with conflicting ideological, political, and economic systems and overlapping spheres of influence who did not finally resort to arms. In the Cold War, however, two such powers found themselves armed with weapons they dared not use. So they found other ways to settle their differences, other ways to fight their war.

Not only did nuclear weapons compel the superpowers to fight their war on the battlefield of economics and politics, it also constrained other wars as well. The superpowers cooperated to ensure that their client states did not draw them into armed confrontation, as in the Arab-Israeli War of 1973. And they negotiated how they would fight with their own and other client states, as in Vietnam and Afghanistan. There was plenty of shooting in the Cold War, but it was not the shooting that would have transpired had there been no nuclear weapons.

At first glance, the huge nuclear arsenals that guaranteed the peace and shaped the fighting looked like products of the old industrial system. Thousands of warheads were mounted on thousands of platforms—missiles, planes, ships, guns—all cranked out by a vast industrial infrastructure. But the key technologies, the technologies mastered by the superpowers and by only a handful of other states, were buried deep inside the monstrous machines. One technology was the controlled release of neutrons into a critical mass of fissionable material, which later provided the energy necessary to fuse still another mass of fusible material. The energy released by these processes could be designed to suit the purpose: heat, shock, radiation. The power of the atom, in short, could be manipulated for purposes of war. This was a technology beyond the reach of lesser nations.

Ironically, the infrastructure necessary to produce this technology is inversely proportional to the size of the technology itself. One classic example is the K-25 plant at Oak Ridge, Tennessee. Here the Manhattan Project separated bomb-grade uranium 235 from naturally occurring uranium 238 by gaseous diffusion. The material being manipulated was infinitesimal—molecules of uranium hexafloride. And the material produced was hardly much larger: the process of which K-25 was a part could produce only 7.2 grams of bomb-grade material a day by January 1945. But the plant itself was enormous—four stories high, one-half mile long, one-fifth mile wide, enclosing 42.6 acres of floor space at a cost of $500 million.[19] And behind this monumental architecture were arrayed the legions of researchers and workers in universities and laboratories around the country who were developing the theories and conducting the experiments to prove the concept of this facility. Behind these researchers was the infrastructure of the Office of Scientific Research and Development that mobilized the personnel and material resources for this vast undertaking. And at the base of this infrastructure was the educational system that prepared these researchers for their work and the economic and industrial system that developed the expertise and the resources to build such a structure as K-25 in the first place. The focus of the technology was minute, but it required a lens of unprecedented scale. The Manhattan Project was, in its time, the largest single technological undertaking in modern history. It is rivaled only by such projects as the pyramids, the medieval cathedrals, and the Great Wall of China, all of which took decades or centuries to complete.

So too with the delivery systems designed for nuclear weapons. The missile was by far the most important, for it was the first one against which no reliable defense could be developed. At the heart of the missile was a guidance system that was also dependent on subatomic particles—in this case the electrons that moved through solid-state devices to direct the weapons on their way. The rocket had been around since the Chinese introduced it in the thirteenth century; it had been used in war since Congreve's rockets in the early nineteenth century. But these were unguided missiles. It was micro-electronics, the technology of solid-state physics, that provided the mechanisms to guide these weapons with useful accuracy.[20] This technology took on its modern guise with the discovery of the transistor in 1948.

It is a nice question whether the defining technology here is the computer or solid-state physics. The computer has roots as deep as Greek calculating machines.[21] It began its modern evolution with the mechanical devices of Gottfried Wilhelm Leibniz in the seventeenth century and Charles Babbage in the nineteenth century. By World War II, huge electrical devices such as the ENIAC machine at the University of Pennsylvania were doing war work, calculating ballistic tables and even solving complex mathematical problems for the Manhattan Project. None of these devices, however, held any promise of guiding missiles; they were simply too large and heavy.

The brains built into our modern weapons systems were made possible only when the trajectory of electrical, digital computers crossed the trajectory of solid-state physics.[22] The conductive properties of the class of materials known as semiconductors was grasped as early as 1876; what was not understood was why the curious materials conducted electricity differently under different circumstances. That riddle was finally solved in the 1940s by a team of researchers at Bell Labs. Looking for an electrical version of the mechanical telephone switching devices then in use, the researchers discovered that they could control the flow of electrons in a semiconducting material such as silicon by "doping" it with other atoms and applying an electrical charge. In 1948 they introduced the point-contact transistor, forerunner of all subsequent solid-state devices.

At first, the implications of this discovery were lost on the commercial world.

The U.S. military, especially the Air Force, provided the research and development funds that helped turn this technological breakthrough into useful products. The main locus of activity was Silicon Valley, where William Shockley, the Nobel Laureate and co-inventor of the transistor, set up a commercial enterprise to develop the potential of these new devices. Seed money from the military soon bore fruit, and a hot-house atmosphere of entrepreneurial development led to self-sustaining growth of the industry.

When this trajectory met that of digital computation and produced the first integrated circuit, the modern computer was born. Replacing the vast array of tubes that manipulated the electrons of the World War II ENIAC were tiny circuits etched on silicon chips. These were smaller, lighter, and cooler than the tubes they replaced; soon they were also cheaper. Placed inside missiles and other military equipment, they proved able to direct the weapons' activities in accord with detailed programs.

From the first-generation computers of the ENIAC type, through the second generation of transistor-driven, hand-held calculators, the field quickly proceeded through a third generation of integrated circuits and a fourth generation of microprocessors. For the last decade, the computer world has been flirting with what the Japanese called the fifth generation, supercomputers capable of achieving artificial intelligence—"machines who think."[23]

This has been—along with nuclear fission and fusion—the defining technology of nuclear warfare. Of course, countless other technologies go into our modern strategic weapons: fuels, synthetic materials, exotic alloys, systems integration, aerodynamics, structures, engines, pumps, valves. The list could be extended indefinitely. But the hearts and brains of these weapons are the nuclear devices themselves and the microelectronics that control them; the blood of these beasts are the electrons and neutrons that course through their circuitry and their explosives.

What about conventional war, then? Surely that is still dominated by the tanks and planes and ships which are very much like the weapons of World War II. At one level, this is true. The revolution in conventional war came more slowly. Viewed from afar, the Gulf War looked like the Battle of the Bulge. Beneath the surface, however, this was a wizard war, won before it began by electrons and photons and the ability to control them. The difference is that the route from the Bulge to the Gulf was indirect; it took a major detour through Vietnam.

As was the case in nuclear war, the United States took the lead in this evolution. Following World War II, all the services institutionalized research and development, and all of them fought for a piece of the nuclear mission.[24] Into this channel was directed much of the new thinking and the new technology. Conventional war, at least conventional war on land, was envisaged as a continuation of industrialized war. The Soviet Union now held the huge advantage in mass. Tank warfare on the plains of Europe, should it come, would pit the Red Army and its impressive armor forces against an outnumbered NATO alliance. The first American response to this threat was not so much new conventional technology as nuclear weapons. Through the Truman Administration and much of the Eisenhower Administration, the United

States relied on the threat of strategic warfare to deter conventional Soviet attack in Europe. Late in the Eisenhower administration it began to deploy tactical nuclear weapons as well.

Otherwise it relied on the land warfare technology of World War II, improved, surely, but hardly revolutionized. This technology served in Korea. When Vietnam entangled the United States in another major ground war, it was deployed there as well. Only when it failed in that setting, did the United States turn to a scale of technology that defined the second half of the twentieth century—not neutrons, this time, but electrons.

Faced with an unconventional, elusive, and intractable enemy in Vietnam, the United States began to develop and deploy a new array of military technology. These weapons and devices ranged from sensors and night-vision equipment to precision-guided munitions. In the course of this war a new style of fighting emerged, what Paul Dixon called "the electronic battlefield."[25] The new devices served as power multipliers. They gave the soldier enhanced knowledge: knowledge of where the enemy was, how to discriminate the enemy from the environment in which he hid, and how to deliver munitions on the enemy and only the enemy. From the bludgeon with which the United States had won the world wars, now emerged the scalpel. Unable and unwilling to destroy the sea in which the Communist fishes swam, the United States moved instead to finding the fishes, driving them into schools, picking them out one at a time from their environment.

Needless to say, the new technique lacked maturity. Sensing devices and smart bombs had their greatest impact late in the war, when political forces had already taken control. Postwar reports emanating from the Vietcong and the North Vietnamese confirm that the U.S. military was experiencing considerable success on the battlefield in the later years of the war, but it was not enough to turn the political tide.[26] Still, the new technology had demonstrated its power. America had to withdraw from Vietnam, but the military services did not abandon the new line of development begun there.

The experience of Vietnam formed the basis of AirLand Battle. Here the technology of knowledge, what has come to be known as C^3I, formed the basis of a new kind of warfare. Instead of meeting mass with more mass, the formula of the world wars, the United States chose to meet mass with precision informed by technology based in electrons. Sensing devices on the ground, in the air, and in space would track enemy dispositions and provide information in real time to the theater commander. Computers at the disposal of the commander would generate tactics to attack the enemy where he was weakest, identify vulnerable nodes of logistics, transportation, and command and dispose his own resources to achieve maximum battlefield effectiveness. The plan would go out to unit commanders over the most sophisticated communications network ever seen on a battlefield. Individual units would receive their orders ready-tailored for their particular role. And weapons would be guided to their targets with surgical precision. Perhaps the rhetoric occasionally outran the capability, but the concept was nonetheless sound. It won the Gulf War.[27]

One key to success in AirLand Battle and in the Gulf War where it played out was the need to deny the enemy the same capabilities that were to be deployed against him. Denial was predicated on quality, not quantity: fast computers, more reliable communications, better electronic countermeasures, more powerful satellites. The first targets in the attack on Iraq were the eyes and ears of the Iraqi military—the radars to guide antiaircraft defenses, the communication nodes that controlled the forces in the field. The successful attack that launched the ground offensive worked because of superior U.S. intelligence and communications and because the Iraqi forces were by then blind and deaf.[28]

In some cases, the United States simply deployed technology completely unavailable to the enemy. For example, the global positioning system (GPS), a satellite-based navigation network, allowed ground unit commanders down to the squad level to move across a trackless battlefield with precision measured in tens of feet. GPS stole the desert from its denizens.

Of course, electrons did not win the Gulf War. Though they may have been defining, they were not deterministic. Both sides relied on troops on the ground. Both used critical technologies that were not dependent on computers and solid-state electronics. The trucks and tanks and planes and ships still carried and delivered the people and equipment that are the stuff of war. Food and fuel still fed the moving parts that make up the modern machine of war. But this was not like the Battle of the Bulge. That was a battle of brute force. This was a battle of knowledge and information. The Germans knew what hit them; the Iraqis did not.[29]

Low intensity conflict is more complicated. It too has a defining technology— the same one as nuclear war and conventional war. But in low intensity conflict, femtotechnology works at cross purposes. It favors both the insurgent and the counterinsurgent.

Consider first the counterinsurgent. The "electronic battlefield" of Vietnam was invented specifically for his purposes. The key was intelligence. The United States had more than enough military power in Vietnam to destroy the enemy; if only we could find the enemy. The Vietcong, however, knew that their success depended on avoiding set-piece battles with the American war machine. They pursued a different strategy. That strategy was developed by Mao Tse-tung and his colleagues in the mountains of Hunan Province. Though Mao had been driven to that remote region by Chiang Kai-shek, he developed his theory of "people's war" to fight another, more intractable enemy—the Japanese. Japan had invaded northern China in 1931, extending its control through much of eastern China after the "China incident" of 1937. From his mountain refuge, Mao contemplated how a backward, agricultural nation such as China might successfully fight against a modern, industrialized state such as Japan. He gave his answer in a series of essays, the most important of which is "On Protracted War."[30] It has proved to be a blueprint for insurrectionary movements around the world.

Successful people's war has three components. First, the insurgents need a sanctuary, a safe haven beyond the reach of the superior conventional power of the enemy. Second, the insurgents must retire before the conventional force of the enemy and attack where the enemy is weak, operating behind enemy lines from a succession of temporary, mobile, and expendable bases. Third, the insurgents must seek a political resolution of a war of attrition, relying on domestic pressure within the enemy's own country and international public opinion to force the enemy to come to terms. Mao had his Clausewitz through Lenin; he knew that war was an extension of politics.

Mao's formula fell short of rigorous application in China; the Japanese were defeated by the United States and its allies, and the civil war that followed between Chiang and Mao was, for the most part, the third stage of a process allowed to proceed from strategic defense, through stalemate, to strategic offense. Vietnam, however, followed the formula with great success. As the French had done before them, the Americans first deployed the industrial arsenal of World War II to defeat the Vietcong. Unable to fix the enemy, the United States turned increasingly to the electronic battlefield, an array of sensors and a first generation of new smart weapons to find the enemy and attack him with speed and precision before he could disperse. Failing that, the United States also tried to isolate the enemy from his base, and finally to attack the enemy's sanctuaries in Cambodia and elsewhere. These new techniques were having some success before the politics of the war overtook other events—just as Mao had predicted they would.

Out of its Vietnam experience the United States developed not only the arsenal of AirLand Battle, but also the arsenal of counterinsurgency. The equipment with which Americans armed the government of El Salvador to fight the contra rebels was the harvest of this line of development: star scopes, airborne sensors, improved communications, signals interception, etc. The goal, of course, was still to bring superior force to bear on the enemy, but in counterinsurgency that capability was determined by knowing who and where the enemy was.

Unfortunately, the microelectronics revolution that fed this new arsenal also armed the insurgent. Unlike the pattern of large-scale industrial production that fueled warfare in the first half of the twentieth century, the femtotechnology of the second half of the century puts more and more power in the hands of the individual. Only industrialized states could produce and deploy the tanks and planes and ships that won the world wars. However, insurgents of modest means can buy a Stinger missile on the black market and bring down a jumbo jet at the end of any airport runway. The threat of the future is a guerrilla with a computer—and perhaps a nuclear weapon.

The irony of the present situation is that both the insurgent and the counterinsurgent are being driven to the same technology. When Mao first invented People War's, its specific goal was to overcome the advantage enjoyed by an industrialized state deploying the massive technology of the world wars. It worked so well in Vietnam that the United States began to move away from war of industrial production and toward war of femtotechnology—especially war of electrons and photons.

That technology, however, can serve the insurgent as well as the counterinsurgent. And it is harder to keep out of the international marketplace than the military hardware that it replaces.

The industrialized states still enjoy an advantage in this competition. An enormous technological infrastructure is necessary to develop and produce the femtotechnology of war. Just as the Manhattan Project sat atop a vast economic, industrial, research, and education infrastructure, so too do the new technologies of electrons and photons rest upon a sprawling institutional base. The guidance mechanism that could steer a home-made cruise missile is filled with chips that only a handful of states produce. Secure satellite communications go to those states that can launch and maintain satellites. The speed and security of fiber-optic communications and computing will be available to those states that have the technological and financial resources to develop and deploy such systems.

Today, the microelectronics chip foundry is the analog of the K-25 plant at Oak Ridge in World War II. By the end of the 1980s, a "world-class" foundry for very large-scale integrated (VLSI) circuit chips cost $200 million, an order of magnitude increase in cost since 1975.[31] By 1993 it was estimated that the fabrication facilities for ultra large-scale integration (ULSI) circuits, those in the range of 1 million to 1 billion bits per chip, would exceed $1 billion.[32]

The problem is that the new technology is increasingly difficult to control. The technique of building a nuclear weapon is public knowledge; the only real trick is finding enough fissionable material on the international black market. The chips that can control the wizard weapons of today's battlefield are available commercially. So too is fiber-optic communication. While only a few states can produce such technologies, most states have access to them in the international arms bazaar. Parts of that bazaar are open to the public, parts are under ground. The overall effect is to diffuse the femtotechnology of war to an ever wider circle of players.

Furthermore, the femtotechnology that undergirds the military might of the industrialized states is vulnerable to types of countermeasures that did not plague the owners of the industrial technology of the world wars. Computer systems are only as good as their security; one smart insurgent hacker with a $2,000 work station and a modem can do as much military damage to the United States as a whole regiment of guerrillas in the field. Instant and secure communications can lull a military establishment into the complacent assumption that they will always be up. An insurgent with a listening device from Radio Shack sitting outside a military base is likely to glean more information than the furtive interloper of an earlier day pawing through files in the night. A resourceful terrorist with a GPS receiver can land a cruise missile on the Pentagon.

Femtotechnology defines low intensity conflict less clearly than it defines strategic warfare. Indeed, strategic warfare as we have come to know it in the second half of the twentieth century did not even exist before the femtotechnology of nuclear fission and fusion. Just as nuclear war has dominated and shaped all other kinds of warfare since World War II, so too has femtotechnology had its greatest influence in that realm and spread more slowly into conventional war and finally low intensity conflict. If this analysis is correct, it may well suggest that the pattern of war in the immediate future will be toward a greater permeation of all forms of warfare with femtotechnology.

Students of modern technology will not fail to notice that the pattern being suggested here echoes one that has shaped civilian life as well. It began with neutrons in the 1950s and 1960s. Commercial nuclear power promised a technological revolution of great economic consequence: "electricity too cheap to meter," in the memorable hyperbole of one enthusiast.[33] In the United States, at least, this revolution fizzled in the late 1970s, though approximately 20 percent of the nation's electricity is now generated by nuclear power, compared to a world average of 17 percent. In some other countries, most notably France, the revolution thrives, accounting for 73 percent of that nation's power.[34]

A more lasting impact on civilian life was achieved by solid-state electronics and computers. Some people call this the computer revolution, others the microelectronics revolution. James R. Beniger calls it "the control revolution," arguing that the greatest change introduced by these new technologies is the change in management of human affairs.[35]

Perhaps the most accurate label is the "information revolution," the source of the National Information Infrastructure, more popularly known as the "information highway."[36] If knowledge is power, then information is power. Robert Reich, the Harvard political economist and secretary of commerce in the Clinton administration, even argues that in the emerging world economy it does not matter where goods are produced, only where knowledge is produced. The country that generates the most new knowledge will prosper; the goods and services that result from that knowledge can be produced anywhere.[37]

This is akin to the "third wave" argument made by Alvin and Heidi Toffler.[38] They believe that mankind has evolved through three stages or waves: the first was agricultural, the second industrial, and the third is the culture of knowledge. Smokestack industries are irrelevant to real power in this age. Someone will have to manufacture the products we consume, but this is no more essential to America's prosperity than its agricultural productivity is. England did not need agricultural self-sufficiency to have an industrial revolution; America does not need industrial might to have an information revolution.

The Tofflers have actually written a book on warfare in the era of the third wave.[39] Their argument is similar to the one advanced here, save that they look across the entire spectrum of new technologies, instead of focusing primarily on the femtotechnology that forms the core of my analysis. As with many futurologists, the Tofflers seem never to have met a potential technology they did not love. Their book, therefore, is filled indiscriminately with the probable and the unlikely, science and science fiction. Still, their underlying argument has merit; the world is entering a postindustrial phase in which information will be the desiderata of power, and this change will shape warfare just as fully as it shapes the rest of society.

Evidence abounds that the American military is well aware of this trend. For example, the National Critical Technologies Panel was created in 1990 to identify those technologies essential to the military and economic security of the United States. That list of technologies suggests where the country is and where it is heading:

1. *semiconductor materials and microelectronic circuits*
2. *software engineering*
3. *high performance computing*
4. *machine intelligence and robotics*
5. *simulation and modeling*
6. *photonics*
7. *sensitive radar*
8. *passive sensors*
9. *signal and image processing*
10. *signature control*
11. weapon system environment
12. *data fusion*
13. <u>computational fluid dynamics</u>
14. air-breathing propulsion
15. pulsed power
16. hypervelocity projectiles and propulsion
17. high energy-density materials
18. composite materials
19. <u>superconductivity</u>
20. biotechnology
21. <u>flexible manufacturing</u>[40]

Those in italics are straight femtotechnology, that is, the technology of electrons and photons. Those underlined have a significant femtotechnology component. The pattern is clear. Femtotechnology leads and dominates the list of critical technologies. Many of these are in place and working today; others are on their way to development. Just a decade ago, research in information technology consumed about one-quarter of the budget of the Defense Advanced Research Projects Agency; today it consumes more than half. Howard Frank, Director of ARPA's Computer Systems Technology Office says "it is *the* core technology."[41]

To prepare for this brave new world, the Information Resources Management College of the National Defense University began offering in the fall of 1994 a senior-level course of study on the information component of national power. It will cover everything from the "Global Information Society" to wargaming and simulation to the achievement of "information advantage" over the enemy.[42] The military leaders of tomorrow will have to be prepared for this new dimension of war.

Furthermore, the direction in which science and technology are tending, at least in the short term, is clear. A National Academy of Sciences Panel on the Future of Atomic, Molecular, and Optical Sciences has recently concluded that the manipulation of microscopic particles, molecules, and even individual atoms holds enormous potential for development and exploitation. It recommends that the nation's resources be invested in these fields, including the manipulation of charged particles and light.[43]

Is it, then, fair to conclude from this necessarily brief examination that femtotechnology is the defining technology of modern war? I would say yes, but I cannot prove it. "Defining technology" is not a fact or even a conclusion. It is a label, a model, a category of thought. It represents not the truth about a subject but a way of thinking about a subject. Its utility lies in the insight it might generate.

For this particular rubric to have any explanatory power, four caveats must be kept in mind. First, femtotechnology is an imprecise term to encompass the set of technologies I mean to highlight. The term works, I hope, if it is taken in the special sense of the manipulation of neutron, electrons, and photons.

Second, femtotechnology has not *determined* the outcome of modern war. Many other technologies play important roles—materials, propulsion, biotechnology. All of these have shaped warfare in important ways, but the outcome of war is determined by many other variables as well—people and chance, for example, to name only two.

Third, femtotechnology cannot even be said to dominate the battlefield. It does not win wars by itself. Rather it is the power behind the power. It is the least common denominator of those military technologies that have dominated warfare since 1945. In this sense it is analogous to Walter Millis's internal combustion engine in World War II.

Fourth, as a cultural icon it is not as satisfying as, say, the steam engine in the nineteenth century. There was a presence and tangibility to the steam engine that is lacking in femtotechnology. We can barely envision neutrons and electrons scurrying about, let alone conceptualize how photons work. They are mysterious phenomena, outside the realm of our experience and even our intuition. We can grasp the mushroom cloud and the graphics on our computer screen and the voice at the other end of our fiber-optic telephone call, but the particles that make those things happen live in a world we cannot enter.

Still, I think these very small things capture the essence of modern war more fully than any other single technology. One commentator on the Gulf War said that "control of the battlefield means control of the electromagnetic spectrum."[44] That has been true for some time now. It is likely to be even more true in the future.

Notes

1. J. David Bolter, *Turing's Man: Western Culture in the Computer Age* (Chapel Hill: University of North Carolina Press, 1984), 11.

2. Karl Marx, *The Poverty of Philosophy* (New York: International Publishers, n.d.), 92.

3. Donald MacKenzie, "Marx and the Machine," *Technology and Culture* 25 (July 1984).

4. The issue is explored most recently in Merritt Roe Smith and Leo Marx, eds., *Does Technology Drive History? The Dilemma of Technological Determinism* (Cambridge, Mass.: MIT Press, 1994).

5. Jacques Ellul, *The Technological Society* (New York: Knopf, 1964); Lewis Mumford, *The Myth of the Machine*, 2 vols. (New York: Harcourt, Brace, Jovanovich, 1967–70).

6. Langdon Winner, *Autonomous Technology: Technics-out-of-Control as a Theme in Political Thought* (Cambridge, Mass.: MIT Press, 1977).

7. Alex Roland, "Theories and Models of Technological Change: Semantics and Substance," *Science, Technology, & Human Values* 17 (Winter 1992): 83.

8. William H. McNeill, *The Rise of the West: A History of the Human Community* (Chicago: University of Chicago Press, 1963), 104–6; Robert Drews, *The End of the Bronze Age: Changes in Warfare and the Catastrophe ca. 1200 B.C.* (Princeton, N.J.: Princeton University Press, 1993).

9. Yigael Yadin, *The Art of Warfare in Biblical Lands in the Light of Archaeological Study*, 2 vols. (New York: McGraw-Hill, 1963), 1:16–24.

10. For Lynn White, Jr.'s. controversial hypothesis about the stirrup, see his *Medieval Technology and Social Change* (Oxford: Oxford University Press, 1962), 1–38. Appraisals of the attacks on this theory may be found in Philip Contamine, *War in the Middle Ages*, tr. Michael Jones (Oxford: Basil Blackwell, 1984), 179–84; and Kelly DeVries, *Medieval Military Technology* (Lewiston, N.Y.: Broadview, 1992), 95–110.

11. Daniel J. Kevles, *The Physicists: The History of a Scientific Community in Modern America* (New York: Vintage Books, 1979), 137, 302–23.

12. Walter Millis, *Arms and Men: A Study in American Military History* (1956; New Brunswick, N.J.: Rutgers University Press, 1984), 283.

13. B. C. Crandall and James Lewis, eds., *Nanotechnology: Research and Perspectives: Papers from the First Foresight Conference in Nanotechnology* (Cambridge, Mass.: MIT Press, 1992). This field traces its roots to a 1960 paper by physicist Richard Feynman, "There's Plenty of Room at the Bottom: An Invitation to Enter a New Field of Physics," reprinted in ibid. See also K. Eric Drexler, *Engines of Creation: The Coming Era of Nanotechnology* (1986; New York: Doubleday, 1990).

14. K. Eric Drexler, "Appendix A: Machines of Inner Space," in Crandall and Lewis, eds., *Nanotechnology*, 326.

15. K. Eric Drexler and Chris Paterson with Gayle Pergamit, *Unbounding the Future: The Nanotechnology Revolution* (New York: Morrow, 1991), 85.

16. Thomas P. Hughes, *American Genesis: A Century of Invention and Technological Enthusiasm, 1870–1970* (New York: Viking, 1989); see also idem, *Networks of Power: Electrification in Western Society, 1880–1930* (Baltimore, Md.: Johns Hopkins University Press, 1983).

17. Michael Carver, *War since 1945* (London: Ashfield, 1990), 270–71.

18. Patrick Brogan, *The Fighting Never Stopped: A Comprehensive Guide to World Conflict since 1945* (New York: Vintage Books, 1990), vii–xvii.

19. Richard Rhodes, *The Making of the Atomic Bomb* (New York: Simon & Schuster, 1986), 494, 496, 601.

20. Donald MacKenzie, *Inventing Accuracy: A Historical Sociology of Nuclear Missile Guidance* (Cambridge, Mass.: MIT Press, 1990). The German V-2 rocket of World War II was guided by an inertial/electronic system that accounted for its notoriously poor accuracy.

21. Good general histories are Herman H. Goldstine, *The Computer: From Pascal to von Neumann* (1972; Princeton, N.J.: Princeton University Press, 1993); Michael R. Williams, *A History of Computing Technology* (Englewood Cliffs, N.J.: Prentice-Hall, 1985).

22. See Hans Queisser, *The Conquest of the Microchip*, tr. Diane Crawford-Burkhardt (Cambridge, Mass.: Harvard University Press, 1988).

23. Edward A. Feigenbaum and Pamela McCorduck, *The Fifth Generation: Artificial Intelligence and Japan's Computer Challenge to the World* (Reading, Mass.: Addison-Wesley, 1983); Pamela McCorduck, *Machines Who Think* (San Francisco: Freeman, 1979).

24. No single history captures this story. Parts of the research and development story appear in Nick A. Komons, *Science and the Air Force: A History of the Air Force Office of Scientific Research* (Arlington, Va.: Office of Aerospace Research, 1966); Michael H. Gorn, *Harnessing the Genie: Science and Technology Forecasting for the Air Force, 1944–1986* (Washington, D.C.: Office of Air Force History, 1988); Thomas Sturm, *The USAF Scientific Advisory Board:*

Its First Twenty Years, 1944–1964 (Washington, D.C.: Office of Air Force History, 1986); and Harvey Sapolsky, *Science and the Navy: The History of the Office of Naval Research* (Princeton, N.J.: Princeton University Press, 1990).

25. Paul Dixon, *The Electronic Battlefield* (London: Morian Brothers, 1976); Frank Barnaby, *The Automated Battlefield* (New York: Free Press, 1986).

26. William S. Turley, *The Second Indochina War: A Short Political and Military History, 1954–1975* (Boulder, Colo.: Westview, 1986); Jayne S. Werner and Luu Doan Huynh, *The Vietnam War: Vietnamese and American Perspectives* (Armonk, N.Y.: Sharpe, 1993).

27. John Pimlott and Stephen Bedsey, *The Gulf War Assessed* (London: Arms & Armour, 1992); see esp. pp. 63–79 on the evolution of AirLand Battle. The authors conclude that "electronic superiority over the enemy has become the most important single factor in determining victory" (pp. 271–72).

28. Gen. John Galvin, Supreme Allied Commander Europe, said immediately after the war that Gen. Norman "Schwartzkopf was able to dismantle the electromagnetic spectrum [so that] he effectively closed Saddam's eyes and ears." Quoted in Bruce W. Watson, Bruce George, Peter Touras, and B. L. Cyr, *Military Lessons of the Gulf War*, ed. Bruce W. Watson (Novato, Calif.: Presidio, 1993), 157.

29. Eliot A. Cohen, author of a comprehensive assessment of air power in the Gulf War has said that "the most profound change in military technology . . . [to be demonstrated in the Gulf War] was the vast increase in usable and communicable information." Eliot A. Cohen, "The Mystique of U.S. Air Power," *Foreign Affairs* (January/February 1994): 112.

30. Mao Tse-tung, "On the Protracted War," in *Selected Works,* 5 vols. (New York: International Publishers, 1954), 2:157–243.

31. Fred Warshofsky, *The Chip War: The Battle for the World of Tomorrow* (New York: Scribner's, 1989), 50.

32. Anthony Ralston and Edwin D. Reilly, eds., *Encyclopedia of Computer Science,* 3rd ed. (New York: Van Nostrand Reinhold, 1993), 693.

33. Lewis Strauss, chairman of the Atomic Energy Commission, 1954, quoted in George T. Mazuzan and J. Samuel Walker, *Controlling the Atom: The Beginnings of Nuclear Regulation, 1946–1962* (Berkeley: University of California Press, 1984), 77.

34. United Nations Department of Economic and Social Information and Policy Analysis, *1991 Energy Statistics Yearbook* (New York: United Nations, 1993), 414.

35. James R. Beniger, *The Control Revolution: Technological and Economic Origins of the Information Society* (Cambridge, Mass.: Harvard University Press, 1986). See also JoAnne Yates, *Control through Communications: The Rise of System in American Management* (Baltimore, Md.: Johns Hopkins University Press, 1989).

36. U.S. Congress, Congressional Budget Office, *Promoting High Performance Computing and Communications* (Washington, D.C.: CBO, 1993).

37. Robert Reich, *The Work of Nations: Preparing Ourselves for 21st-Century Capitalism* (New York: Knopf, 1991). Reich asserts that in the twenty-first century, "there will be no *national* products or technologies, no national corporations, no national industries" (p. 3).

38. Alvin and Heidi Toffler, *The Third Wave* (New York: Bantam, 1980).

39. Alvin and Heidi Toffler, *War and Anti-War: Survival at the Dawn of the 21st Century* (Boston: Little, Brown, 1993).

40. Board on Army Science and Technology, Commission on Engineering and Technical Systems, National Research Council, *Star 21: Strategic Technologies for the Army of the Twenty-first Century* (Washington: National Academy Press, 1992), 277–80.

41. Howard Frank, "ARPA's Information Technology Strategy," opening remarks for the session on "Information Technology" at ARPA's Seventeenth Systems and Technology Symposium, San Francisco, 25 October 1994; emphasis in original.

42. Draft syllabus provided by Daniel T. Kuehl, Professor of Military Strategy at the National Defense University.

43. National Academy of Sciences, Panel on the Future of Atomic, Molecular, and Optical Sciences, Board on Physics and Astronomy, Commission on Physical Sciences, Mathematics, and Applications, *Atomic, Molecular, and Optical Science: An Investment in the Future* (Washington, D.C.: National Academy Press, 1994).

44. David C. Isby, "Electronic Warfare Lessons," in Watson, George, Tsouras, and Cyr, *Military Lessons of the Gulf War*, 164.

Soldiers and Civilians Confronting Future War: Lev Tolstoy, Jan Bloch, and Their Russian Military Critics

Jacob W. Kipp

The relationship among military science, the social sciences, and forecasting in military affairs was one of the central features of the Soviet military system and continues to be of considerable interest to the Russian military.[1] Soviet military theorists summed up their understanding of the relationship between forecasting and military science quite explicitly: "Military science is the science of future war."[2] In his recent work on the changing nature of armed conflict over the next twenty to twenty-five years, General Makhmut Akhmetovich Gareev, the former chief of the Director of Military Science of the Soviet General Staff and the president of the newly founded Academy of Military Sciences, speaks of the exercise of military foresight as a necessary but difficult and frustrating activity.[3] Gareev began his military career as a Red Army cavalryman in the late 1930s, served throughout the Great Patriotic War, rose to prominence as a commander, staff officer, and advisor to foreign armies during the Cold War, and brings a unique perspective to foresight in military affairs. The chief imperative behind the exercise of foresight is to push conservative, bureaucratic military institutions to address the fact that the next war will be different from the last. The forecaster seeks to grasp the direction of change in all aspects, from the causes to the nature of armed conflict. To Gareev, foresight is that activity where military science and military art join, where theory and practice are linked. The forecaster seeks to examine "law-governed patterns" (*zakonomernosti*) and trends dialectically. But complete success in this endeavor is quite unlikely:

> History knows many sagacious predictions regarding separate aspects of future war; however, to foresee correctly the nature of new armed conflict in its entirety has practically never been achieved.[4]

This makes the task of the military forecaster, like the labor of Sisyphus, one of unending toil and no reward. Yet, the task is necessary since foresight is a basic ingredient in the successful resolution of a host of problems not only associated with defense policy but also "with the goal of preventing armed conflicts and wars."[5] Gareev argues that it is much better to make mistakes in military forecasting than to fall back on the assumption that "to peek at the future of military affairs is impossible."[6] Studying the problem of future war is a matter of numerous, repeated attempts from diverse perspectives in seeking a forecast with fewer errors and more insights. General Gareev, although retired, is in a particularly good position to appre-

ciate the problems of integrating forecasting in military-political and military-technical areas under a common methodological approach fitting the requirements of military science, the social sciences, and the natural sciences after their emancipation
from Marxism-Leninism and the veil of total secrecy that surrounded military issues
under the Soviet system.

Christopher Bellamy, one of the handful of Western military analysts who had
taken note of Soviet military forecasts of the Revolution in Military Affairs, drew
attention to the new problem confronting Russian forecasters: the need to address
both political-military and military-technical changes in order to understand the full
implications of the revolution. Bellamy drew attention to the renewed role of civilian
experts with social science credentials in Russian/Soviet defense thinking and properly linked it to the tsarist experience with civilian experts. He focused on the contributions of Jan Bloch and his six-volume study of future war and compared Bloch's
contributions with those of Andrei A. Kokoshin, then a senior researcher at the
Academy of Sciences Institute of the United States and Canada, who in May 1992
became first deputy minister of defense of Russia.[7] Kokoshin did his most important
early work on the U.S. national security system and its methods of forecasting.
During the final years of the Cold War he took an active part in the Soviet efforts to
undermine the Strategic Defense Initiative. During *perestroyka* he became one of the
most important voices for an alternative political-military posture for the Soviet Union,
writing in collaboration with General V. V. Larionov and General V. N. Lobov.[8] One of
the key arguments advanced by Kokoshin and Larionov during this period was the
relevance of professional military judgment that was outside party control, especially the writings of A. A. Svechin, tsarist *genshtabist* and Soviet *voyenspets*.
Kokoshin supported Boris Yeltsin during the August Putsch of 1991 and was actively involved in seeking a security arrangement during the transition from union to
commonwealth. In the Ministry of Defense, where he is the ranking civilian, he has
been involved in military research and development, procurement, and foreign military sales.

Bellamy's linkage of Bloch and Kokoshin raises the larger question of continuities and changes in the context of Russian civil-military relations and in the specific
area of military forecasting over the last century. The very first issue of the newest
military journal of the Russian Armed Forces, *Armeyskiy Sbornik*, raised just this
issue in an article titled, "Does Russia Need an Army?" It answered that question
with a resounding yes. The author took the question of the place of the army in
Russian society to 1900 and noted that extremists had undermined the position of the
armed forces on several occasions, at the turn of the century, following the Russo-
Japanese War, and during *perestroyka.* The author went on to make a case for a
world in which the use of force to protect national interests is a necessity of national
policy, given the threats posed by conflicting national interests. At that stage the
means of achieving state goals begin to play the main role. To decipher the "genetic
code" of war or armed conflict means to determine when and why in the resolution of
disputed issues first priority has been given to the use of military power.[9]

There are, of course, fundamental continuities in the debate over this issue—whether the object is future war, the prevention of future war or peacekeeping/peace-enforcement. War, that is, "the continuation of politics by other, violent, means," remains central to both. But war as a chameleon has changed color to adapt to its new environment, and therein lies one of the central tensions between military science and the social sciences at the present time. Furthermore, the subject of the debate is also unchanged. It remains the state as a sovereign institution which possesses the power and authority to mobilize society and create, maintain, and employ the instruments of war. The nature of the state is, however, in flux. The Soviet Union no longer exists, but the Russian Federation is only now taking shape, and a host of contradictions over its character are unresolved. Moreover, the society from which the state draws its resources is subject to radical changes involving the collapse of old institutions and social arrangements and the slow, painful process of the emergence of new ones. In short, Russia and the other successor states are living through a profound revolutionary process affecting both state and society.

One of the most central issues confronting Russian reformers is the legacy of the militarization that penetrated every aspect of Soviet society. Forecasting in military affairs has, therefore, become both more difficult and more important in its potential impact.

The continuity in subject and object in military forecasting extends back beyond the Soviet period, and an examination of some of the tensions created by the current revolutionary changes in military art and the state are best evaluated in the context of the debate over military theory and social science as it developed in the last decades of tsarist Russia.

Lev Tolstoy and the Sociology of People's War: Mass War and the End of the Great Captains

An appropriate starting point for this discussion is the publication of Count Lev Tolstoy's (1828–1910) *War and Peace* in 1869. Having written some of the best essays on the psychological impact of war on combatants in his "Sevastopol Tales," Tolstoy turned to the writing of *War and Peace* in 1862, one year after the emancipation of the gentry's serfs. He took part in the emancipation as a justice of the peace and saw the fate of Russia's peasantry as the issue at the very heart of the empire's further development. Tolstoy devoted significant effort to the reform of popular education, running a school at Yasnaya Polyana for that purpose of popular enlightenment. "In education the main thing is equality and freedom."[10] In short, Tolstoy wrote in a postwar period, when the military reforms associated with the creation of a modern mass army, led by professional officers, was getting under way. Tolstoy saw the issue of war in a much broader context and tied it to the larger issue of socioeconomic reform in his own time. The very design of *War and Peace,* combining a family tale with a narrative of the great affairs of that era, culminating with Napoleon's invasion of Russia, bespoke the need for some linking device to portray the epic qualities inherent in the daily affairs of men caught up in great transformations.

Tolstoy wrote and published *War and Peace* over several years and in the process recast the work to become one of the great epic works of Russian fiction. In spite of the objections of the critics, he included at the very heart of the novel an exposition of his philosophy of history and sought to relate that philosophy to the core issue of the book: the problem of freedom and necessity. Literary critics, generally, were annoyed with the philosophizing and tended to dismiss it. But, as Isaiah Berlin has pointed out, those discourses are central to the novel and provide a detailed deposition on that philosophy.[11] They are to provide guidance for the reader in understanding and comprehending the flow of events.

One of the major points regarding the relationship between Tolstoy's literary works and his philosophy, according to Mark Aldanov, is the riddle of science and its contradictions. Tolstoy rejected natural science in its Newtonian and Darwinian forms when applied to human society because they could not explain cause and effect, even as he called for applied science to improve people's daily lives. He scorned the German strategists of *War and Peace,* Pful' and Weierother, whose science amounted to nothing more than "die erste Kolonne marschiert . . . die zweite Kolonne marschiert" (the first column marches . . . the second column marches). Tolstoy attacked them and the officers whom they represented for promoting a military science that in its formalism had lost all touch with the reality of war, which Tolstoy depicted in all its destruction and awesome beauty. Tolstoy, who sought to combine rationalism with fatalism, understood that common soldiers must act, follow orders, accept their fate, and above all, not think.[12] Those in command, particularly Napoleon, who seemed to be directing events, were unable to control those social phenomena and, in fact, were more objects acted upon than subjects willfully shaping the world around them. The laws in this case were not matters of prediction but trends.

Tolstoy and Mass War: Combat Experience

Tolstoy, himself a combat veteran of the struggle with Shamil in the Caucasus and a defender of Sevastopol during the Crimean War, broke with those who saw war as a rational phenomenon of simple cause and effect and rather focused his observations on the role of freedom and necessity in human affairs.[13] His account of the siege of Sevastopol, where he served with various artillery batteries and had a good opportunity to see the mindless brutality of Nicholas I's army, reflected the horrors of protracted siege warfare and was pessimistic and honest even in the face of the censor's scissors.[14] The Crimean War, in which the senior commanders on both sides were aging veterans of the Napoleonic Wars, had two Janus-like faces: one pointed to the future where technology would transform strategy and tactics. New technologies—the rifle, steamship, shell-firing gun, galvanic mine, and ironclad battery—pointed vaguely towards the future. But the other, the applied tactics, training, and strategic leadership bore all the marks of epigones trying badly to execute the

Napoleonic art of victory. Like many of his fellow officers, Tolstoy treated his troops like serf-automatons, who had to be beaten into obedience. While his initial writing on the war in Crimea was full of patriotic appeal, his sketch of the situation in Sevastopol in May 1855, as the struggle was drawing to its inevitable conclusion and a Russian defeat, was too honest in its pessimism to escape butchery by government censors.[15] Having observed war fought by the epigones of Napoleon Bonaparte, the Duke of Wellington, and Mikhail Illarionovich Kutuzov, he belittled the ability of "great captains" to impose their will on war, by which he meant wars fought by nations in arms. Napoleon was no free actor, and his "genius" was no answer to the social forces that set nations in motion. Great events were the product of multiple "accidents" (*sluchainosti*) behind which stood powerful social forces. Only by understanding the social processes might one escape the trap of assuming freedom of action, when, in fact, one was forced to act by the necessity imposed by those forces. Theory must be armed with experience, and in this case sociological insight was to guide the interpretation of experience.[16]

> What is the cause of historical events?—Power [*vlast'*]. What is power? Power is the totality of wills, transferred to one person. Under what circumstances is the will of the masses transferred to one person?—Under the conditions of the expression by the person of the will of all people. That is, power is power. That is, power is a word, the meaning of which is unknown.

> If the area of human knowledge was confined to just abstract thought, then having critically considered that definition of power which is given by *science*, humanity would have come to the conclusion that power is nothing more than a word and in reality does not exist. But, in addition to abstract thought, for the cognition of a phenomenon man is also armed with experience which he uses to verify the results of his thought. And experience says that power is not a word but a phenomenon that really exists.[17]

The Nation in Arms and the Professional Army

Tolstoy's concern with the role of nations in arms and his critique of the role of "great captains" in war drew criticism from General Mikhail Dragomirov (1830–1905), a figure who dominated Russian thinking about tactics during the last decades of the nineteenth century. His essay appeared in 1868 with the appearance of the four-volume *Voina i mir* (War and Peace). Tolstoy had heard about Dragomirov's review, and A. A. Bers informed him that the essay had "circulated among all the Grand Dukes' circles."[18] Dragomirov noted Tolstoy's ability to describe "military sciences and types," called attention to his characterization of regimental life and values, and praised his description of combat in all its complexity and confusion.[19] Dragomirov, however, rejected Tolstoy's historical fatalism and questioned Tolstoy's assertions on a number of larger theoretical questions:

> Is some sort of theory in military art possible? What is the significance of the commander-in-chief in the army? What causes set in motion the world movement of 1812?[20]

Dragomirov singles out Prince Andrei's attack on the concept of military genius and the existence of a theory of military art for special attention. Dragomirov accuses Prince Andrei and Tolstoy of confusing science and theory in their attack on military genius.

> Mixing the understandings of science and of theory, Prince Andrei strives to show that in military affairs there is neither science, nor theory and, *consequently,* (!) there can not be military genius.[21]

But it is clear from this discussion that Dragomirov's own focus is on tactics, that is, the employment of forces in battle, as opposed to strategy, the actions which bring forces into contact.

Dragomirov then draws a distinction between science and theory based on the difference between exact sciences and applied arts. Military affairs belonged to the applied arts, like poetry, music, and painting.

> At the present time no one would get it in his head to claim that *military science* is possible. It is as senseless as to call sciences: poetry, painting, and music. But it does not follow from this that there could be no theory of military art, just as it exists in any applied art. Theory in these latter arts does not make a Raphael or a Beethoven or a Goethe, but it does give them the technique of the art, without which they could not achieve that height, which they achieved. The theory of military affairs does not claim to prepare either Napoleons or even Timokhins [a character from *War and Peace*]. But it reenforces our understanding of the "condition of troops and of terrain. It points to the examples of creative thought in military affairs and, consequently, makes easier the work of those who are endowed by nature with the gift of military skills."[22]

The theory of military art could not be a science precisely because it was an applied art and had to "admit its powerlessness to investigate the third and most terrible given in military affairs—the given of accidents." This part of war was inaccessible to science. There existed no metaphysical foundation for empirical exploration; rather the criteria for the appreciation of military art rested on an aesthetic foundation. "The theory of military art does not give formulas on how to create Austerlitzes, Friedlands, Wagrams, the Swiss campaigns of 1799, Königgrätzes; but it does present these masterpieces of military creativity (*tvorchestvo*) to the military person for study, just as the painter, musician, and poet study the masterpieces, each in his own specialty: not in order that these [works] could literally be copied but in order that they [the artists] might absorb their spirit."[23] Moreover, military art was different from other arts in that on the opposing side was another master trying to practice the same art to frustrate those actions, impose his will on the first, and achieve victory. Dragomirov accused Tolstoy through the character of Prince Andrei of being a dilettante who had not studied the military classics, that is, the writings of Marshal de Saxe, Lloyd, and Frederick the Great, rejecting the old in favor of what was new, modern and fashionable.[24]

Dragomirov's critique of Andrei's reading of "fashionable authors," as opposed to military classics, according to Boris Eikhenbaum, a leading literary scholar and formalist of the 1920s, was directed at Tolstoy's own eclecticism. Tolstoy in developing his concept of this novel sought to create an epic by drawing together a tale of family relations against the backdrop of the titanic struggle between Napoleonic France and Imperial Russia. In developing his concept he relied on the ideas and observations of a diverse group of authors who were widely read in the 1860s, including Homer, Goethe, Trollope, Proudhon, Joseph de Maistre, S. Urusov, and M. Pogodin. While the first three provided Tolstoy with models for his epic, the last four's writings offered insights into the specific historical events that made up the grand theme of the novel. Tolstoy drew on Proudhon's observations on "war and peace" and his conception of Napoleon; adapted Joseph de Maistre's account of the response of the Russian court to the campaign; used Urusov's combination of differential analysis and observations on the campaigns of 1812–13; and absorbed Pogodin's historical-philosophical speculations and Slavophile sympathies in support of his "archaic school."[25]

On military matters it was the ideas of Sergei Semenovich Urusov (1827–97), with whom Tolstoy had forged a friendship during the Siege of Sevastopol, that had influence. Tolstoy later described Urusov as "a very brave officer, a great eccentric, and one of the best chess players in Europe at the time." Urusov once proposed during the bloody battles at Sevastopol that the fate of the hotly contested fifth bastion be resolved by challenging the English to a chess match for the bastion. He would be the Russian champion, and the English would name their own.[26] Urusov, an artillerist by training and experience, was reputed to be an outstanding mathematician, who sought to apply insights from chess and higher mathematics to the study of military art. He published a number of works in both areas and used applied mathematics to solve chess problems. Urusov sought to interest military officers in using applied mathematics, including probability theory, to solve actual strategic and tactical problems relating to mass and maneuver.[27] With Tolstoy he shared a Slavophile's view of Russia and Russian virtues. Tolstoy drew on Urusov's military theory in his discussion of 1812. Those views were infused with a search for law-governed patterns that might explain the course and outcome of wars, campaigns, and battles. In his volume devoted to this topic, *A Survey of the Campaigns of 1812 and 1813, Military-Mathematical Problems, and Concerning the Railroads,* which also appeared in 1868, Urusov noted his debt to Tolstoy in raising the issue of historical cause versus law-governed tendencies and directed his attention to its application to the study of future war:

> The author's views and his discussion of the causes of the War of 1812 inspired the idea of looking for historical laws, primarily laws of war, with the aid of mathematical analysis. One of the problems in the book's First Supplement (concerning the movement of troops by railroad) was the reason for writing a separate article on the transport of troops (the Second Supplement). Until now, questions of this sort have been solved empirically, and therefore my calculations are the first attempt to surround this part of strategy in a formula of exact science. My article reveals the unusual similarity and even in certain cases the identity that exists between the problems of troop movements and chess problems.[28]

Urusov rejected an empirical, subjective approach to causes and historical laws in favor of objective laws. He had in mind an objective path "to discovering historical

laws according to which scholars have studied in experimental laws [*nauki opytnye*]."
Urusov proposed to bring mathematics into the study of history via "*the mathematical theory of probability*" just as it was being used by insurers and in games.[29]
Urusov reasons that war was "the most simple of historical events" and therefore this
made the search for the laws of war more practical. This was because armies were
ordered systems. "Armies are the most unfree societies: located under civil laws,
connected from above by military discipline, they are formed, moved, reduced and
generally cease to exit with such regularity, that from prehistoric times *strategy*—the
science of war—was born."[30] The study of objective laws stood in contrast with the
study of subjective laws. The former emphasized the realm of necessity, the latter
looked to the world of freedom. But the subjective path could not find support in
mathematics. Urusov described his own work as "the first attempt to find and study
the fatalistic laws of war."[31]

By applying mathematical analysis to questions of strategy Urusov came to the
conclusion that differential calculus could be used to address the role of speed and
impact in warfare. "The force of the aggressor is expressed as *vis viva of the mass*, i.e.,
as the product of half the mass times the square of velocity."[32] In a letter to Tolstoy
in May 1868 Urusov had titled his book *A Collection on the Laws of War.*[33] Urusov
reduced the question of Napoleon's genius to the law-governed pattern influencing
the course and outcome of battles and campaigns, which he defined as the "basic
laws of strategy" and "expressed in *speed and impact, i.e., in velocity and mass.*"[34]

Technological progress was transforming the context of maneuver within a campaign. Under Napoleon, Marshal Louis Alexandre Berthier could rely on keen-
wittedness (*smetlivost'*) and habit (*privychka*) to organize the march deployment of
the corps so that they marched separately to combine in a timely fashion on the
battlefield.[35] Indeed, Urusov asserted that one of the most important acts of a great
captain was the selection of such a "keen-witted person" to oversee "the quartering
of the troops and the designation of the march routes for the units." Writing two
years after Helmuth von Moltke's successes in employing railroads to speed the
concentration of Prussian forces on exterior lines against the Habsburg's army, Urusov
stated that the problem of strategic maneuver had become more complex and required the use of applied mathematics. Railroad maneuvers for the deployment of
armies in theaters of war thus took on revolutionary import since they offered the
possibility of increased velocity in deployment and sustained mass through enhanced logistics. The faulty exploitation of a state's rail network would have catastrophic consequences. For this reason he proposed that mobilization criteria be
taken into account in the building of railroads and addressed the problem of optimizing the Russian rail system to maximize "*the degree of centralization and military
strength of the network.*"[36]

Urusov proposed to use mathematics to guide railroad construction in order to
increase the rational use of the railroads as a system or network. He acknowledged
that railroad development was not a simple matter of strategic choice but a complex
social, economic, and political process:

A plan of railroads must be subordinated and usually is subordinated to commercial, political, financial and other conditions. The issue of forming via railroads the main communications of the state and the points designated by the government, just as with the selection of the centers of administration, store houses and iron works, can be reduced to a geometric task that corresponds with the problem *of least distances and of the stability of equilibrium.*[37]

Urusov noted this approach had already been discussed by an author in a journal of communications in 1827. Urusov sought to extend the use of applied mathematics to the problem of the optimization on the basis of throughput rates between one station and another. Railroad maneuver was not a matter of empirical experience and sharp assessments, as march-maneuver had been under Napoleon. Rail maneuver now involved much more complex deployments, provided the existing lines were organized to support such deployment. "One must take this task as the basis for and in the evaluation of a railroad project in military terms."[38] He treated the problem of optimizing rail movement as one similar to chess combinations under Warnsdorf's rule "for moving a knight around the chess board." Urusov modified the equation to fit the movement of a king and spoke of its applicability to military and commercial movements.[39] The question of successful mobilization and strategic deployment had moved beyond the competence of the keen-witted guided by habit, to be subjected to method, rigor, and mathematical precision. The key ingredient in the military exploitation of railroads was effective, rational management.

At the same time Urusov's observations on 1812 began the process of creating a Kutuzov legend, turning a "cunning court liar" into a commander who deceived Napoleon.[40] How had Russia defeated the great Napoleon without winning a single battle and losing its capital? He cited favorably Kutuzov's reply to the question of one of his relatives about how he hoped to "beat" Napoleon: "Beat him? No! I hope to fool him."[41] Urusov reduced Kutuzov's "strategy" to the formula of exhausting the enemy army by imposing distance on its maneuvers without offering decisive battle, being willing to abandon the field after a heavy, but indecisive battle at Borodino and saving his army to fight again even if it meant giving up his capital.[42] Kutuzov used pursuit, partisan war, and winter to break the enemy, once Napoleon's army had gone beyond its culminating point without destroying the Russian army in the field.

In his study of the campaign of 1813, Urusov turned his attention to the study of "the laws of engagements," that is, tactics.[43] In this case Kutuzov's approach in 1812, avoiding a general engagement with Napoleon when possible, became an objective law:

When engaged with Napoleon, fall back, avoiding a fight, when engaged against his marshals one must attack; in the event of an unavoidable battle with Napoleon, stress only your defense; the individuals will fulfill their duty if you keep in mind the wise principle of *laissez faire.* It is very easy to generalize this principle, to disseminate it to all the generals; we plan to present these tactics in the form of a mathematical law.[44]

For Urusov, Kutuzov was first of all a strategist, who sought consistently to weaken the Grand Army even as he tried to keep the Russian Army in the field. At Borodino he

achieved a tactical stalemate and then abandoned the field, leaving Moscow unde-
fended to save that army. His march-maneuver to the southeast put his army in a
position to threaten Napoleon's supply lines to the West, setting the stage for
Napoleon's withdrawal and the Russian pursuit in the fall and winter of 1812.[45] The
explicit linkage between strategy as the art of maneuvering forces to combat, and
tactics being the conduct of combat, broke down in Urusov's hands. Successful
strategy did not lead inevitably to the general engagement and annihilation of the
enemy there. The focus shifted from tactics to the campaign, which took on new
dimensions. In Tolstoy's hands Kutuzov's insight of merit became his understanding
of the Russian people, whose endurance, faith, and patriotism became the instrument
of Napoleon's defeat. Thus, a distinctly Russian interpretation of 1812 was born. In it,
the relationship between the mass army and people's war became something distinctly Russian. Rationality was a false guide to action since causes of people's actions led to unintended but foreseeable consequences. As *War and Peace* appeared
in serial form, the critics began to attack the historical and philosophical underpin-
nings of the work. Regarding the initial objections of other major literary figures,
Tolstoy simply avoided confrontation, even in the face of their bitter attacks. As he
wrote A. A. Fet in June 1866, he refused to be drawn into a fight with Turgenev over
the latter's criticisms of his philosophy of history, even when Turgenev accused
Tolstoy of using contemporary epigones as his models for Kutuzov and Bagration.[46]

By the time Tolstoy had published his completed masterpiece in 1869, military
reform was well underway in Russia. D. A. Miliutin had created a more effective
administrative system based on military districts and reformed the system of military
education. The creation of a mass army, based on a universal conscription obligation, was only five years from enactment. The army had been rearmed with rifled
weapons and was experimenting with breechloaders. Military professionalism was
advancing under the guidance of the Main Staff. And under General Geinrikh Leer
the study of Napoleonic strategy was reaching its high point at the Nikolaev Academy of the General Staff.

Thus, the reviews of *War and Peace*, which appeared in the military press are of
particular interest for what they say about the military's response to Tolstoy's observations on Russian military history, military theory, and the laws explaining the course
and outcome of campaigns and battles. Military historians accused Tolstoy of putting together negative anecdotes and scandals to present a negative picture of higher
Russian society and the army. A. Norov complained bitterly of this tendency and
described the book as demonstrating "an absence of any sort of sense in military
actions and even an absence of any sort of military valor, on which our army always
rightly prided itself." Tolstoy had reduced glorious 1812 to a mere "soap bubble."[47]
A. Vatmer, a "critic-specialist," addressed the fourth volume of the novel, where
Tolstoy laid out his observations on the role of "great captains," the utter uselessness of war plans, the limits of causal explanations of events, and the negative
portrait of Napoleon. Vatmer pointed out that the opinions expressed by Price Andrei
and those of the narrator were in agreement and mutually reenforcing. Vatmer dis-

agreed with Tolstoy on each point. He agreed that "really it is impossible to foresee beforehand all chances." But the point was to make every effort to reduce the impact of negative chance and to seize opportunities as they arise. Where Tolstoy pictured a Napoleon who only appeared to control his destiny and was acted on by forces beyond his control, Vatmer presented Napoleon at Borodino as a commander making rational judgments with incomplete information, providing broad direction that his deputy commanders would act on from their understanding of the situation. Such general instructions could not address unforeseen development as was the case with Prince Poniatowski's corps, which Napoleon had ordered to flank the Russian left wing, and which ran into General Tuchkov's corps and was drawn into heavy fighting for a village. Vatmer admitted that Napoleon had not been in his best form. He criticized Napoleon's decision not to commit his Guards to achieve decisive victory. Vatmer also rejected Tolstoy's presentation of Kutuzov as the antipode of Napoleon, that is, a commander who understood the limits of rational control in a complex battle and sought to manipulate the psychological mechanisms within the army that give men the spirit to fight and die.[48] Vatmer rejected Tolstoy's image and suggested that Kutuzov had tried to lead and direct the battle, had made decisions, and, thanks to oversights of his staff and the death of his commander of artillery, Gen. Aleksandr Ivanovich Kutaysov, who fell trying to retake the Kurgan battery, failed to make use of his advantage in field artillery, keeping more than 300 guns in reserve at a critical moment in the battle.[49] His focus was the general engagement. Vatmer spoke of the spirit of a people as the force driving nations and armies and rejected Tolstoy's attention to the psychology of the combatant, who in the wake of a day's heavy fighting might just loose the will to continue the struggle. Vatmer said that armed bands might act in that fashion but not "disciplined troops."[50] In short, the larger world of mass war remained beyond his purview. Vatmer explained his sharp attack on Tolstoy's military theory and history in terms of the author's position as a writer of influence and the dangerous potential consequences of his ideas going unchallenged:

> We have hoped that no one, including the author himself, will accuse us of careless analysis. Perhaps, in the heat of the moment we too passionately disputed the author's false thoughts and conclusions. But this came about as a result of the fact that they were expressed by a writer, who has on our public a fascinating influence, by an artist, with whom there is not now any one to compare in our fatherland. They were expressed with disgusting self-confidence and defiant intolerance.[51]

Vatmer suggested that Tolstoy not waste "his talent, efforts, and time on a matter, alien to his brilliant gift."[52]

Tolstoy responded to these criticisms by agreeing with some, rejecting others, and complaining that many readers simply did not understand the system that lay at the basis of his grand design. In response to specific criticisms on historical details by N. Lachinov, a military historian, writing under the initials, N. L., Tolstoy wrote the editor of *Russian Invalid,* the War Ministry's newspaper, thanking the author for his comments and saying that he wanted to contact him directly to discuss other points of dispute.[53] On the question of his historical ideas and philosophy Tolstoy was adamant. In an entry in his diary for 2 February 1870 he complained:

> I can hear the critics: "the sleigh ride at Christmas, Bagration's attack, the hunt, the dinner, the dancing—all this is good; but his theory of history, his philosophy is bad; it is tasteless and cheerless."[54]

Tolstoy went on to relate a parable of a cook preparing a meal and in the process tossing to dogs in the yard the refuse—bones and blood—from his preparations. With the appearance of these items the dogs labeled him a great cook. When the cook later tossed other items—egg shells and artichoke peels—the dogs complained that he had become a bad cook. He concluded: "But the cook continued to cook dinner, and those for whom it was cooked ate it up."[55] Tolstoy's dinner did not appeal to Russia's leading contemporary soldier-theorists. Some were still under the thrall of Napoleon's genius as the epitome of military art. They were unready to address the issue of the psychology of the mass army and people's war.

General Leer did not sup at Tolstoy's table. Leer championed the idea of a military science and spoke of strategy as "the science of military actions" and tactics as "the science of battle" and wrote approvingly of a "critical-historical methodology" which would promote both historical and theoretical understanding and "setting it [military science] on solid, positive ground."[56] For Leer, Napoleon was the ideal "great captain" (*polkovodets*) who grasped and applied the "unchanging laws," which are the bases of "the theory of military art." Leer went on to observe: "The number of laws, serving as the basis of military art is not great. They are immortal, self-evident, unchanging. Only their application *depending on circumstance* becomes infinite." Thus, there are no formulas to fit each unique situation.[57]

Leer addressed railroads in his discussion of march-maneuvers and described them as "*the most advantageous, most modern mobile warehouses*" and suggested that one or two railroads would have solved Napoleon's logistic problems in 1812 and thereby have made that impossible enterprise quite feasible.[58] But Leer confined the impact of railroads to the preparatory period of a campaign and excluded the possibility of railroads serving as the operational line of a campaign. On the question of theater preparations in case of a defensive war Leer discussed the increased speed of concentration that railroads offered and discussed at length the problem of defense of the Polish salient and the western borderlands. Railroads offered the possibility of overcoming the division of the theater into two parts by making possible the construction of a line that linked the existing network from Dinaburg through Vitebsk with a north-south line parallel the Dnieper and joining Vitebsk with Kiev. Leer suggested that economic factors and strategic considerations both had to be taken into account and pointed out that such a line to the east of the Pripyat Marshes could run through major towns—Mogilev, Rogachev, and Chernigov:

> Economic conditions would be improved since this railroad would run through populated and important commercial points. In strategic relations the negative influence of the Pripyat Marshes would be significantly reduced in the sense that it

divides the western borderland space into parts. Finally, security would to a significant degree be achieved because the road would be covered by the Dnieper and the Pripyat Marshes, not taking into account the support, which the totality of defensive measures, concentrated in the Polish Kingdom itself, could give to it.[59]

Whereas, Leer spoke of "immutable laws," Dragomirov acknowledged profound tensions between the practical side of military affairs in which matériel dominated and the more "sublime" side in which the psychological dominated. But war was also changing, and Dragomirov identified the following factors in that transformation:

> (1) the modernization of firearms; (2) the expansion of railroads; (3) the tremendous rise in the numerical strength of armies, and as a result of this rise, short terms of service. The first two factors are purely material and as such they can not be the cause of a change in the spiritual nature of the person, and consequently, of the *partie divine de la guerre.*

Miliutin in late 1870 proposed the creation of a mass army, based on a universal military obligation for all estates. Tolstoy revealed the conservative social basis of his views on a mass army. In a letter to Urusov, Tolstoy came down squarely against copying Moltke's Prussian system. Long-term service created competent troops at much less cost, so it made more sense to retain the old Russian system: "The advantage of this solution is that the only thing you have to do is nothing, to not destroy the type of old Russian soldier that has given so much glory to Russian arms, and so not to try anything that is untested."[60] Tolstoy evoked the memory of the defenders of Sevastopol to make his case.

When Miliutin's reform imposing a universal military obligation on all males was adopted in 1874, it fell to Dragomirov to seek a way to train conscripts to make them into competent soldiers. Dragomirov's focus was on the impact of such material changes upon this soldier, and he singled out shifts in the size of the battlefield, its lethality, and the very different character of the young soldier in mass armies.[61]

Battles won and battles lost still decided the fates of empires. Dragomirov, who would play a leading role in forcing the Danube in 1877 and went on to be the preeminent tactical authority in the Russian Army down to the end of the century, championed preparing this soldier for the reality of combat and not preparing an automaton who feared his officer more than he did the enemy.[62] Here Dragomirov was already commenting on the second transformation of warfare that had made the latter half of the century the age of Moltke—the impact of industrialization on the means of war. In this regard, Dragomirov was a step beyond Tolstoy, the gentry-agrarian with his attention focused on the peasant. For Tolstoy, material progress had its own unintended and often negative consequences, far beyond the participants' ability to comprehend. Thus, Tolstoy wrote of fashion and the connection created between the rich seeking to establish their distinction by dress and the proletariat who inherited the old and unfashionable clothes discarded by their betters. So too Tolstoy saw railroads bringing unforeseen and unwanted consequences. Like the trains in *Anna Karenina,* they brought chaos and corruption, disorder and death. Following his own spiritual path over the decade after the publication of

Voina i mir, Tolstoy came to a pacifism that rejected modern battles and wars in the aftermath of the Russo-Turkish War of 1877–78.

Jan Bloch and Modern War: The Industrialization of War

The other great civilian student of war, although he too ultimately embraced pacifism, was the very antipode of Count Tolstoy, the Russian aristocrat. Jan Bogomil (Gotlieb) Bloch (1836–1902), a.k.a. Jean de Bloch and Ivan S. Bliokh, the son of Jewish factory owner from Radom, whose business failed in the aftermath of the November Insurrection in Poland, went on to become one of the "kings of the railroads" and a prominent advisor to the Russian Ministry of Finance. Several of his employees in the railroad business rose to the position of minister of finance, that is, I. A. Vyshnegradskii and S. Iu. Witte. Bloch made technological change and industrialization the focus of his pioneering work on the study of "future war."[63] Count Tolstoy belonged to one of the most ancient and honored gentry families in Russia, Bloch was a Polish Jew and "un parvenu intellectuel," an outsider who by his talent became both rich and influential.[64] While Tolstoy had been a soldier and seen combat, Bloch was a complete outsider to military affairs. While Tolstoy's pacifism emerged during the Russo-Turkish War, Bloch's intensive interest in the problem of modern war was only just beginning. Bloch's famous six-volume work, *Budushchaya voina* (Future War) appeared in 1898, the same year as the republication of Dragomirov's critique of Tolstoy's *War and Peace.* Bloch is remembered in the Anglophone West as a pacifist who questioned whether a general European war under modern conditions was possible, largely as a result of the decision to confine the translation of the work to his volume of conclusions and to give the work the title, *Is War Now Impossible?* His introductory essay to the English edition, which addresses the risks that a war of mass armies and modern means would pose to the states of Europe, is not found in the Russian edition which begins simply with "the description of the mechanism of war" and is a summary of the preceding five volumes.[65]

Bloch's presentation of his description of modern war, which begins with the impact of fire on the modern battlefield, and discusses the impact of the magazine rifle, smokeless powder, machine guns, and quick-firing artillery, actually conceals his central point, which can be stated simply: the combination of mass armies, based on universal conscription, strategic mobility created by rail maneuver, and the revolution in fire power had moved war from a struggle for a single point into a linear battle of trying to turn enemy flanks. However, the density of forces, the ability of a defender to counter maneuver, and the lethality of fire would lead to repeated battles, culmination for the attacking force, and the generation of new forces by the defender, leading to stalemate and attrition warfare. Mass armies and industrialization had reduced the factor of chance to deciding battles but not wars involving multiple armies in continental theaters. Industrial war would consume society until the strain of war broke the social fabric and brought revolution in its wake. Commanders who sought decisive engagement were deluding themselves. War would become protracted and lead to the exhaustion of the contending states:

> The struggle [between the German and Russian armies] in the end could lead to nothing less but both sides being exhausted before either could achieve its objectives. No other result can be foreseen, even allowing that one side would make more mistakes than the other and that chance would play some role. In view of the numerousness of the troops on both sides and in such basic conditions, in which this colossal war would be conducted, such chance phenomena could not change the final result, which would involve namely the mutual exhaustion of the governments and the factual uselessness of the war.[66]

Bloch expected modern war to bring with it social breakdown and revolution as a consequence of the economic strain imposed by the war. What is interesting in this regard in that Bloch linked the strain to class struggle, pitting disgruntled workers against their social betters. The model for this was the Paris Commune, but the context was the fate of Warsaw in a war fought by Germany and Russia over the Polish Kingdom. Bloch thus linked the fate of front and rear in making his case against the utility of war as an instrument of rational policy.[67] He assumed that statesmen, once they understood this risk, would seek to avoid war. By publishing his volumes he sought to arm civilian authorities and society at large with the means to assess the utility of war as an instrument of policy.[68]

Bloch's critics accused him of being a "utopian," "cosmopolitan," "dilettante," and worse, "a civilian." Colonel Simansky, the author of an extended review of *Future War,* said that Bloch could not fully comprehend "the reality of military affairs, the goal and bases of military art" because he was a civilian and did not love military affairs. "Such an understanding comes only by love of military affairs."[69] Bloch's critics spoke of the inevitability of war as a social phenomenon and, while admitting the horror of war, proclaimed that it had a noble side. "War develops in a person the very highest and best side of his intellect and heart."[70] Invoking social Darwinism, these critics judged war to be inevitable, since it was the way in which national energies could be mobilized to promote and protect national interests. As General Leer wrote, "War is a quite natural phenomenon in the life of nations. Struggle lies at the basis of all existence."[71] Bloch dismissed such criticism by soldiers as the ravings of romantic utopians, who refused to grasp the nature of modern war, by which he meant general war among the European powers.[72] Bloch's work, which appeared at the time of the calling of the Hague Peace Conference of 1898, was seen by many as a powerful statement for disarmament and international arbitration of conflicts.[73] Nicholas II had suggested the conference to the other powers of Europe to stop the progressive development of existing armaments and reduce the burden of defense for the states of Europe. In this cause Bloch could invoke no less an authority than Count Miliutin, Russia's preeminent military reformer and minister of defense under Alexander II. Miliutin wrote Bloch on the eve of the Hague Conference:

> The main object of your work [on the Future War] has been to draw a picture, faithful but terrible, of that war which in a future more or less near will ruin Europe

in order to allow recent inventions to be utilized. From that very reason your book would have an immense and beneficent effect if it could influence the directing spheres, the men who shape the policy of States and, above all other, the Delegates to the Conference at the Hague. This, however, is unfortunately not to be hoped for; the appalling consequences which may be expected to follow the catastrophe are not capable of turning back the obstinate fanatics of militarism from the road which they have mapped out for themselves.[74]

His critics also claimed that Bloch underestimated Russia's strategic superiority and that Bloch was wrong in his prediction of an economic crisis in case of a protracted war, wrong on the costs of such a war, and wrong on the number of losses.[75] Most of all, Bloch's critics questioned his ability to comprehend Russia's greatness. Simansky spoke of that "greatest superiority" which autocracy bestowed on the Russians over other nations. The "adoration of the autocrat" was the embodiment of what Tolstoy had called power. "This unity of thought with the divine sovereign creates indestructible power." Being a Polonized Jew and an outsider, Bloch, according to Simansky, could not understand those national virtues, which arose from "this flesh of its flesh."[76] Finally, Bloch's call for international arbitration of disputes to reduce the impact of militarism in European statecraft was misplaced. Militarism came from only one source: Imperial Germany and its allies.[77] Of Bloch's warning about war bringing revolution and the advance of socialism, his critics claimed that he had just missed a new cause of war, "civil war" (*mezhduusobnaya voina*).[78]

Modern critics have noted that Bloch relied heavily on professional military expertise and studies and failed to anticipate the impact of indirect artillery fire, ignored the possibility of linking internal combustion and armor into a war-fighting vehicle to cover group maneuver, and did not foresee the appearance of the airplane as an instrument of war.[79] Bellamy accuses Bloch of only extrapolating on the accepted judgments of military professionals. "His failure was to apply any imagination or new perspective—precisely the traits which supposedly make civilian analysts useful."[80] Yet, many contemporary Russian officers wanted to dismiss Bloch precisely because he was a civilian. Gen. A. K. Puzyrevsky, the chief of staff of the Warsaw Military District, raised just that objection. In answer to that, Boleslaw Prus pointed out that Bloch was trying to inform public opinion about issues that were of critical importance. Bloch was one of those self-made men, who used their wealth for the public good and suffered unfair attacks for it.[81]

Moreover, it does seem strange to attack Bloch for not anticipating technology which would only make its appearance on the battlefield almost two decades after Bloch wrote and well after his death. The standards for modern military foresight, based on the time required for weapons development, procurement, training, and integration into the force, point to time frames of fifteen to thirty years.[82] By these standards Bloch's technical, economic and political forecasts about modern war hold up quite well. Bellamy also makes the point that Bloch was not an isolated amateur, and this, in fact, deserves close examination. Michael Howard, while more positive on Bloch's insights, concludes that Bloch missed the point on war in Eastern Europe, since war on the Eastern Front in World War I did not bog down in stalemate but retained an element of maneuver.[83] But that *Gummikrieg* (rubber war) involved operational successes and not strategic victory. Only the internal collapse of the Russian state and society in the revolutionary upheaval of 1917, which Bloch did foresee, brought about an end to that war and a descent into a bloody civil war.

Bloch was not a prophet, but he did develop a methodology for applying strategic foresight to war. Bloch's observations on the transformation of battle, that is, from a tactical perspective, were informed by the judgments of contemporary military professionals, whose expertise Bloch acknowledged. Bloch's relationship with the professional military was, however, more complex than just his soliciting the experts' advice. At the time of his research and writing, the Imperial Russian Army was itself in the process of debating the need to change basic field regulations to take into account the appearance of smokeless powder, the magazine rifle, and quick-firing artillery. Bloch's views were in direct opposition to those of the aging General Dragomirov, whose approach remained one of what a leading modern scholar of the Imperial Army has called "incrementalism." Bloch's views coincided with those who supported a fundamental reform in the field regulations to take into account the increased significance of fire.[84]

Regarding the strategic transformation of war brought about by the development of the railroad and the appearance of the telegraph, Bloch's own expertise was significant and professionally founded. With the outbreak of the Russo-Turkish War in 1877 Bloch, as president of the South-Western Railroad Company, found his firm involved in the deployment and supply of Russian troops in the theater of military actions.[85] His conversations with senior commanders during the mobilization and deployment for that campaign left a strong impression on him. He became convinced that while they understood the technical advantages which railroads bestowed in the initial period of war, the commanders failed to appreciate the fact that modern war would not lend itself to a single decisive battle, but would become protracted, thanks in large measures to strengthening of the defense by the very industrialization of war that had gotten the troops there in the first place. Strategy had become a matter of campaign planning, successive operations, and logistics. Bloch made repeated references to these issues and stressed the economic and social crises that mass, industrial war would bring.[86] No one understood better the importance of the railroad for both deployment and logistical support of armies and no one better appreciated the linkage between front and rear that industrialization had forged. As Bloch observed, general staffs that prepared plans for mobilization, concentration, and supply of forces, while ignoring the linkage between the theater campaign and the national economy, were missing a key aspect of successful prosecution of a war "the necessity of anticipating the means of protecting all functions of the social organism against possible disruptions." Progressive military thinkers had to work out questions of military supply, railroad utilization and such with civilian experts in peace time, as was then being done in Prussia.[87]

While Bloch was a self-made man and an autodidact, his approach to research was professional and modern. In 1874 at the 6th Congress of the Representatives of Russian Railroads, Bloch and others proposed the creation of a pension fund for railroad workers. The congress charged Bloch to provide a scientific study of a financially sound mechanism for funding the proposed pension fund in time for the next congress. In this, his first research project, Bloch relied on his deputy in St. Petersburg, Vyshnegradskii, and recruited another railroad official, one Maliszewski, who had a bent for mathematics and used probability theory and applied mathematics to provide an actuarial basis on which the retirement fund could function.[88] For this project Bloch received the Order of St. Vladimir and the rank of State Counselor. On the basis of the data collected for the pension study, Bloch and Vyshnegradskii undertook in 1875 a study of the impact of the railroad development on the Russian economy. This study, *On Russian Railroad, Influences, and Expenditures of Exploitation, Costs of Transport and Movement of Goods,* was presented to the Ministry of Finance in the form of memoranda.[89]

In 1879, Bloch proposed the creation of a library for merchants and a commercial-industrial statistical bureau in Warsaw and contributed 6,000 rubles to its establishment.[90] To undertake his studies he used his "Statistical Bureau," located at the corner of Krolewska and Marszalkowska Streets in Warsaw. He staffed the bureau with specialists, researchers, and writers, organized into special scientific subbureaus. They included historians, sociologists, economists, novelists, and translators.[91] Bloch paid willingly for talent. Boleslaw Prus, a promising journalist with the newspaper, *Kurier Warszawski,* was recruited for an annual salary of 900 rubles.[92] For his study on the economic impact of war on society, Bloch maintained a special research bureau in St. Petersburg to collect material on the economic impact of war on industry, trade, the money supply, credit, and state revenues.[93]

Bloch stood at the center of these activities—reading, editing, and discussing every detail with his team. In the midst of one of these efforts he became a man possessed: "He could forget about food, sleep, walks, the advice of his doctor."[94] Bloch was a positivist and employed "a strictly inductive, scientific method based on the thorough study of the subject, on facts, which were not selected for each individual case but rather grouped according to a certain system: conclusions were drawn from such facts, which were assembled and tested."[95] Bloch consistently applied one rule to each of his research projects: statistical relevance of his data or the application of the law of large numbers: "to make judgments only concerning those questions, which directly or indirectly could be proven by numbers, following the well-known dictum, *'Il n'y a rein d'aussi brutal que les chiffres.'"[96] Finally, Bloch was an entrepreneur, who fully understood the role of advertising in the market place of ideas. He made certain that his works were immediately translated into the major languages of Europe; he sent copies to men of power and influence, took part in scientific meetings and congresses, and cultivated ties with the press.[97]

Future War was the last of Bloch's great scientific enterprises, which extended back over two decades, but it was not his first commentary on the relationship between economic development and military power. The nature of these preceding studies, directed and funded by the "railroad king" and banker, had deep roots in Bloch's own experiences. According to his biographer, Jan Bloch's most vivid childhood memories were attached to his early years in Radom and his father's struggle with bankruptcy and debts as a consequence of his factory's failure after the November Insurrection.[98] Bloch came from the "Jewish proletariat" of provincial Poland and by hard work, dedication, and self-discipline raised himself to a position of wealth and prominence.[99] But Bloch grew up in a Polish Kingdom and Russian Empire, where instinctive forms of Judeophobia and a common state policy of seeking to regulate Jewish life and to render Jews "harmless to the rest of the population," were joined with even more reactionary policies.[100] Bloch had no reason to like the Jewish policy of the *Nikolaevshchina,* Nicholas I's attempt to strengthen autocracy and impose military discipline on society. For Jews this system involved repression, regulation, and forced assimilation under the policy of "recruitment and conversion."[101] While Bloch became one of those successful Jewish businessmen, who later converted to Catholicism to achieve acceptance in Warsaw society, Bloch's first deviation from the faith of his fathers was his conversion to Calvinism in 1851, shortly after his arrival in Warsaw. Calvinism's rationalism and attention to the external virtues that were the manifestations of internal salvation seem to have guided Bloch in this choice.[102] It also was the beginning of Bloch's connections with Austrian and German banking institutions, ties cemented by the marriages of his sisters.

With this source of capital Bloch proved to be in a highly favorable position to compete for state-subsidized railroad construction in Russia, which took off in the 1860s under Minister of Finance M. Kh. Reutern. Down to 1860 two fundamentally different approaches to railroad development had dominated in the Polish Kingdom and Russia proper. In Russian Poland foreign capital, a mature banking system, and market considerations had led to the construction of main lines tying Warsaw to Vienna and Central Europe. In 1857 these lines were turned over to private, joint-stock companies. In Russia military considerations, that is, mobilization and deployment of forces in time of war and suppression of insurrections, and fear of social unrest had placed railroad construction in the hands of a conservative, capital-starved government.[103]

In the early 1860s these two policies collided in the Polish Kingdom. As early as 1849 the Field Marshal Ivan Fedorovich Paskevich had used the Warsaw-Vienna Railroad to move 781,172 poods (one pood equals 36 pounds) of supplies and 18,518 men and their equipment. Paskevich had noted in his report to Nicholas I that this railroad would be even more useful "when it is in service with the completed railroad linking both capitals [St. Petersburg and Moscow] with the western borders of the empire."[104] Railroads constructed on the left bank of the Vistula had European gauge, while those on the right bank, by government decree were to have the wider Russian gauge.[105] Thus, by the early 1860s there was a fundamental contradiction in state policy toward railroad construction. On the one hand, market forces were leading toward the building of railroads to support economic development and, on the other, government policy continued to treat railway construction as a matter of "strategic goals," with the intent of turning the Polish Kingdom into an "armed camp" for the deployment of large armies in case of "European political confrontations."[106]

This situation changed under Reutern, who encouraged private enterprise and led to a period of robust, dynamic, if corrupt, capital accumulation.[107] Reutern, a liberal bureaucrat and close ally of the Grand Duke Konstantin Nikolaevich, judged private enterprise to be the most effective road for economic development and technological innovation. An alliance between state interests and private profit was the best guarantee of national power. Even before the Crimean War the Grand Duke, as president of the Imperial Russian Geographic Society, supported demographic, economic and statistical studies.[108] Examining the naval arms race set off by the introduction of screw-propelled ships-of-the-line in the late 1840s, Reutern wrote that Britain had won the race thanks to the Admiralty's use of a decentralized administration to mobilize public works and private enterprise to national purposes, while France had relied on centralized, bureaucratic controls and state facilities.[109] Following the Crimean War, Reutern traveled abroad investigating administrative policy, budget process, banking institutions, and economic strategy in Prussia, France, England, and the United States and became deeply interested in the process of economic transformation and industrialization transpiring in the West. He advocated a radical reconstruction of state banking institutions to encourage the formation of capital and its investment in industry. Confronted by a worldwide industrial crisis and economic depression following the Crimean War, Reutern sought its roots in a crisis of currency, credit and finance.[110] In looking for a solution, Reutern put a premium on entrepreneurship and venture capitalism on the model he had seen practiced in the United States. Reutern likened the American willingness "to put everything on a single card," the "go ahead" and "never mind" to a steamboat race on the Ohio River where captains tied down the safety valve and ignored sand bars to race ahead. It had created an economic miracle. Reutern cited the growth of Chicago from a small village into a major city in a few short years, the construction of 22,000 miles of railroads, and the growth of American banks.[111] His objective was to create an investment climate for private capital by practicing a balanced state budget, *glasnost* in the area of state finances, and a coherent national policy in the areas of finance and credit under the direction of the Ministry of Finance. Reutern favored the emancipation of the gentry's serfs as part of the Russia's modernization:

> Steam, electricity, chemistry, and engineering do not have a nationality. Their results belong, therefore, to him who wishes to use them. It is not enough that the government or educated classes, witnessing the use of instruments and inventions abroad, should introduce them among us. . . . All economic improvements, even railroads, are of use only in proportion to the development of free labor. Personal profit makes it possible to accept without excess ceremony the belief that the fruits of intensive labor belong absolutely to the worker—here is what has inhibited in the clever Russian nation those considerable economic qualities, with which it has been blessed.[112]

Reutern rejected the idea of the well-ordered police state, where the autocracy and its bureaucratic agents sought to guide and control state and estate-based society. In fact, serfdom stood in the way of uniting town and village into a national market. Economic transformation, which was essential to Russian power and influence, would not come from the estates of old Russia but from new social classes of free laborers and entrepreneurs. Railroads figured prominently in Reutern's vision of a modernized Russia since they would promote trade, especially the grain trade, and speed economic growth.[113] Alexander II appointed Reutern minister of finance in early 1862, just as private railroad construction was taking off. Reutern held that position until 1878. Railroad construction during this period reached a feverish pace, going from less than 1,500 versts total in 1860 to nearly 10,000 versts of new construction in the 1860s and over 11,000 new versts in the 1870s.[114]

Great fortunes were made during this railroad boom. The most prominent entrepreneurs in Russian railroad construction were the great bankers of Warsaw. The institution for funding their operations was the Polish Bank, which, according to Russian officials, formed a "state within the state" and was dominated by "Jew-capitalists."[115] Small traders operated outside the bank building exchanging money, and the rotunda of the bank itself served as the Warsaw Stock Exchange.[116] While nominally under military rule following the November Insurrection of 1830, the Polish Kingdom enjoyed a period of economic development, thanks to the Polish Bank's role in promoting investment. In short, the Polish Kingdom was relatively advanced in economic terms compared to the rest of the Empire.

The large role of Warsaw's bankers and entrepreneurs in the railroad boom of the 1860s and 1870s invites an examination of the comparative climate for investment and development in the Polish Kingdom and the Russian Empire proper. As Colleen A. Dunlavy has pointed out in her study of early railroad development in the United States and Prussia, "the overall structure of political institutions" contributed to radically different railroad systems.[117] In the Russian case we face three dimensions of comparative history: First, this involved the later Russian process in comparison with the earlier experiences in Prussia and the United States, which Russian officials and entrepreneurs knew about and discussed. Second, there were distinct differences in institutional arrangements with the Polish Kingdom and the Empire itself, with the former dominated by market considerations and linkages to Western economies and the latter constrained by the demands of national defense in the initial period. Finally, we have the prism of institutional competition during the Russian railroad boom, pitting the vision of the Ministry of Finances and its entrepreneurial allies against that of the Ministry of Ways of Communication and Ministry of War in a contest over state-supported private railroad development versus state-built and owned railroads.

In the late 1850s and early 1860s many Russian officials, particularly those associated with the military, distrusted this market dominated by Jewish merchants and Polish nobles and complained of the corruption and political motivations that led to wild fluctuations in price with every procurement action. At auctions for state contracts in forage for animals or grain for troops, contractors manipulated prices artificially, thanks to the intervention of "more or less influential persons" and bribed officials to overlook the delivery of poor-quality products. "Who does not know to what extent the self-interested mind of the Jew in view of the pressure [for profit] is capable." These same officials suspected the Polish nobility via the Polish Agricultural Society was conspiring in a secret boycott of the government's forage and grain auctions to deny the Russian military supplies.[118] But the policy of the Russian administration in the Polish Kingdom down to January 1863 was to seek some sort of compromise with Polish society and to promote the economic integration of the Kingdom with Russia. Conservative Russian civil and military officials in St. Petersburg and Warsaw looked back to the Paskevich era of martial law as a time of order and discipline when Jewish corruption and Polish nationalism had been held in check by military power. For many Russian officers Polish nationalism and the threat of revolution were explicitly connected. They were uneasy with the processes of social and economic transformation which their government encouraged in the Polish Kingdom and in Russia proper in an attempt to overcome economic backwardness and to create a new foundation for national power.

In military terms the Polish Kingdom was the forward bastion of Russian defense in Europe. In political terms, it was the most combustible material for bringing revolution to the Empire. In economic terms, it was one of the most advanced regions of the Empire with significant ties to the economies of Central and Western Europe. These considerations had a persistent, often contradictory, impact on Russian policy, leading to a vacillation between repression and reform, coercion and co-option. In October 1862 the Council of Ministers discussed the advisability of building a railroad connecting Kiev and Odessa. A. V. Golovnin, the minister of education and a close ally of the Grand Duke Konstantin Nikolaevich, wrote that opponents of the railroad said it should not be built unless Kiev was linked to Moscow. Otherwise the entire region would be Polonized. Prince Gorchakov warned that delay in building the railroad would have dangerous political consequences. He stated that if the region did not get a railroad "it would become an enemy of Russia." Unable to reach an agreement and unhappy with the terms offered by an English banking house for construction, the Council agreed to wait for an offer from the House of Rothschild.[119] The next item of business considered by the Council of Ministers at that same session made a direct connection between the politics of reform and the threat of unrest in the Empire's western borderlands. It involved the petition of the newly elected marshal of the Podolsk Assembly of Nobility to have the Podolsk Gubernia joined to the Polish Kingdom. The petition was rejected, the marshal placed under investigation by the Senate, and the election annulled. This was a moderate response in comparison with those advocated by some "reformist" ministers. D. A. Miliutin, the minister of war, called for the exiling of those who signed the petition and the abridgement of the "Podolsk Szlachta's right of assembly." P. A. Valuev, the minister of internal affairs, proposed that Podolsk Gubernia be abolished, its territory divided among other gubernias, and the Podolsk Assembly of Szlachta be liquidated.[120]

In this context Jan Bloch profited from the tsarist government's experiment with reform and entered into one of the great business rivalries of the era with Leopold Kronenberg, whose ties with foreign banks ran through the Rothchilds' banks in Paris and Brussels. Bloch relied on Herman Epstein to get contracts for railroad construction in St. Petersburg for his various joint-stock companies. To get conces-sions and state subsidies for construction of lines, the entrepreneurs had to bribe officials and court figures. This raised the cost of laying track. But concessions for rubles/verst of construction were liberal and a 5 percent return on capital was com-mon. Some builders, in order to cut costs or meet deadlines, cut corners on construc-tion or failed to invest in maintenance. Later, describing in his memoirs what he called the excesses of the Reutern era, that is, state subsidies and bribes to private railroad builders, Minister of Finance Witte, who had worked for Bloch's Southwestern Rail-road and emerged as a champion of state-financed railroad construction in the 1890s, attributed these excesses to the "railroad kings," who were "Jews or by origin Jews."[121] It was not corruption and speculation that brought about the end of Reutern's policy, but war. And war in the Balkans was to have a profound impact on Bloch's views of the relationship among technical, economic, and political factors shaping warfare. The Russo-Turkish War of 1877–78 was Russia's first railroad war, and its conse-quences shaped much of Bloch's later thinking.

Military Strategy and Railroad Maneuver: The Russo-Turkish War of 1877–78

This conflict provided the first test of Russia's ability to engage in large-scale railroad maneuvers for the mobilization and deployment of a field army in a theater of war. In 1859 the War Ministry had carried out a partial deployment of several corps along the Austrian border during the Franco-Austrian War relying on preindustrial means. The deployment proved protracted and confused. In 1863 the War Ministry relied on rail maneuvers to suppress the Polish Insurrection of that year. But since this was a struggle against partisan bands and not an opposing army, it left little impression on the question of using rail maneuver to support mobilization and de-ployment in a theater of war in the case of conflict with a foreign power. Obruchev had been quite right about the inability of the War Ministry to predict the actual strategic rail requirements of the Russian state in case of a war in an unanticipated theater. A combination of Serbian adventures, Russian volunteers, led by Gen.-Maj. M. G. Chernyaev, the "Lion of Tashkent," the "Bulgarian Horrors" committed on the local population by Turkish irregulars, and patriotic sentiment in Russian society pushed the tsarist government towards a war for which it was ill prepared diplomati-cally and politically.

In the spring of 1876 General N. N. Obruchev delivered a series of lectures at the Nikolaev Academy of the General Staff on a possible Balkan Campaign, and in May he turned these into an essay that became part of the discussion within the govern-ment over Russia's response to the Balkan crisis. Obruchev focused on the need for a Russian Army to move rapidly into the theater and defeat the Turks before they could receive foreign assistance. This depended on rapid mobilization and deploy-ment of forces to the theater. Obruchev emphasized the forcing of the Danube as the culminating point of the campaign.[122] Turkish control of the lower Danube and the Black Sea complicated the problem of maneuver and logistical support for the Russian Army in theater. Thanks to the demilitarization of the Black Sea after the Crimean War, Russia, although it had secured the right to remilitarization in 1871, lacked a Black Sea Fleet that would provide command of the sea as had been the case in earlier wars between just Turkey and Russia.

For this reason Obruchev looked to a forced crossing of the Danube in mid-course in the area of Sistovo. This choice meant that a Russian army would have to advance a considerable distance across Romania to reach the crossing point. In the fall of 1876 the Russian government used a partial mobilization on its border with Romania to threaten the Ottomans with war unless there were concessions. The railroad mobilization and deployment of forces was ordered in early November and completed by mid-December. And this coercive diplomacy brought concessions from the Sultan. However, in the early spring the Ottomans went back on their pledges and the Emperor made the decision for war, having gained Romania's support.

General Obruchev's campaign plan called for an all-out concentration for a crossing in the mid-Danube and an immediate drive on Constantinople. Russian financial weakness and diplomatic isolation demanded such a course of action and the appearance of Turkish weakness seemed to confirm the probability of successful execution. Obruchev's plan called for the concentration of six-and-one-half corps south of the Danube to drive through the Balkan Mountains and on to Constantinople. To achieve mass, Obruchev relied on weak screening forces and demonstrations to hold the Turkish armies on the lower Danube and in the Western Balkan Mountains in check. This was the epitome of Napoleonic military art and required a man of genius, a Bonaparte or a Moltke, to oversee its execution. William Fuller has described Miliutin and Obruchev as "technologists" in their military reforms and modernization but suggests that the Obruchev's war plan was infused with what he calls the spirit of the "Magician," that is, an appeal to the special qualities of the Russian soldier in overcoming all barriers and hardships in the path to victory.[123] But the key ingredients in Obruchev's plan were surprise, tempo, and maneuver. Russia was not fighting a major European power.

In fact, the execution of Obruchev's plan was placed in the hands of Grand Duke Nikolai Nikolaevich. The Grand Duke and his staff revised the plan and committed more forces to cover the flanks and left the march on Constantinople to Gen. Iosef Vladimirovich Gurko's detachments of ten-plus battalions. A. Svechin judged Obruchev's plan a considerable gamble, given the low level of tactical training of the Russian Army and the poor execution by the Grand Duke and his staff.[124] General N. N. Nepokoichitskiy, the chief of staff of the Active Army, was no Moltke. M. Gazenkampf, who served on that staff, noted that political motives had figured in the rejection of Obruchev for the post and described Nepokoichitskiy as a military bureaucrat committed to paper shuffling.[125] This was not a magician like Dragomirov, but a chinovnik in uniform out of the worst side of Nicholas I's army. In the face of a plan requiring audacity and speed, the Russian command acted conservatively.

In Romania the Russian military had to confront a host of problems that made rail maneuver difficult and time-consuming, and this influenced the development of Russian strategy. Obruchev's notes to the Emperor and Miliutin in early 1877 served as the basis for Russian strategy in that war. He noted the dangers of an unsuccessful war in the Balkans but made the case for war as a better course than concessions and compromises made out of weakness. Obruchev designed a strategy for lightning war, that is, victory in a single campaign. It accepted Reutern's analysis of Russia's fiscal and economic difficulties and noted the risk of foreign intervention by a coalition of hostile powers:

> We have no choice. We are not free to pose the question; is it possible or impossible to end the war in one campaign? *It must be ended in one campaign*, as we do not have the resources for a second, and , moreover, because then we would have to fight not only Turkey but with all of those who are only waiting for our exhaustion. It is necessary quickly to put an end to the matter and suppress the Turks, while we still preserve the entire extent of our strength and have not revealed our weakness.[126]

The success of such a campaign depended on the rapid deployment of Russian forces into the depths of theater, and that depended in part on the effective utilization of Romania's existing network of railroads operating within the theater. Romania, although nominally a vassal of the Porte, found every reason to side with Russia to gain complete independence. On 4 April Russia concluded an agreement with the Romanian government to move its forces through Romanian territory and on 12 April declared war on the Porte.

While the political arrangements proved easy to negotiate, the exploitation of the Romanian rail system to speed movement to the Danube proved disastrous. In Romania, of the four main lines with a length of 1,241 km, only one line, that from Ungena to Jassy, was wide-gauged, and that was a minor line with only four engines and fifty cars.[127] Thus, where the gauge changed to European standard, a monumental log jam of cars developed once the war began. There Russian freight had to be unloaded and then reloaded on Romanian cars for movement into the Balkan theater.[128] Russian officers inspecting the Romanian railroads complained of poorly laid track, badly placed right-of-ways, inadequate infrastructure, and anti-Russian railroad personnel, especially Austrians and Poles. In short, the Romanian railroad system, much of which had been built by Stroussberg at a high profit, proved totally inadequate for strategic rail maneuver.[129] Russian attempts to begin repair and modernization of the Romanian roads in the winter of 1877 met with opposition within the leadership of Romania's railroad administration. In late March, less than a month before the declaration of war, the chief of staff of the Active Army noted that it was impossible to prepare the Romanian railroads in a timely fashion for intense exploitation to move Russian troops.[130] This situation had a significant impact on the course and outcome of the campaign of 1877. Only part of the Russian I Corps made the deployment into Romania by rail maneuver, traveling from Ungena to Bucharest. The other three corps marched to the Danube and with further reenforcements began the crossing of the river only in late May.

Count von Pfeil, a Prussian officer in Russian service in the Balkans, described his travel by rail in the fall of 1877 from Moscow to the Russian border as uneventful—he complained of some lost luggage and boring traveling companions but reserved his disdain for the Romanian railroads. Von Pfeil found the main Romanian line south of Bucharest unsafe at any speed:

> Thanks to Herr Stroussberg's forethought, the line had been constructed on such moist ground between two halting stations, that the railway embankment could not stand the enormous traffic of the past month, and several trains had already had accidents owing to driving too fast. Earth and stones were constantly being deposited there, but without any good effect, so the track was really very dangerous, and the train could only proceed very slowly for a full hour—slower than a man can walk.[131]

The strategic implications of a slow rail maneuver to the Danube three months after the deployment of the Russian Army to the Russo-Romanian border were not lost on General Obruchev. Delays in the crossing of the Danube had cost valuable time, which the Turks had put to good use to concentrate troops in the theater. Obruchev wrote: "We have lost much time in the delay in crossing." He went on to point out that in the meantime the situation had shifted unfavorably. "What could have been done a year ago by divisions, and at present requires corps, soon will demand whole armies."[132]

Obruchev, however, did not let this development affect the political objectives of the Russian campaign. Indeed, he made them even more sweeping: "in order to achieve decisive results, the goal of our strategic actions must be more than ever Constantinople itself. On the shore of the Bosphorus we can really smash Turkish ownership and get a sound peace that once and for all resolves our conflict with them over the Balkan Christians."[133] Russian troops forced the Danube successfully in May, but their advance into the Balkan Mountains bogged down in the face of stubborn Turkish resistance at Plevna and Shipka Pass. In late July Minister of War D. A. Miliutin wrote Alexander II informing him that Turkish resistance had proven more effective than expected. He attributed this to foreign military assistance. Tactically, he admitted to errors, especially the conduct of repeated frontal attacks in the face of an opponent armed with modern rifles. Strategically, Miliutin concluded: "it is obvious that we can no longer count on one rapid, decisive move through the Balkans to create panic and fear in the enemy's troops, and people and expect to be in a few weeks time under the walls of his very capital where we will dictate peace terms to him."[134] A. Svechin, Russia's most talented strategic thinker of the twentieth century, categorized Obruchev's war plan as radically flawed because of the disconnect between "the incomplete nature of Miliutin's reforms and the terrible condition of our operational thought with the [plan's] arch-annihilation tendencies."[135] As Svechin's account of the campaign makes clear, Obruchev gambled on a lightning campaign of swift maneuver to disconcert the enemy and reach Constantinople before the Turks could rally or the other powers could intervene.[136] Failing to achieve that result, Russia found itself in a protracted war that it could win on the battlefield but could neither afford in an economic, fiscal, or political sense nor win in terms of a final settlement. Reutern's warning had proven right. Bloch shared Reutern's pessimism over the risks of war and defended Reutern's position publicly. A popular war thus turned into an albatross around the neck of an unpopular government and had profound domestic consequences.

Railroads, Reform, and the Crisis of Autocracy

On the issue of Russia's economic situation and the war, Bloch not only supported Reutern and the Ministry of Finance but also put the current economic situation in context and noted that the fiscal policies, private railroad construction, and the other great reforms had created the infrastructure for further economic development. Indeed, in late 1877, when the questions of Russia's economic situation and the state's fiscal condition were under a cloud because of the war, Bloch published a series of articles on the situation in the national economy, defending railroad construction as critical to economic development and prosperity.

Bloch compared Russia's current situation with its economic backwardness prior to the Crimean War. Without reform and private enterprise Russia was doomed to a declining position in the world as it lost its ability to compete. Railroad construction with its huge capital costs had been worth the price because it was the basis of a modern national economy, integrated into world trade.

Reutern's fiscal policies, which had sought to reduce annual budget deficits, reform the banking system, and stabilize credit, had faced major difficulties as a result of the disruptions associated with the emancipation of the serfs and the Polish Insurrection, and it proved difficult to get access to foreign capital markets for the stocks and bonds of private railroad companies, which were considered risky investments. Only in the mid-1860s had state guarantees on the return of capital encouraged foreign capital, particularly the Berlin stock market, to invest in Russian railroad projects.[137] Bloch called attention to the general economic crisis affecting Europe's stock markets in the mid-1870s and warned of the serious dislocations to state finances and the national economy that the war in the Balkans had already caused. But this crisis would be "only a temporary interruption" because Russia's new economic infrastructure and reforms had laid the basis for a rapid recovery of trade and industry.[138]

Bloch's five-volume work, *The Influence of Railroads on the Economic Development of Russia,* continued with this theme and ended with a direct challenge to those opposing private, as opposed to state-owned, railroad construction. In this detailed study of Russian railroad construction from its beginning to 1878, Bloch linked the Great Reforms and private railroad construction. The issue was not one of railway construction but of national direction: building roads that served troop mobilization and deployment at the expense of the civilian economy, or supporting construction of economically sound railroads that linked Russia's internal economic development and foreign markets:

> Such are the basic features of the fundamental improvement, as the present eco-
> nomic situation of Russia represents in comparison with the past. The present war,
> bringing about a break in this gradual improvement, thanks chiefly to this, i.e., to
> the reforms and the construction of a network of 20,000 versts of railroad, can not
> have the same negative effect on our finances and general economic situation as was
> brought about by the war of 1853–1856 as a consequence of the period of stagna-
> tion.[139]

To Bloch, reform and enterprise were the tools for the sustained development of
national might, "not in stagnation, not in isolation from other peoples, and not in the
confinement of the conditions of internal life."[140] One of the main contributors to
stagnation was excessive military spending and the associated deficits, which represented dire threats to financial markets, a sound currency, and a prudent fiscal policy.
His four-volume study of Russian state finances in the nineteenth century, which
appeared in 1882, made these very points.[141]

Russia had to adopt a "peaceful, lawful but daring movement forward along the
common-European path."[142] War did not, as conservative myth held, lead to social
peace. Writing of the Russo-Turkish War of 1877–78 and Reutern's threat to resign in
1876 because of the effect that war would have on the Russian economy, Bloch noted
that the war had itself been caused "in part namely by the sick languor of society,
which is condemned to complete inaction by the falsity surrounding it." Bloch noted
that "if anyone thought that by this way it would be possible to divert those evil
forces and to satisfy society's thirst for action, then this proposition proved a grave
mistake. The results of the war brought to society disappointment in all relations,
military, civil, and financial."[143] A war won only after a long campaign brought gains
that were lost at the conference table in Berlin. And trust between government and
society widened into a gap that threatened the autocracy itself.

The hard-won gains of the 1860s, which had created the climate for Bloch's own
success and opened opportunities for Jews within the Kingdom and Empire, were in
danger. A brief wave of renewed reform in the face of revolutionary terrorism collapsed after the assassination of Alexander II, to be replaced by reaction,
counterreform, and political anti-Semitism under Alexander III.

Bloch and the Jewish Question

Bloch's subsequent projects, leading up to *Budushchaya voina,* have an intellectual coherence that gives special meaning to that work. Bloch was moved by a
series of events which suggested to him the real risks of war for state and society.
Bloch began as a liberal with a faith in progress and human emancipation that would
come from enterprise. He had been one of those who had profited by the atmosphere
of toleration that had gone hand-in-hand with the Great Reforms. But after the Russo-
Turkish War it occurred to Bloch that dark forces were threatening that vision of
progress. The crisis of autocracy which began with foreign policy setbacks at the
Congress of Berlin, extended to rural unrest, and urban strikes, and culminated in a
terrorist campaign against the government and the Tsar himself, reached its peak

with the assassination of Alexander II on 1 March 1881 (O.S.).[144] The twin problems of social unrest and official anti-Semitism were becoming linked during the decade of counterreforms in the reign of Alexander III.

Bloch was intimately connected with the development of Warsaw in the years after the January Insurrection of 1863. During that period when "organic work" replaced revolutionary romanticism among patriotic Poles, the rivalry between the Bloch and Kronenberg banking houses was fought out on many fronts. In addition to a bitter struggle over railway concessions, their competition also extended to the area of civic improvements, culture, education, and philanthropy. Bloch supported his scientific work by building personal contacts with the leading universities of Europe, recruiting writers and scholars to serve as his researchers, providing funds for the sustained, systematic collection of data, and personally overseeing the completed research. In the age of the artisan-scholar working alone, Bloch built a powerful, independent think-tank that he could aim at the most immediate problem of concern to him, whether it was the feasibility of building a modern sewage system for Warsaw or reviewing the state of Russia's finances in the nineteenth century. For his many services to the Empire, Alexander II ennobled Bloch giving him the coat of arms of "Ogonczyk Odmienny" (the Distinct), because Ogonczyk had been an old Polish heraldic symbol, and to differentiate its new holder from one Jan Bloch, a sixteenth-century Jewish convert from Oswiecim [Auschwitz] parish.[145]

The decades following the January Insurrection were ones of remarkable economic progress in the Polish Kingdom in spite of the tsarist government's attempts at iron rule and Russification. Popular agitation for reform and self-rule in the early 1860s had created a climate of solidarity and tolerance among Jews and Poles in a common struggle for self-government and civic rights. Poles and Jews, as Rabbi Izaak Kramsztyk declared in March 1861 in a Warsaw Synagogue, were brothers of different faiths setting out "on the path of enlightenment and civilization, on the path of science and toleration."[146]

The Jewish Question arose in the context of a relaxation of Russian rule in Poland. Following the November Insurrection of 1831 and for the next quarter century, Nicholas I had relied on Field Marshal Paskevich, the Russian Army, fortresses, and martial law to maintain Russian rule.[147] By the early 1860s reformers in the Russian government had come to question the effectiveness of this militaristic regime:

> The [Paskevich] system, which rejects any thought of legality . . . and has as its only objective the creation of an external pacification, does not bring into existence any proper, lawful government. With the passage of time it has created more evil than good for it has stirred up in the whole country a hatred toward the government, which will live on from generation to generation. It has caused disrespect toward it [the government]. Even beyond that it has destroyed the very instruments on which a government must depend—law, respect for law, and order.[148]

In the face of popular demonstrations, Alexander II's government embarked on a reform program in Poland. This effort, which culminated in the viceroyship of Grand Duke Konstantin Nikolaevich in 1862–63, sought to undercut the revolutionary thrust of Polish nationalism by a program of social and economic reforms. Concessions in

Poland were supposed to give Russia time for a deeper program of reforms at home and to provide a sound basis for international cooperation between Russia and France's Second Empire. The program failed to win the support of the Polish Szlachta, that is, the whites, and pushed the Polish radicals, that is, reds, into more terrorist acts and finally open rebellion.[149]

The reforms undertaken by Count Aleksander Wielopolski and a reformist Russian administration in the Polish Kingdom, including the granting of legal equality in June 1862, however, provided sufficient opportunity to spur Jewish entrepreneurship in the economic development of the Kingdom, which took off after the Insurrection.[150] The cosmopolitan character of Warsaw was much in evidence among the financial, industrial, and commercial leadership of the city, where German and Jewish businessmen had prominence, much to the dismay of the Polish nobility, who now saw their "natural" position as leaders of society being challenged.[151] No matter what they tried to do in this new, dynamic world of capital, the dislocated nobility found one common source for their failures, as Prus suggested: "Oh! I could write a whole poem about our disordered life. But even here I can not move for the Jews have more money than me."[152] To combat this trend, Bloch, the Catholic convert, used his position with the tsarist government to promote the cause of Jewish rights and lent his support to a policy of assimilation. Already in 1877 Bloch had raised the Jewish Question during an audience with Alexander II devoted to financial issues. Bloch had spoken of the positive influence of the Jews on the national economy to the displeasure of his Sovereign.[153]

The economic depression of the mid-1870s and the downturn after the Russo-Turkish War of 1877–78, combined with local unrest, revolutionary terrorism, and governmental crisis in Russia, undercut that climate of optimism. In the summer of 1881, after the assassination of Alexander II by terrorists, a wave of anti-Semitic pogroms swept across South Russia, extending into the Polish Kingdom.[154] On Christmas Day, 1881, the crowd leaving the Holy Cross Church in Warsaw panicked and about thirty persons were trampled to death. In a matter of hours rumors spread throughout the poorer districts of the city that the Jews had caused the deaths. Two days of violence and looting followed during which hundreds of Jews were hurt, scores killed, and over one million rubles in property destroyed. Boleslaw Prus described the pogrom as an "orgy of violence." For three days the police and garrison took no actions to end the violence.

For Bloch these anti-Semitic pogroms and the tsarist government's reactionary policies represented a dire threat to the process of reform, enlightenment, and enterprise in which he believed. Alexander III and his advisors, who included Konstanin Pobedonostsev, the tsarist tutor and advocate of solving the Jewish Question by thirds: one-third convert, one-third emigrate, and one-third die off, viewed the Jews as implacable enemies of the Russian idea. Repression and pogroms within Russia increased the pressure for migration by tens of thousands of Jews out of the Pale of Settlement and into the Polish Kingdom. The appearance of these refugees increased social discontent among those artisans and workers who saw Jews as competitors for employment and as merchants who exploited them. Tsarist officials were quite willing to use the Jews as scapegoats for all social ills and thereby manipulate this modern form of political anti-Semitism.[155] As Andrei Subbotin noted, Russian anti-Semitism in the late 1870s and early 1880s went through three stages—progressive, populist, and conservative—but all had in common a desire to protect the "nation"—meaning the peasantry—from the ravages of capitalist exploitation.[156]

The wave of pogroms gave way to overt anti-Semitism in the much of the national press. The Russian government seeking to mobilize this instrument to its own agenda of counterreform now began to look for ways to curtail the rights of the Jewish population. Governor-General I. V. Gurko proposed in February 1885 to take back the civic equality granted Jews in 1862. To counter this threat the Warsaw Jewish Commission undertook the collection of information that would establish the important and positive role of the Jewish community in the life of the Kingdom. Jan Bloch and Henryk Natanson presented a petition (*Memorial*) in defense of the Jewish community and the rights granted it in May 1862. The petition was leaked to the press and became the focal point of the Jewish Question in public opinion. The ultra-conservative, anti-Semitic publication, *Niwa* denounced the *Memorial* in an article entitled, "Not by that Path."[157] In the ensuing struggle, Bloch undertook a massive study of the position of Jews in the Polish Kingdom and with a solid argument, based on social and economic statistics, took on the counterreformers' main argument that all social ills could be explained by one simple fact: "Jews bring only one thing—evil!" Bloch used his court connections to reach Minister of Finance Vyshnegradskii, his former employee, and Alexander III delayed action on the proposal. The appointment of Sergei Witte, "an even stronger enemy of wild nationalism," as minister of finance put an end to the project.[158]

Between 1884 and 1891 Bloch collected materials on the conditions of Jews in the Polish Kingdom and Russia proper. Bloch remained committed to an open society and assimilation and suggested that the successful assimilation of the Jews in Germany was a model to be applied in Poland and the rest of the Russian Empire. Bloch set out to prove his point by assembling these materials into a massive socioeconomic study of the Jewish population in the Polish Kingdom and Russia. This five-volume work that ran to over 2,000 pages took as its point of departure the recommendations of the government's own Pahlen Commission, which had studied the Jewish Question and had concluded that voluntary assimilation was the best long-term answer. In his study Bloch addressed the fate of the mass of the Jewish population in the Pale of Settlement, noting their marginalization and growing poverty. He drew a picture of the Jews of the Pale of Settlement as mostly small traders and artisans, who eked out a subsistence existence of low income and long hours. This he attributed to the discrimination that Jews faced as a community. He took issue with those who blamed the Jews for a decline in public morals among the peasantry, associated with increasing use of alcohol and rising crime rates. The Jews were charged with the manufacturing, smuggling, and illegal sale of spirits. Yet, the gubernias of the Pale of Settlement had a much lower crime rate and level of alcohol consumption than those outside. Bloch noted that in Western Europe, where Jewish assimilation on the basis of civic equality had been going on since the French Revolution, such problems were hardly in evidence. "All this leads to the conclusion that if such a moral decline of Jews exists then it is caused to a certain degree by the very conditions imposed upon them."[159] Feudal institutions in their struggle to retain their power over society were responsible for the discriminations under which the Jewish population still labored.

Bloch also noted the emergence of modern anti-Semitism in the West, calling attention to the linkage between socialist and anti-socialist impulses in modern anti-Semitism. On the one hand, socialists sought to create the utopia of a classless society, and their propaganda called for a revolutionary struggle against capitalism which the dark masses in their semiconsciousness took to be an invitation to kill and rob Jews whom they held as the representatives of the "unjustly rich." On the other hand, conservative forces, ranging from what Bloch described as Bismarck's "state socialism" to the aristocratic, ultramontane version of Christian Socialism, looked back to a conservative utopia of a society of orders where each man knew his place and sought to blame all of modern society's dislocations and ills on the Jews. They used their own anticapitalist propaganda to assume the mantle of protector over the lower classes. "All anti-Semites are enemies of innovation and of that which is called the *esprit moderne*."[160]

Before the study could be distributed a fire at the printing plant destroyed all but twenty-five copies of the five-volume sets, which Bloch bestowed on a few close friends, making it into a very rare and difficult to locate publication.[161] Andrei Subbotin, a leading Russian economist and specialist on the Jewish Question, saw Bloch's work as a point of departure for the serious study of the issue. He agreed with Bloch that the only answer was rapidly to expand the rights of Jews to engage in productive activities. He warned against "the conservative course of treatment" for this problem because it always seemed to prescribe "surgical operations," designed to cut off and isolate the Jewish population.[162] Bloch himself seems to have recognized this danger, for in the last decade of his life, according to N. Sokolow, he became involved in projects for Jewish emigration and colonization and he moved from a skeptical view of Zionism to one of sympathy.[163]

Bloch followed this study with others on the credit crisis and agriculture, land question and rural debt in Poland, and on the socioeconomic situation in Galicia, all of which addressed the economic problems associated with backwardness.[164] Boleslaw Prus, who took part in the study of the socioeconomic situation in Galicia, spoke of that Austrian province as "not being a society and already uncivilized or only just becoming civilized," where there was little work, literacy, prosperity, and fellowship. Popular education, land reform, industrialization, and improved public hygiene were the best solutions to this crisis. But Prus feared that anarchist propaganda under such conditions could enflame the "entire proletariat" into violent revolt.[165] The idea that rural poverty might create the conditions for popular revolt ran directly counter to one of Bloch's central arguments concerning Russia's ability to fight a protracted war; that is, peasant subsistence agriculture was less likely to be affected by the economic dislocations of modern war than would be modern commercial farming with its dependence on financing, chemical fertilizers, and so on.[166]

War and Revolution

Thus, when examining Bloch's *Future War* it is critical to understand the social context under which he saw a general European War being fought. From the optimist of the 1860s and 1870s with his faith in progress, Bloch had turned into a pessimist. By 1893 Bloch spoke of a "premonition of war" in the air across Europe and warned that the slightest miscalculation could rapidly shift the scales of war and peace toward war. A general European war would require the complete mobilization of national economies, and only those states who had taken such measures were likely to withstand the strain placed on finance and economic life by a protracted conflict. Russia, with a backward state edifice and economy, was not in a position to carry out the prewar measures necessary to ensure effective mobilization of the national economy. This entailed foreseeing those areas of disruption in the social organism that would lead to crises threatening internal order.[167] Bloch's pacifism was not based on some belief that men had changed for the better, but rather that prudent statesmen had to understand that modern war would unleash socioeconomic collapse and revolution precisely in those areas of Eastern Europe where reform and enlightenment had failed to sweep away the old regime and its militarism. Bloch died in 1902 and did not live to see the first appearance of his twin demons—war and revolution. In one of his last articles, published posthumously, Bloch addressed the issue of chance, which Dragomirov had raised against Tolstoy's "dilettantism." The role of chance in warfare had changed because of the expansion in the scale of war. This was a quantitative leap thanks to mass armies, modern weapons, and industrialization. The new scale had brought the factor of probability into the equation of war:

> If war does break out, the human factor and the factor of accident will undoubtedly play a considerable part. But there are reasons why they cannot play so decisive a part as they have often in the past. The law of averages applies in human affairs. If two men toss pennies a half dozen times, one may possibly win all six. But if they toss a hundred times there is no human possibility that one will win all hundred; it is a hundred to one that the winnings will be equally divided. This may be taken as a parable to illustrate the part which accident is likely to play in the prolonged and complex warfare of the future.[168]

Jan Bloch left his reader with war as an engine beyond human control, in which the disconnect between means and ends was so self-evident that rational statesmen would not take the risk of relying on force of arms to resolve even questions of vital interests but would depend on other institutions of international arbitration. But Bloch had already witnessed the terrible consequences of wars that got beyond the control of the statesmen who embarked upon them. Militarism had not been dethroned. It still might unleash those forces which it did not understand and could not control. Modern war and revolution were thus linked to the assault on an open society and threatened to unleash a virulent anti-Semitism.

Conclusion

Bloch died of a heart attack on 7 January 1902, having willed a large portion of his fortune to philanthropic enterprises, ranging from the Charity Institution of the Polish Kingdom, the Warsaw Municipal Charity, the Warsaw Politechnicum, to the Permanent International Bureau of Peace and its Museum of War and Peace in Berne, Switzerland. Bloch did not live to see the Russo-Japanese War or the Revolution of 1905.

In his hope that statesmen would see the folly of war, Bloch proved sadly incorrect. Nicholas II, the sovereign responsible for calling the Hague Peace Conference, following a policy of imperial aggrandizement and racist overconfidence managed to drag Russia into a war with the Japanese Empire. And that war became the prism through which Russians judged Bloch and his method.

Regarding the dispute which pitted Urusov and Bloch as champions of applied mathematics against their conservative military critics, who objected to applying mathematical methods to the art of war, there were already signs that some reform-minded officers grasped the utility of parts of Urusov and Bloch's method in approaching modern war. There were already Russian officers who were adapting one part of their methodology, the use of applied mathematics and probability theory to the study of military art, even as the Russo-Japanese War was beginning. They used these applications of probability theory to manage the factor of chance on the more complex modern battlefield and to find solutions to pressing logistical problems created by protracted combat, that is, the average consumption of shells by artillery batteries and of ammunition by infantry units.[169] During World War I, M. Osipov used coupled differential equations to bring together advanced statistical techniques and historical data to study the influence of numerical strength of opposing forces on their losses, making him one in the forefront of analyzing and understanding combat dynamics.[170]

Regarding the dispute between Bloch and the "incrementalists," the postwar judgments of Russian military theorists who were veterans of Manchuria, came down on neither side. As professional soldiers, they could not agree with him on his pacifism—war was too possible—nor did they agree with the proposition that the defense would dominate the offense, but they did agree that war had changed profoundly. Lieutenant Colonel A. A. Neznamov (1872–1928), reflecting on the Russian defeats in the Far East wrote in 1909 that the problem had been: "We did not understand modern War."[171] Neznamov agreed with Bloch that the Achilles heel of modern war was the linkage between tactics and strategy, and Neznamov, therefore, focused on the conduct of operations. He also acknowledged the difficult problem of troop control on the modern battlefield and proposed as an answer to that the creation of military doctrine. In its absence, "these diverse views go to war with us, and thus unity of thought will cost rivers of blood and perhaps can be purchased only at the cost of a serious failure. *But still it is necessary.*"[172]

Bloch proved a much better forecaster when it came to the campaign, which proved protracted and indecisive. Some officers were even willing to challenge the cult of the offensive at the strategic level. A. A. Svechin wrote in 1913 that France, given its geostrategic situation, should forego initial offensive operations and that Russia should not undertake offensive operations until it had completed its mobilization.[173]

Bloch proved correct in his assessment of the negative impact of fiscal and economic dislocations of such a war on Russia. But he proved quite wrong in his assessment of Russia's invulnerability to revolution. The linkage between army and society which the autocracy invoked broke under the weight of military failure and protracted war in a region far removed from the Russian heartland. It produced urban and rural unrest and finally revolution in January 1905. Socialism, which Bloch, like many others, had seen as the illness of the West, had raised its head in Russia. The first Social-Democratic Party in Russia held its first congress in 1898. Lev Tolstoy, living through those events, returned to the great question that he had raised in *War and Peace*, the relationship between people and the sovereign power:

> A revolution is taking place in Russia, and all the world is following it with eager attention, guessing, and trying to foresee whither it is tending and to what it will bring the Russian people.
>
> To guess at and to foresee this may be interesting and important to outside spectators watching the Russian Revolution, but for us Russians, who are living in this revolution and making it, the chief interest lies not in guessing what is going to happen, but in defining as clearly and firmly as possible what we must do in these immensely important, terrible and dangerous times in which we live.
>
> Every revolution is a change of a people's relation toward power [*vlast'*].
>
> Such a change is now taking place in Russia and we, the whole Russian people, are accomplishing it.
>
> Therefore to know how we can and should change our relation toward power, we must understand the nature of power: what it consists of, how it arose, and how best to treat it.[174]

The final resolution of that struggle over power would only come in the wake of another great, protracted war and revolutionary upheaval.

The tragic outcome of that process would be the triumph of an isolated, socialist state, directed by revolutionaries bent on building a socialist society and conducting class struggle to bring about the world revolution. Surrounded by hostile powers and fighting a bloody civil war, Lenin and his Bolsheviks militarized Marxism on the way to building a totalitarian, warfare state.[175] In the process they would smash the Russia of Tolstoy's villages and of Bloch's cosmopolitan cities and middle-class values. And Warsaw and St. Petersburg, the keystones of that world, would become the quintessential symbols of total war in all its barbarity.

Under Stalin the Soviet Union became an order adapted to the demands of total war but imposing so great a burden on society that it stifled technological innovation, economic development, and social welfare. Dogmatism and stagnation precluded its entry into the ranks of information societies. As one of the foremost Soviet forecasters, N. N. Moiseyev noted, the centralized, planned economy that won the Great Patriotic War, could not adapt to new requirements.[176] The very process of reforming it, which meant demilitarizing it, brought about a fatal crisis and the end of the Soviet system.

After three-quarters of a century that state collapsed. Russia and the other successor-states have embarked on a new experiment, embracing democracy and a market economy. Once again, the issue of power stands at the very heart of any forecast of Russia's future. Once again, the question of forecasting future war embraces a host of political-military and military-technical questions. Once again, questions of the nature of the threats and dangers confronting Russia are issues of debate. And, once again, soldiers and civilians are struggling with these issues. In a time of revolutionary change and severe social dislocation such questions invite a wide range of answers. This time, however, the world shares a profound interest in those answers, if this Russia is to find its proper place in the world. Forecasting in military affairs can be about those methods to maintain a sufficient military force to prevent war, to protect legitimate Russian interests, and to contribute to regional stability and global peace. Or it can be about preparations for war in the name of revenge, the imposition of Russia's will on others and imperial expansion. And those who wish Russia well may hope that the values which Tolstoy and Bloch embraced, that is, respect for the nation, the rule of law, and the engines of material progress and peace, will find their champions. At the same time, those friends of Russia must be concerned by the signs of a resurgent populist nationalism that embraces extreme nationalism, statism, imperialism, and militarism, as represented by Vladimir Zhirinovsky, his Liberal-Democratic Party and others.[177]

Notes

1. Jacob W. Kipp, *Foresight and Forecasting: The Russian and Soviet Military Experience* (College Station, TX: Center for Strategic Technology Stratech Studies, 1988).

2. I. E. Shavrov and M. I. Galkin, eds., *Metodologiya voyenno-nauchnogo poznaniya* (Moscow: Voyenizdat, 1977), 64.

3. Makhmut Akhmetovich Gareev, *Esli zavtra voina? . .* (Moscow: Vladar, 1995).

4. Ibid., 5.

5. Ibid.

6. Ibid.

7. Christopher Bellamy, "'Civilian Experts' and Russian Defence Thinking: The Renewed Relevance of Jan Bloch," *RUSI* (April 1992): 50–55.

8. Timothy L. Thomas, "Soviet Military Theoretician A. A. Kokoshin," *Journal of Soviet Military Studies* 5 (March 1992), and Jacob W. Kipp, "General of the Army V. N. Lobov: One of Gorbachev's *Genshtabisty*," *Journal of Soviet Military Studies* 2 (September 1989).

9. V. Cheban, "Nuzhna li armiya Rossii?" *Armeyskiy Sbornik: Zhurnal dlya voyennykh professionalov*, no. 1 (July 1994), 15.

10. R. F. Christian, ed., *Tolstoy's Diaries, 1847–1894* (London: Athlone, 1985), 157.

11. Isaiah Berlin, "The Hedgehog and the Fox," in Isaiah Berlin, *Russian Thinkers,* ed. Henry Hardy and Aileen Kelly (London: Hogarth, 1978), 22–80. Berlin focuses on the intellectual and literary sources of Tolstoy's philosophy.

12. Mark Aldanov, *Zagadka Tolstogo* (Providence: Brown University Press, 1969), 7ff. Tolstoy's Pful' was Prussian General Karl Ludwig August Pfül (1757–1826).

13. Ernest J. Simmons, *Lev Tolstoy* ((New York: Vintage, 1960), 1:82–112.

14. Leo Tolstoy, *Sevastopol* (Ann Arbor: University of Michigan Press, 1968), 113–229.

15. Simmons, *Lev Tolstoy,* 113–37.

16. Lev Tolstoy, *Voina i mir* (Moscow: Khudozhestvennaya Literatura, 1983), 2:607ff.

17. Ibid., 623–24.

18. L. N. Tolstoy, *Polnoe sobranie sochinenii,* 90 vols. (Moscow: Gosudarstvennoe Izdatel'stvo Khudozhestvennoi Literatury, 1928–64), 61:204.

19. M. Dragomirov, "Razbor *Voiny i mira,*" in *Ocherki* (Kiev: N. Ya. Ogloblin, 1898).

20. Ibid., 39.

21. Ibid., 46–47.

22. Ibid., 48.

23. Ibid., 49.

24. Ibid., 49–50.

25. Boris Eikhenbaum, *Tolstoi in the Sixties,* tr. Duffield White (Ann Arbor, Mich.: Ardis, 1982), 175ff.

26. Ibid., 210.

27. S. Urusov, *Obzor kampanii 1812–1813. Voyenno-matematicheskie zadachi i o zheleznykh dorogakh* (Moscow: Tipografiya V. A. Got'e, 1868), 119–55. See also S. S. Urusov, *Differentsial'nye raznostnye uravneniya* (Moscow, 1863); *O reshenii problemy konya (v shakhmatakh)* (Moscow, 1865); and *Ob integriruyushchem mnozhitele raznostnykh i differentsial'nykh uravnenii* (Moscow, 1865).

28. Ibid., iii.

29. Ibid., ix–x.

30. Ibid., x–xi.

31. Ibid., xix–xx.

32. Eikhenbaum, *Tolstoi in the Sixties,* 220.

33. Tolstoy, *Polnoe sobranie sochinenii,* 61:208.

34. Urusov, *Obzor,* 47–63.

35. Ibid., 158.

36. Ibid., 165–66.

37. Ibid., 156.

38. Ibid., 159.

39. Ibid., 188–93.

40. Ibid., 38–46.

41. Ibid., 22.

42. Ibid.

43. Ibid., 78–97.

44. Ibid., 22, and Eikhenbaum, *Tolstoi in the Sixties,* 226.

45. Ibid., 34–37.

46. L. N. Tolstoy, *Perepiska s russkimi pisatelyami* (Moscow: Khudozhestvennaya Literatura, 1978), 379–81.

47. A. Norov, "Voina i mir (1805–1812). S istoricheskoi tochki zreniya i po vospominaniyam sovremennikov. (Po povodu sochineniya grafa L. N. Tolstago: *Voina i mir,*" *Voyennyi sbornik* 64 (December 1868), *neof.,* 189–90.

48. Ibid., 120–21.

49. Ibid., 126.

50. Ibid., 131–32.

51. Ibid., 134.

52. Ibid.

53. Letter to the Editor of *The Russian Invalid*, Moscow, 11 Apr. 1869, in R. F. Christian, ed. and trans., *Tolstoy's Letters, 1828–1879* (New York: Scribner's, 1978), 1:218–19.

54. *Tolstoy's Diaries, 1847–1894*, 1:187.

55. Ibid.

56. G. Leer, *Opyt kritiko-istoricheskago izsledovaniya zakonov iskusstva vedeniya voiny (Polozhitel'naya strategiya)* (St. Petersburg: Tipografiya Tovarishchestva "Obshchestvennaya pol'za," 1871), 1–4.

57. Ibid., 19–22.

58. Ibid., 162–63.

59. Ibid., 268–69.

60. Tolstoy, *Polnoe sobranie sochinenii*, 61:254.

61. One of the best accounts of this new army in the field comes from V. M. Garshin, who served with the army in the Balkans. His short story, "From the Memoirs of Private Ivanov," depicts the new relationship between officers and men who are no longer serfs in uniform and not yet citizen-soldiers. See V. M. Garshin, "Iz vospominii ryadovogo Ivanova," in *Krasnyi tsevtok* (Kiev: Izdatel'stvo khudozhestvennoi literatury "Dnipro," 1986), 113–55.

62. M. Dragomirov, "Zametki o Napoleone," in *Ocherki*, 182–86.

63. I. S. Bliokh, *Budushchaya voina v tekhnicheskom, ekonomicheskom i politicheskom otnosheniyakh*, 6 vols. (St. Petersburg: Tipografiya I A. Efrona, 1898).

64. Ryszard Kolodziejczyk, *Jan Bloch (1836–1902)* (Warsaw: Panstwowy Institut Wydawniczy, 1983), 300.

65. I. S. Bloch, *Is War Now Impossible? Being an Abridgement of the War of the Future in Its Technical, Economic and Political Relations* (London: Gregg Revivals, 1991). The false impression provided by this volume regarding Bloch's views on modern war still reverberates in contemporary scholarship. Thus, Bruce D. Porter in his excellent study on war and the rise of the modern state places Bloch among "the choir of intellectuals assuring the world of the absolute incompatibility of the military spirit and industrial society and of the inevitable obsolescence of war." As we shall see Bloch was by no means so optimistic and had no blind faith in human progress. See Bruce D. Porter, *War and the Rise of the State: The Military Foundations of Modern Politics*(New York: Free Press, 1994), 161.

66. I. S. Bliokh, *Obshchie vyvody iz sochineniya "Budushchaya voina v tekhnicheskom, politicheskom i ekonomicheskom otnosheniyakh" I S. Bliokh*, (St. Petersburg: Tipografiya I. Efrona, 1898), 112–13.

67. Ibid., 192.

68. Hans Delbrück, who combined the serious study of military history with what would today be called defense analysis, grasped this point in Bloch's work. In a review written in 1899 of the three-volume German edition, Delbrück critiqued the work as "unscholarly," that is, a historical, but praised the practical value of the mechanism that Bloch had employed to look at war and its possible consequences. As an historian he dismissed the idea of technological determinism as a check on war. New weapons in the past had created their own crisis of arms. The wars of Frederick the Great had been wars of attrition and those of Napoleon wars of annihilation. But on the core point of the possibility of a general war setting off economic collapse and social crisis, Delbrück described Bloch's argument and then said that only "practice," that is, the actual fighting of such a conflict would determine the truth of Bloch's assumptions. Within a decade Delbrück was coming close to accepting Bloch's position on the disutility of war for Germany. See Has Delbrück, "Zunkunftskrieg und Zunkunftsfriede," *Prussiche Jahrbücher* 96 (April–June 1899): 207–9; and Arden Bucholz, *Hans Delbrueck and the German Military Establishment: War Images in Conflict* (Iowa City: University of Iowa Press, 1985), 73–75.

69. Panteleimon Nikolaevich Simanskiy, *Otvet G. Bliokhu na ego trud "Budushchaya voina v tekhnicheskom, ekonomicheskom, i politicheskom otnosheniyakh"* (Moscow: A. N. Levinson, 1898), 7–8.

70. Ibid., 26.

71. Ibid., 13.

72. William T. Stead, "Conversations with the Author," in Bloch, *Is War Now Impossible?*, iv.

73. L. Slonimskii, "Vooruzhennyi mir i proekty razoruzheniya," *Vestnik Evropy* 33 (October 1898).

74. Jean de Bloch, "The Wars of the Future," *Contemporary Review*, 80 (September 1901): 315–16.

75. Simanskiy, *Otvet G. Bliokhu*, 60–61.

76. Ibid., 56–57.

77. Ibid., 68.

78. Ibid., 19–20.

79. Bellamy, "'Civilian Experts,'" 54–55.

80. Ibid., 55.

81. B. Prus, *Kroniki*, 20 vols. (Warsaw: Panstwowy Institut Wydawniczy, 1951–1970), 16: 443–44.

82. In this regard it is important to remember that Bloch presented his initial argument not in 1898 but in a series of articles that began to appear in March 1892. See Jan Bloch, "Przyszla wojna, jej ekonomiczne przyczyny i skutki," *Biblioteka Warszawska* 1 (March 1893): 576ff.

83. Michael Howard, "Men against Fire: The Doctrine of the Offensive in 1914," in Peter Paret, ed., *Makers of Modern Strategy: From Machiavelli to the Nuclear Age* (Princeton, N.J.: Princeton University Press, 1986).

84. Bruce W. Menning, *Bayonets before Bullets: The Imperial Russian Army, 1861–1914* (Bloomington: Indiana University Press, 1992), 139–43.

85. Kolodziejczyk, *Jan Bloch*, 219.

86. Bliokh, *Budushchaya voina*, 2:56–82.

87. I. Bliokh, "Ekonomicheskiya zatrudneniya v sredneevropeyskikh gosudarstvakh v sluchae voiny," *Russkiy vestnik* (April 1893): 316.

88. Kolodziejczyk, *Jan Bloch*, 187–89.

89. Ibid., 189–90.

90. Ibid., 196.

91. Ibid., 190–91.

92. Prus, *Kroniki*, 2:388, 688.

93. Kolodziejczyk, *Jan Bloch*, 301–2.

94. Andrei Pavlovich Subbotin, *Evreyskiy vopros v ego pravil'nom osveshchenii v svyazi s trudami (svyazi s trudami I. S. Bliokha)*, (St. Petersburg: Tipografiya A. E. Landau, 1903), 1–2.

95. Ibid., 4.

96. Ibid., 5.

97. Kolodziejczyk, *Jan Bloch*, 202–6.

98. Ibid., 30–35.

99. Subbotin, *Evreyskiy vopros*, 1–2.

100. John Doyle Klier, *Russia Gathers Her Jews: The Origins of the Jewish Question in Russia, 1772–1825* (London: Northern Illinois University Press, 1986), 171–79.

101. Emmanuil Flisfish, *Kantonisty* (Tel-Aviv: Effect, 1980), 91ff.

102. Kolodziejczyk, *Jan Bloch*, 36–37.

103. V. S. Virginskii, *Vozniknovenie zheleznykh dorog v Rossii do nachala 40-kh godov XIX veka* (Moscow: Gosudarstvennoe Transportno-Zheleznodorozhnoe Izdatel'stvo, 1949), 131.

104. "Raport Namestnika Tsarstva Pol'skago ego Imperatorskomu Velichestvu," *Sbornik Imperatorskago Rossiyskago Istoricheskago obshchestva*, no. 98 (1896): 609–10.

228 Tooling for War

105. Archiwum Glowne Akt Dawnych, Protokoly Rady Administracyjnej, no. 147 (4/16 June 1862), 794–98.
106. Ryszard Kolodziejczyk, ed., Gospodarka i Finanse Krolestwa Polskiego przed Powstaniem Styczniowym: Raport Josefa Bossakowskiego z 1862 r. dla Ministra Finansow M. Ch. Reuterna (Warsaw: Panstwowe Wydawnictwo Naukowe, 1969), 231–32.
107. Jack W. Kipp, "M. Kh. Reutern on the Russian State and Economy: A Liberal Bureaucrat during the Crimean Era," Journal of Modern History 47 (September 1975).
108. W. Bruce Lincoln, In the Vanguard of Reform: Russia's Enlightened Bureaucrats, 1825–1861 (DeKalb: Northern Illinois University Press, 1982), 93–99.
109. M. Kh. Reutern, "Opyt kratkago sravnitel'nago izsledovaniya morskikh byudzhetov angliiskago i frantsuzskago," Morskoy sbornik, no. 9 (September 1854).
110. Gosudarstvennaya Publichnaya Biblioteka im. Saltykova-Shchedrina, [hereafter cited as GPB], fond 208 (A. V. Golovnin), delo 65/11-12 "Zapiska v. kn. Konstantinu Nikolaevichu o prichinakh promyshlennogo krizisa vo vsekh stranakh," (Paris, 4/16 Nov. 1857).
111. Ibid., 8–10.
112. M. Kh. Reutern, "Vliyanie ekonomicheskago kharaktera naroda na obrazovanie kapitalov," Morskoy sbornik 46 (April 1860): 65–66.
113. Ibid., 68–70.
114. Kolodziejczyk, Jan Bloch, 43–44.
115. Kolodziejczyk, ed., Gospodarka, 384.
116. Ibid., 365.
117. Colleen A. Dunlavy, Politics and Industrialization: Early Railroads in the United States and Prussia (Princeton, N.J.: Princeton University, 1994), 4.
118. "Zapiski senatora N. P. Sinel'nikova," Istoricheskiy vestnik 40 (April 1895): 47–48.
119. Tsentral'nyy Gosudarstvennyy Arkhiv Voyenno-Morskogo Flota, fond 224, opis' 2, delo 22.
120. Ibid.
121. S. Iu. Witte, Vospominaniya, 2 vols. (Moscow: Izdatel'stvo Sotsial'no-Ekonomicheskoy Literatury, 1960), 1:115–17.
122. N. N. Rostunov, ed., Russko-turetskaya voina 1877–1878 (Moscow: Voyenizdat, 1977), 56.
123. William C. Fuller, Jr., Strategy and Power in Russia, 1600–1914 (New York: Free Press, 1992), 300–317.
124. A. A. Svechin, Evolyutsiya voyennogo iskusstva (Moscow: Gosudarstvennoe Izdatel'stvo, 1928), 2:365–66.
125. M. A. Gazenkampf, Moi dnevnik 1877–1878 (St. Petersburg: V. Berzovskiy, 1908), 15.
126. Sbornik materialov po russko-turetskoi voine 1877–1878 gg. na Balkanskom poluostrove, vol. 22, "prilozhenie" (St. Petersburg: Voyennaya Tipografiya, 1899), 365.
127. Ibid., 19:142.
128. Ibid., 142–44.
129. Ibid., 145–48.
130. Ibid., 135.
131. Richard Graf von Pfeil, Experiences of a Prussian Officer in the Russian Service during the Turkish War of 1877–1878, tr. C. W. Bowdler (London: Edward Stanford, 1893), 26.
132. Sbornik materialov po russko-turetskoi voine 1877–1878, 19:365.
133. Osoboye pribavleniye k Opisaniyu russko-turetskoi voiny 1877–1878 na Balkanskom poluostrove, 4:29.
134. Sbornik materialov po russko-turetskoi voine 1877–1878, 26:153.
135. Svechin, Evolyutsiya voyennogo iskusstva, 2:9.
136. Ibid., 337–411.
137. I. Bliokh, "Ekonomicheskoe sostoyanie Rossii v proshlom i nastoyashchem," Vestnik Evropy 12 (September 1877).

138. Ibid. 12 (December 1877): 735.

139. I. S. Bliokh, *Vliyanie zheleznykh dorog na ekonomicheskoye sostoyanie Rossii* (St. Petersburg: Tipografiya M. S. Vol'fa, 1877–80), 5:164–65.

140. Ibid., 165.

141. I. S. Bliokh, *Finansy Rossii XIX stoletiya,* 4 vols. (St. Petersburg: Tipografiya M. M. Stasyulevicha, 1882).

142. Ibid., 1:9.

143. Ibid., 25.

144. P. A. Zayonchkovskiy, *Krizis samoderzhaviya na rubezhe 1870–1880 godov* (Moscow: Izdatel'stvo Moskovskogo Universiteta, 1964).

145. Marian Fuks, *Żydzi w Warszawie* (Poznan: Sorus, 1992), 128.

146. A. Eisenbach, D. Fainhauz, and A. Wein, eds., *Żydzi a Powstanie Styczniowe: Materialy i Dokumenty* (Warsaw: Panstwowe Wydawnictwo Naukowe, 1963), 14.

147. Jacob W. Kipp, "Policing Paskevich's Poland: The Corps of Gendarmes and Polish Society," Bela K. Kiraly, ed., *East Central European Society in the Era of Revolutions, 1775–1856* (New York: Brooklyn College Press, 1984), 4:200–215.

148. GPB, fond 208, delo 37/8 "Zapiska neizvestnogo o deyatel'nosti v. k. Konstantina Nikolaevicha v kachestve namestnika v Pol'she (Sentyabr' 1863 g.)."

149. Irena Koberdowa, *Wielki Ksiaze Konstanty w Warszawie, 1862–1863* (Panstwowe Wydawnictwo Naukowe, 1962), 215–60.

150. A. M. Skalkowski, *Aleksander Wielopolski w Swietle Archiwow Rodzinnych (1861–1877)* (Poznan: Poznanskie Towarzystwo Przyjaciol Nauk, 1947), 3:136.

151. Kolodziejczyk, *Jan Bloch,* 137–65.

152. Prus, *Kroniki,* 9:324.

153. Fuks, *Zydzi w Warszawie,* 128.

154. I. S. Bliokh, *Sravnenie material'nago byta i nravstvennago sostoyaniya naseleniya v cherte osedlosti evreyev i vne eya,* (St. Petersburg: Tipografiya I. A. Efrona, 1891), 2:8–22.

155. Kolodziejczyk, *Jan Bloch,* 175–77.

156. Subbotin, *Evreyskiy vopros,* 17.

157. "Nie tedy droga," *Niwa,* (1886), 825.

158. Kolodziejczyk, *Jan Bloch,* 178.

159. Bliokh, *Sravnenie,* 2:322–23.

160. Ibid., 21.

161. Subbotin, *Evreyskiy vopros,* 2–3.

162. Ibid., 63.

163. N. Sokolow, "The Late M. Jean de Bloch," *Jewish Chronicle,* 24 Jan. 1902, 11.

164. Kolodziejczyk, *Jan Bloch,* 193, 197–98.

165. Prus, *Kroniki,* 19:328–29.

166. Bliokh, *Budushchaya voina,* 4:260–79.

167. I. Bliokh, "Budushchaya voina, eya ekonomicheskiya prichiny i posledstviya," *Russkiy vestnik* 224 (February 1893).

168. Jean de Bloch, "South Africa and Europe," *North American Review* 174 (April 1902), cited in Jean de Bloch, *Selected Articles by M. Jean de Bloch,* (Ft. Leavenworth, Kans.: U.S. Army Command and General Staff College, Combat Studies Institute, 1993), 196.

169. Nikolai Volotskiy, "Teoriya veroyatnostei i boyevoye snabzhenie artillerii," *Voyennyy sbornik,* no. 2 (February 1904): 139ff; ibid., no. 3 (March 1904); and "Teoriya veroyatnostei i boyevoye snabzhenie patronami," ibid., no. 11 (November 1904).

170. M. Osipov, "Vliyanie chislennosti srazhayushchikhsya storon na ikh poteri," *Voyenny sbornik,* no. 6 (June 1915), no. 7 (July 1915), no. 8 (August 1915), no. 9 (September 1915), and no. 10, (October 1915). See also idem, *The Influence of the Numerical Strength of Engaged Forces on Their Casualties,* ed. and tr. Robert L. Helmbold and Allan S. Rehm (Bethesda, Md.: U.S. Army Concepts Analysis Agency, 1991).

171. A. Neznamov, *Sovremennaya voina: Deistviya polevoi armii,* 2d ed. (Moscow: Vysshiy Voyennyy Redaktsionyy Sovet, 1922), vi.

172. A. Neznamov, "Na zlobu dnya," *Russkiy invalid,* no. 12 (15 Jan. 1912), 4–5.

173. A. A. Svechin, "Bol'shaya voyennaya programma," *Russkaya mysl',* August 1913.

174. Lev Tolstoy, "The Russian Revolution," in Marc Raeff, *Russian Intellectual History: An Anthology* (New York: Harcourt, Brace & World, 1966), 323–24.

175. Jacob W. Kipp, "Lenin and Clausewitz: The Militarization of Marxism," *Military Affairs* 49 (December 1985).

176. N. N. Moiseyev, *Sotsializm i informatika* (Moscow: Izdatel'stvo politicheskoi literautry, 1988), 132–39.

177. Jacob W. Kipp, "The Zhirinovsky Threat," *Foreign Affairs,* May–June 1994. For a systematic exposition of Zhirinovsky's views, see Vladimir Zhirinovskiy, *Posledniy brosok na iug* (Moscow: Liberal'no-Demokraticheskaya Partiya, 1993).

The Sorcerer's Apprentice: Social Science and the American Military

Sam C. Sarkesian

The American military is undergoing a revolution in military affairs triggered by changes in military technology and the nature of warfare.[1] Others call this "War in the Information Age" or third-wave warfare demanding new conceptual, technical, and organizational skills of military leaders.[2] While authorities may disagree over the nature of warfare, the direction of change, and the consequences, they agree that major changes are occurring. This is a result of the post–Cold War new world order—or disorder, and the broad-based technological developments in society and the military. The same scientific theories, concepts, and methodologies are driving most disciplines in social science and permeating most of academia. At the same time, these developments have deeply penetrated the military system, its leadership structure, and its civilian counterparts. In the process, traditional principles of leadership and command are being eroded—for some, even Carl von Clausewitz is suspect—to the point where it is important to pause, reflect, and critically analyze the consequences.

These prefatory comments lead to the purpose of this essay: an analysis of social science and the "scientific" methodology driving the revolution in military affairs, and its impact on the social science–military equation. Also, what all of this may mean to the American military and the military profession is examined with a final commentary on what needs to be done.

But first, a brief comment is in order about the title of this essay. At first glance the notion of "Sorcerer's Apprentice" may raise questions about the scope of the author's intellectual horizons and the seriousness of this work. As this essay explores the social science–military equation, the seriousness of the sorcerer's apprentice label will become clear. In the conclusions, specific reference is made to the essay's title and what this may mean to social scientists and the military.

The connection between social science and the American military is relatively recent. Although World War I was the starting point, it was not until World War II that social science became an important tool in the study of the military. From that time to the present, social science has become an increasingly important component for military professionals, defense policymakers, and scholars studying the military, as well as military antagonists. This is not to suggest that some components of social science as we know them today were not part of Western civilization from the earliest years. The political writings of Plato and Aristotle during the Greek city-state period reflect early application of social science as do some historical studies such as Thucydides's *The Peloponnesian War.* Indeed, some authorities argue that the history of social science *is* the history of Western civilization. Others point to the Renaissance in fourteenth- and fifteenth-century Europe as the beginning of modern social science. Yet, the serious application of social science to the military is primarily a twentieth-century phenomenon.

Although this essay is on social science and the military, it is important to place this in the context of the broader concept of social science which goes beyond the specific issues of the military. This raises several important questions: What is meant by social science? What academic disciplines are included in social science? What is social science methodology?

Social Science and Society

Social science is the study and examination of how human beings form social organizations and systems, and the consequences of such processes. While there are a variety of approaches to study these processes, there are two major views regarding the "scientific" basis of social science. The empirical and "scientific" view is expressed by one authority who writes, "The main body of social science is empirical and, like the natural sciences, seems to have escaped from the influence of the ancient Greek philosophers."[3] According to other scholars, however, social science is an oxymoron. The application of scientific methods presumes experimentation, data collection, factual analysis, rational inquiry into the meaning of data, and predictability of results. Thus, while the study of society can be based on scientific methods, the vagaries of human behavior and the imponderables that always exist in relationships between human beings and social systems preclude the application of "scientific" methods.[4]

While some social scientists and natural scientists accept the notion of a social "science" and its empirical basis, others challenge such notions. The challenge is based on at least three major propositions.[5] First, social sciences have limited use of controlled experiments. Even though there is a wide range of sophisticated research methods, computer procedures for collecting data, and application of mathematical models, the fact is that such procedures cannot capture the essence of human behavior or provide a laboratory for experimentation in many substantive social science matters: "Few of the phenomena that interest social scientists are amenable to manipulation and control."[6]

Second, inherent in social science research are value judgments of researchers; it is difficult to proceed on a purely scientific basis. While this may be true to a certain degree in the natural sciences, there is a great deal of distance between natural science research, moral issues, and value judgments; it is less difficult to proceed on a clinical and empirical basis.

Third, disciplines in social science lack an overarching conceptual commonality: "Although important theoretical advances have also occurred, the social sciences still lack a commonly accepted unifying theory. Such a theory is essential to a mature science."[7] Universal gravitation, relativity, and quantum mechanics are examples from the physical sciences.

These matters are complicated by questions regarding which academic disciplines are part of social science. According to one source, social sciences include history, geography, economics and business administration, sociology, anthropology, psychology, education, and political science.[8] Some colleges and universities, however, do not include history as a social science but as part of humanities. Moreover, some historians decry the inclusion of their discipline as part of the social sciences. Written more than three decades ago, the conclusions by Thomas C. Cochran are relevant today:

> The inclusion of history as a full-fledged cooperating member of the social science group has not appealed to most historians. They see difficulties in the way of such a union that range from philosophical doubt regarding the possibility of a "social" science, to objections to a new terminology.[9]

In the relatively recent period, "behaviorism" with its emphasis on empiricism and "scientific explanation of behavior,"[10] added to the concept as well as problems associated with social science.

The inclusion of a variety of academic disciplines in social science, some of which have tenuous links with each other, troubles some scholars and academicians. Moreover, any number of subjects within various social science disciplines may be legitimate areas of research in non–social science disciplines:

> Each specialized area of social science stands alone, with its own limited special theories, vocabularies, and measurement units. Social scientists in different disciplines may not understand each others' language and may be unaware of relevant findings in fields other than their own.[11]

In an earlier work Bernard Brodie raised similar questions: "The devotees of a science like economics, which is clearly the most impressive social science in terms of theoretical structure, tend to develop a certain disdain and even arrogance concerning other social science fields."[12]

These questions raise another important one. What is the focus of social science inquiry and methods? Here again there are many questions and approaches. Some argue that social science is not constrained by any formal disciplinary boundaries and that anything that has to do with society and human behavior is logically a social science matter. Others argue that to have any rational perspective, one must identify the substance and components of social science and what is supposed to be researched and studied. The lack of clear research boundaries and disagreements on components do not strengthen the concept of social science: "The history of social science shows a great variety of approaches, and we shall have to note that there are many difficult philosophical problems that are as yet unresolved."[13]

It is also important to recognize the limits of social science. Daniel Bell has argued:

> If I think of social sciences in the last twenty years, it seems to me the major task should be an assessment of failures, or of shortfalls, or (of) promises not realized.

. . . Why is it that if one looks back twenty or thirty years, there were a whole series of extravagant claims that have not been fulfilled?[14]

Similarly, Albert Cherns writes:

We are in danger of overselling ourselves and our sciences. Danger lies in the expectations that have been aroused in governments, among politicians, administrators and members of their public. They have been led, or have led themselves, to believe that the social sciences will solve their problems. As social scientists, then, we are under strong and increasing pressures to deliver goods of whose nature neither we nor our "market" are clearly aware.[15]

Such commentary is particularly relevant to the study of the American military.

Regardless of the persistent concerns about the concept of social science and its components, there is little question about social science contributing to the understanding of society and human behavior. This is true even though social science as we know it today may be in its infancy.[16] There is little need to review these matters here. For a serious assessment, the vast literature within and about social science should be reviewed (see select bibliography at the end of this essay).

The Social Science–Military Equation

The issues raised here have important consequences and implications for social science and the American military. The basic questions have to do with the concept of the military system and the military profession, and their uniqueness and distance from society. Much has been written about these matters from several approaches.[17] Yet there is a common theme that the military profession, its purpose, role, and function separates the military from the general notion of society and social order. Further, it is recognized that the primary purpose of the military is to prevail over the enemies of the state. More specifically it is to win wars. All of the military's institutional structures, functions, organizational composition, and training are aimed to that one purpose: To kill in the name of society and the state—but within the proprieties established by state ideology.

It follows that the military profession must develop the necessary professional skills, ethos, and leadership that insure the proper coordination and control of the military institution in order to succeed in battle. Further, human behavior on the battlefield is incongruous to the expectations of social order and peaceful existence within a nation-state system. This is true even though aspects of social science include the study of dysfunction and conflict. In brief, while the military profession may have some characteristics common to other professions and society, it is a self-contained community with its own sense of order, discipline, and purpose. But, it is important to stress that "contemporary trends do not produce a professional armed force isolated and remote from civilian society. Rather, it is a military establishment that is an integral part of the larger society on which its technological resources depend."[18] Yet separation from society is real.

The military profession is further distinguished from society by two unique characteristics. Its sole client is the state and it is committed to ultimate liability. Professional mind-sets, institutional focus, and individual behavior patterns and effort are driven by these characteristics. Again, this is not to suggest that the military is totally separate from society. For the military to perform its role and succeed, it must be connected to society and reflect society's major governing principles, yet remain at some distance from society.

Some social scientists carry this separation to questionable extremes. For example, in a section titled "The War Machine as a Parasitic System," Anatol Rapoport writes:

> There is ample evidence of energetic resistance on the part of the war machine to all attempts to curb its growth, of its insatiable appetite, and above all of its total maniacal preoccupation with efficiency of destruction . . . there are hardly any human beings with a psychological makeup as hostile to what are normally regarded as human traits.[19]

The Two Dimensions of Social Science Research

The social science–military equation can be viewed from two dimensions. First, research is concentrated on the relationship of the military to society, including civil-military relations, the political-social composition of the military, the extent to which the military reflects society, the nature and characteristics of conflict, economic issues, and weapons technology. This is a policy dimension; much of this research has been the basis for developing defense policy and strategy. The second is the focus on human behavior and groups within the military, the nature and characteristics of the military profession, and how all of this affects the military's capability in achieving its primary purpose. This is the inner dimension dealing with internal composition, human motivation, and leadership of the military.

The second dimension of human behavior is a more difficult undertaking not only because of the complex issues involved, but also because it sharpens social science issues and exacerbates the methodological problems discussed earlier. While some attention will be given to the first dimension, policy and of societal context, it is the second with its inner behavioral implications that is the primary focus of this essay. It is also this dimension that reveals the limits of social science and strengthens the critiques of Bell and Cherns. This is not to suggest that there is a clear delineation between the two dimensions. For example, a study of Clausewitz reveals imponderables that are part of the "fog of war" in terms of strategic issues and political-military considerations.[20] But it is also true that the fog of war shapes the immediate battle arena and those involved in combat.

There are other studies on the broader military issues such as civil-military relations and race relations—first (policy) dimension, that also include the second (inner) dimension. For example, although published several decades ago, the works of Janowitz (sociology) and Huntington (political science) examine civil-military relations, but are important in studying both the policy and inner dimensions. Both

works still command attention in studying the military.[21] Another example of social
science that includes both dimensions is the research by Leo Bogart on race and the
U.S. Army. This was based on surveys in which tools of social science were applied
in designing a major national policy.[22] Also social science studies of war and peace
cannot be totally separated from considerations of human behavior and the battlefield environment. These include recent works by Richard Betts, Richard Clutterbuck,
and Donald Snow, as well as earlier works by scholars such as Bernard Brodie and
Russell F. Weigley.[23]

The Second (Inner) Dimension

In turning to the second dimension, a seminal work is Samuel A. Stouffer and
others, *The American Soldier.*[24] One passage is particularly relevant:

> The best single predictor of combat behavior is the simple fact of institutionalized
> role: knowing that a man is a soldier rather than a civilian. The soldier role is a
> vehicle for getting a man into the position in which he has to fight or take the
> institutionally sanctioned consequences.[25]

The work by Stouffer and others was an extension of the research done by social
scientists during World War II to include personnel selection, morale, individual attitudes, and military cohesion. Many other works appeared later that helped establish
the groundwork for contemporary social science research and the military (see the
Select Bibliography). Although it is beyond the scope of this paper to review such
works, it is useful to point to some references in order to demonstrate the depth and
scope of the research.

The volumes by Samuel P. Huntington, *The Soldier and the State,* and Morris
Janowitz, *The Professional Soldier,* were noted earlier. To reiterate, these volumes
remain important social science works on the military and established concepts of
civil-military relations that remain reference points today. The volume by Janowitz
and Roger W. Little, *Sociology and the Military Establishment,* specifically addresses the impact of sociology and examines the military in sociological terms.[26] On
primary groups in the military, for example, the authors write:

> The aspect of military organization that has received the most attention from social
> scientists has been the role of primary groups in maintaining organizational effec-
> tiveness. By primary groups sociologists mean those small groupings in which
> social behavior is governed by face-to-face relations.[27]

And as Janowitz points out:

> The history of military organization and the military professional is generally
> written from the point of view of changes in weapons systems. But from the
> perspective of the social scientists and especially sociology, military organization
> is thought of in different terms and at a different level of abstraction.[28]

Sociologist Charles C. Moskos's volume on the enlisted man is another important work on the inner dimensions of the military.[29] Based on interviews and surveys,

Moskos covers such subjects as race relations, civil-military relations, and the enlisted culture and encompasses subjects within the first (policy) and second (inner) dimensions. Other important works include Edward Shils and Morris Janowitz's insightful study on the Wehrmacht and its cohesion during World War II.[30]

Russell Weigley's *The American Way of War* is an important historical work examining the way America approaches and conducts wars, and the capability of its military leaders.[31] Similarly, Robert Leckie's two volumes on *The Wars of America* are important historical analyses focusing on the human element in wars.[32] According to the author:

> [The book] is an attempt to show not only how our wars were fought but also why they occurred, as well as to illustrate what this country has gained or lost by appeals to arms. Equally, it is an attempt to portray the men who made and fought these wars. . . . War changes, its materials and its methods change, but the hearts of men do not. That is why Marshal Saxe could say, "The human heart is the starting point in all matters pertaining to war," and that is why this book attempts to come down heavily on the human side.[33]

Equally important, following World War II, military professionals were increasingly drawn into formal study of social science in order to develop a better understanding of the profession, the military, and its relationship to society. Curricula at the various service academies and the involvement of military officers in civilian graduate education set the pace. Further, the creation of associations such as the Inter-University Seminar on Armed Forces and Society (IUS) established by Morris Janowitz provided a forum for military officers, social science academicians, and government officials to discuss and debate various aspects of the military and the military profession. The history of the IUS almost parallels the evolution of social science and the study of the military and its incorporation into military professionalism.[34]

In sum, the experience of World War II and the threats and challenges evolving out of the Cold War shaped social science and the study of the military. Also, following World War II not only did social science methodology and its "scientific" thrust become important in research of the military, it became a part of military professional education. Further, a variety of civilian social scientists became engaged in designing national policy. The "scientific" study of defense policy and the military became commonplace.

The Social Science Syndrome

Over the course of the past several decades social science, specifically the disciplines of economics, political science, psychology, sociology, and some history, has expanded its research methods ranging from participant-observer surveys, extended interviews, historical computational analysis, and case studies (lessons learned) to psychological studies of human motivation and behavior, measurement of skills, and personnel selection processes.[35] A variety of empirical devices have become part of social sciences.[36] These have led to a complex set of theories, concepts, and hypotheses about military organizations, military professionalism, individual attitudes, combat behavior, and after-battle issues. Also, the addition of a variety of computer techniques, to include battle simulation, have added to the growing penetration of the military by social science, and its institutionalization within the military.

But for some, serious questions remain. In an earlier volume Brodie noted some of these:

> The present generation of "civilian strategists" are with markedly few exceptions singularly devoid of history. . . . The intellectual leaders of the movement in the 1960s to force an enormous American and NATO buildup of conventional armaments, in order, they thought, to circumvent use of nuclear weapons, were all systems analysts, with no basis in their training or preoccupations for claiming special political insight of any kind.[37]

In a footnote, Brodie names a number of such strategists, noting that they were all trained as economists.[38]

From another perspective, many social scientists hesitate to accept, and often disdain, those who draw conclusions based on anecdotal material. Yet, participant-observer techniques are in the realm of anecdotal material. Moreover, such material, if properly collected and logically analyzed, provides the raw material for drawing conclusions, particularly on battlefield behavior and combat leadership. Indeed, any number of historical studies and case studies owe their inception to anecdotal material.[39]

These problems and contradictions raise a number of questions about the social science–military equation. Also, the incorporation of social science into military leadership and command issues raises another set of troubling questions. In brief, there is a need to rethink the appropriate balance between social science and the military.

The Sorcerer's Apprentice and Gray Areas

Before those in social science can bask in euphoria on their contributions and accomplishments, and to preclude the military profession from crossing the line between common sense and social science, there is a need to recall the critiques of Bell and Cherns about exaggerated claims and expectations. Indeed, there is a vast "gray" area that not only reveals the limits of social sciences but also the fluidity, imponderables, and uncertainties that are a basic part of the military and the military profession. It is at this point that we turn to the title of this essay, "The Sorcerer's Apprentice."

According to *The American College Dictionary,* a sorcerer is one who exercises supernatural power through evil spirits—one who enchants.[40] An apprentice is a learner, a novice. Putting all of this together leads to the conclusion that the military has become enchanted with social science and further, that it acts as an apprentice, accepting the power of social science as the gifted teacher. This may also be true for many social scientists who presume that their knowledge of the world and their intellectual capabilities outweigh those possessed by military professionals. They further believe that it is logical for social scientists to be gifted teachers. In light of the developments over the past several decades in the relationship between social science and the military, this view is to be expected. But continuation of the sorcerer's apprentice syndrome could lead to the diminution of the quality of military professionalism and disastrous results in battle.

These conclusions are based on three propositions. First, the limits of social science not only must be recognized but made an inherent part of the overall study of the military. While such a notion is often given lip service, it tends to be forgotten once social scientists, both within and outside the military, become enamored with their research and engrossed in pursuing "scientific" efforts. Moreover, the military system and any number of its officials tend to seek rational (substitute empirical) bases for decisions in order to convince legislatures and the public of a particular policy or strategy. This is also true of academicians, both protagonists and antagonists, who use social science (and much of its jargon) to try to legitimize their particular perspectives.

One danger is that officials at the highest levels, including the commander in chief, may substitute a more discernible empirical basis for decisions precluding (or lacking) intuition and a sixth sense in exercising appropriate leadership in dealing with the military. This is usually the case with those who have only a superficial understanding—if any—of the inner substance of the military. The Clinton administration has been criticized on such grounds.[41] Driven by economic (social science) considerations for the most part, the administration also has been criticized for lack of strategic vision and for "strategic arthritis."[42] For others, what is lacking in all of this is a good dose of traditional leadership. Unfortunately, the views of many military professionals, who may know better about such matters and who are knowledgeable of social science and understand the "gray areas," remain subdued under the umbrella of institutional norms and professional ethos.

Second, recent developments of electronic information systems and computer techniques have begun to erode the traditional forms of leadership and decision making. Evolving from social science concepts of data collection and data analysis, commanders in the field can be electronically linked to a variety of information and intelligence sources. In the process not only is there a tendency for information overload, but commanders may well seek more and more information and data before making a decision. This is exacerbated by the development of sophisticated technology that demands technical and managerial skills, and even more data.

Commanders may well fall into the "ready, aim" trap—"ready, aim, aim, aim, aim. . . ." Waiting for the final piece of information before making and implementing a decision, always ready and aiming, but never firing:[43]

> The analytical approach to command has difficulty. It craves for certainty that is not there in warfare, and this craving leads to a requirement for more information, which is itself time consuming. In a confused situation, a commander needs what

Carl von Clausewitz described as "the quick recognition of a truth that the mind would ordinarily miss or would perceive only after long study and reflection."[44]

Intuition, common sense, and a sixth sense are often the key to success in battle and effective leadership. Social science methods, intellectual dogmatism, and computer or electronic-generated data cannot substitute for the art of leadership and command. Similarly, technical procedures and techniques cannot substitute for morale, cohesion, and "close-with-the-enemy" motivation needed in battle. One can wonder, for example, how much social science training and social science research went into training of commanders and motivation of individuals on both sides in the American Civil War. The reading of Michael Shaara's historical novel, *The Killer Angels,* may be revealing in this respect.[45]

Unconventional conflicts, in particular, raise serious questions about the utility of social science research and the American military. The characteristics of such conflicts are as the term denotes, unconventional. And as Vietnam demonstrated, all of the conventional wisdom, social science empiricism, and military technology could not come to grips with the essence of battlefield behavior or in shaping the outcome of the war.[46] This has taken on a greater dimension in the post–Cold War period, as the U.S. military develops doctrines for Operations Other Than War (OOTW). Such operations have all of the ear-marks of potential unconventional conflicts and are least susceptible to sophisticated military technology and the expecting methodologies of science—social or otherwise.

Third, after all is said and done, social science and its analytical techniques and methods are important to the military professional and those academics studying the military. But this must be qualified by prudence, caution, and recognition that social science is a tool and not a substitute for experience, a sixth sense, and intuition. Experiments at training centers, battle labs, and well-laid plans may mean little in the confusion and disorder that occurs on the battlefield. This is not to suggest that training is unimportant. But it is equally important to recognize that such matters cannot be simply assessed and determined by social science techniques and then transferred into the battle arena. Long before social science became a part of the military and the military profession, there were successful leaders and commanders. These leaders assessed the moral dimension of their men, determined their capabilities, and were able to identify what made a good leader and fighting man.

As General William Sherman of Civil War fame wrote in his memoirs, "There is a soul to an army as well as to the individual man, and no general can accomplish the full work of his army unless he commands the soul of his men, as well as their bodies and legs."[47] Charles Rogers correctly concludes, "In the final analysis, it seems that in peacetime we tie ourselves to decision making by procedures which stifle intuitive decisions on the battlefield."[48] Trying to correct this by serious studies of Clausewitz and Sun Tzu are steps in the right direction.[49] But the fact is that if such studies do not become an inherent part of the professional military's mind-set and world view, with the importance of intuition and operational art stressed, then the military is ready to aim, aim, aim. . . . In the present environment there is danger that the military is falling into that syndrome.

Unfortunately, many accept the notion that the military is just another federal agency or social grouping and that political-social patterns in civilian life are equally applicable to the military. There are also those who fall into the trap of presuming that the electronic battlefield favors those with the most sophisticated technology. These people neglect the importance of the human element and leadership skills in conflicts across the spectrum—many of which cannot be resolved by military technology. Facile analogies between the precision of scientific research and laboratory experiments with human behavior in combat and successful leadership create dangerous illusions of what it takes to succeed in combat. Such analogies and dependence upon scientific answers become a crutch and substitute for individual judgment.

In sum, "If we wish to succeed in maneuver warfare, then we must train and educate our officers in intuitive thought that emphasizes the 'art' in command rather than the 'science.'"[50]

Conclusions

Changes are continual in history. Dramatic social and military changes have occurred in the past—the industrial age and the advent of the nuclear era, for example. Thus, the changes taking place in the post–Cold War period must be considered in the light of history and not as something necessarily unique. Indeed, it may well be that the post–Cold War period and the Revolution in Military Affairs have distinct historical analogs. The transition to gunpowder weaponry in the sixteenth and seventeenth centuries and the more recent period between the world wars are most suggestive of revolutions.

Further, neither the modern American military nor its profession is a monolith. There are differences in perceptions of warfare, contingencies, and relationships with society (compare, for example, special operations, special forces, and mainstream military units; fighter pilots with bomber and transport pilots; submariners with surface-ship sailors). There are differences in intellectual focus and mind-sets, even though this is within the generally accepted notion of "duty, honor, country" and traditional military professional ethos. To lump all military professionals and the military system into one "parasitic" mold, therefore, is not only incorrect but borders on the disingenuous.

Social science must recognize certain clear distinctions between the military system, its community, and society in general. While there must be a linkage and legitimate relationship between society and the military, it is also clear that the military's unique purpose and characteristics do not allow a total integration with society. This raises serious questions about the wholesale application of social science research and methodology—particularly behaviorist techniques, focused on human behavior and society—to the military. This does not mean that different social science techniques need to be developed—one for society and another for the military. But it does mean that social science research must take into account the differences between the military and the rest of society, and take care on how social science research is interpreted, recognizing the gray areas and limits of social science. Equally important, military professionals must guard against falling into the sorcerer's apprentice trap with all that suggests regarding the denigration of professionalism and possible disastrous results on the battlefield. In sum, leadership and command are more than an analytical social science. The same holds true for motivation and behavior on the battlefield.

American commanders at all operational levels must be engaged in the "art" of killing—tempered by democratic proprieties, and in motivating others to do the same. This cannot be relegated to a quantitative function. It is equally important that those at the highest levels of civilian leadership understand this and how this shapes the purpose of the military and the military profession.

To strengthen the relationship between social science and the military in a prudent and cautious way, one must first acknowledge certain ignorance of social science. That is, the military and social scientists must recognize what social science cannot do. There have been mistakes and misinterpretations, as well as exaggerations as to what social science can accomplish. Once this recognition and acknowledgement are internalized by key social scientists and military intellectuals, then a more realistic interconnection can be developed: That is, social science as a tool can be more realistically integrated into the social science–military equation.

Notes

1. David Jablonsky, *The Owl of Minerva Flies at Twilight: Doctrinal Change and Continuity and the Revolution in Military Affairs* (Carlisle Barracks, Pa.: U.S. Army War College, May 1994), 1. See also Robert Holzer and Stephen C. LeSueur, "Pentagon Plans for 'Revolution'; Technology is Driving Services to Modernize Tactics, Organization," *Army Times*, 6 June 1994, 30.

2. Gen. Gordon R. Sullivan and Col. James M. Dubik, *War in the Information Age* (Carlisle Barracks, Pa.: U.S. Army War College, 6 June 1994), 19. See also Lt. Gen. Frederic J. Brown, USA (ret) *The U.S. Army in Transition II: Land Power in the Information Age* (Washington, D.C.: Brassey's, 1993) and Alvin and Heidi Toffler, *War and Anti-War Survival at the Dawn of the 21st Century* (New York: Little, Brown, 1992).

3. Scott Gordon, *The History and Philosophy of Social Science* (London: Routledge, 1991), 28.

4. Many of these matters are discussed and debated in Karl W. Deutsch, Andrei S. Markovits, and John Platt, eds., *Advances in the Social Sciences, 1900–1980: What, Who, Where, How?* (Lanham, Md.: University Press of America, 1986). See, for example, Karl W. Deutsch, "What Do We Mean by Advances in the Social Sciences?" ibid., 6–9.

5. The ideas for these propositions is based on the detailed discussion in Gordon, *History and Philosophy of Social Science,* 51–56, and Albert Cherns, *Using the Social Sciences* (London: Routledge & Kegan Paul, 1979), 3–14.

6. David Nachimas and Chava Nachimas, *Research Methods in the Social Sciences* (New York: St. Martin's, 1976), 4–5.

7. James Grier Miller, "Social Science and Integration Across Disciplines," in Deutsch, Markovits, and Platt, eds., *Advances in the Social Sciences,* 156.

8. Carl M. White, *Sources of Information in the Social Sciences: A Guide to the Literature* (Chicago: American Library Association, 1973).

9. Thomas C. Cochran, *The Inner Revolution: Essays on the Social Sciences in History* (1964; Gloucester, Mass.: Peter Smith, 1970), 19.

10. Gordon, *History and Philosophy of Social Science,* 53, contains the following: "The contention of behaviorists is not that inner states of consciousness do not exist, but that reference to them is unnecessary in any scientific explanation of behavior. . . . (According to behaviorists) our explanation of a human phenomenon is methodologically similar to our explanation of the connection between wind velocity and bridge failure." See also pp. 4–6.

11. Miller, "Social Science and Integration," 156.

12. Bernard Brodie, *War and Politics* (New York: Macmillan, 1973), 475.

13. Gordon, *History and Philosophy of Social Science,* 2.

14. Daniel Bell, "The Limits of the Social Sciences: A Critique of the Conference," in Deutsch, Markovits, and Platt, eds., *Advances in the Social Sciences,* 314.

15. Cherns, *Using the Social Sciences,* 367.

16. See for example, Dag Anckar and Erkki Berndtson, "Introduction: Toward a Study of the Evolution of Political Science," in Dag Anckar and Erkki Berndtson, eds., "The Evolution of Political Science: Selected Case Studies," *International Political Science Review* 8 (January 1987): 5–7.

17. See for example, Rogert W. Little, ed., *Handbook of Military Institutions* (Beverly Hills, Calif.: Sage, 1971). This volume includes contributions from fifteen authors from the disciplines of political science, sociology, economics, social psychology, history, and social work.

18. Morris Janowitz, "Military Organization," in Little, ed., *Handbook of Military Institutions,* 13.

19. Anatol Rapoport, "Examining the Concept of Advances, Especially in Psychology," in Deutsch, Markovits, and Platt, eds., *Advances in the Social Sciences,* 305.

20. Peter Paret and Michael Howard, eds., *Carl von Clausewitz, On War* (Princeton, N.J.: Princeton University Press, 1976).

21. Morris Janowitz, *The Professional Soldier: A Social and Political Portrait,* rev. ed. (New York: Free Press, 1971), and Samuel P. Huntington, *The Soldier and the State* (Cambridge, Mass.: Harvard University Press, 1957).

22. Leo Bogart, ed., *Social Research and the Desegregation of the U.S. Army: Two Original Field Reports* (Chicago: Markham, 1969).

23. Richard K. Betts, ed., *Conflict After the Cold War: Arguments on Causes of War and Peace* (New York: Macmillan, 1994); Donald M. Snow, *Distant Thunder: Third World Conflict and the New International Order* (New York: St. Martin's, 1993); Richard Clutterbuck, *International Crisis and Conflict* (New York: St. Martin's, 1993); and Russell F. Weigley, *The American Way of War: A History of United States Strategy and Policy* (Bloomington: Indiana University Press, 1977). Many such works relate to the nature and characteristics of the military and the military profession.

24. Samuel A. Stouffer et al., *The American Soldier,* 4 vols. (Princeton, N.J.: Princeton University Press, 1949), vols. 1 and 2.

25. Ibid., 2:101.

26. Morris Janowitz and Roger W. Little, *Sociology and the Military Establishment,* rev. ed. (New York: Russell Sage Foundation, 1965).

27. Ibid., 77.

28. Janowitz, "Military Organization," 13.

29. Charles C. Moskos, Jr., *The American Enlisted Man: The Rank and File in Today's Military* (New York: Russell Sage Foundation, 1970).

30. Edward S. Shils and Morris Janowitz, "Cohesion and Disintegration in the Wehrmacht in World War II," *Public Opinion Quarterly* 12 (Summer 1948). For a more recent work with a different approach and conclusion, see Omer Bartov, *Hitler's Army: Soldiers, Nazis, and War in the Third Reich* (New York: Oxford University Press, 1992).

31. Weigley, *American Way of War.*
32. Robert Leckie, *The Wars of America,* 2 vols. (New York: Harper Perennial, 1993).
33. Ibid., 1:xi.
34. See James Burk, "Morris Janowitz and the Origins of Sociological Research on Armed Forces and Society," *Armed Forces & Society,* No. 2, 1993.
35. See Select Bibliography below.
36. See, for example, Nachimas and Nachimas, *Research Methods.*
37. Brodie, *War and Politics,* 475.
38. Further, Brodie and others also argue that the military is reluctant to change even when evidence is present that such change is necessary. In the more recent period, it was Congress that forced changes on the military regarding the role of the chairman of the Joint Chiefs of Staff and joint duty, and in shaping special operations forces. See Goldwater-Nichols National Defense Reorganization Act, Public Law (PL99-433, passed in October 1986, and the Cohen-Nunn Act (PL99-661), passed in November 1986.
39. See, for example, S. L. A. Marshall, *Men Against Fire* (New York: Morrow, 1947); Michael Herr, *Dispatches* (New York: Avon, 1978); Jonathan Schell, *The Real War: The Classic Reporting on the Vietnam War* (New York: Pantheon, 1987); Al Santoli, *Everything We Had: An Oral History of the Vietnam War by Thirty-three American Soldiers Who Fought It* (New York: Ballantine, 1981); and Bob Woodward, *The Commanders* (New York: Simon & Schuster, 1991).
40. *The American College Dictionary* (New York: Random House, 1955).
41. See, for example, Harry Summers, "White House Climate Spurns Military Loyalty," *Army Times,* 22 Mar. 1993, in which the author writes, "For the first time in American history we have a climate in the White House that openly scorns the military and holds those who have put their lives on the line for their country in total contempt."
42. Ray Moseley, "'Strategic Arthritis' Cripples U.S. Foreign Policy, Think Tank Asserts," *Chicago Tribune,* 24 May 1994, sec. 1, 4. See also Merrill Goozner, "In Asia, Lack of Coherent Strategy May Prove Costly," ibid., 3 July 1994, sec. 4, 1.
43. Thomas E. Cronin, *The Write Stuff: Writing as a Performing and Political Art,* 2d ed. (Englwood Cliffs, N.J.: Prentice-Hall, 1993), 18.
44. Col. Charles T. Rogers, "Intuition: An Imperative of Command," *Military Review* 74 (March 1994): 42.
45. Michael Shaara, *The Killer Angels* (New York: Ballantine, 1974).
46. See, for example, Sam C. Sarkesian, *Unconventional Conflicts in the New Security Era: Lessons from Malaya and Vietnam* (Westport, Conn.: Greenwood, 1993).
47. William Tecumseh Sherman, *Memoirs of General W. T. Sherman* (New York: Library of America, 1990), 879.
48. Rogers, "Intuition," 50.
49. See Paret and Howard, eds., *Carl von Clausewitz,* and Samuel B. Griffith, *Sun Tzu: The Art of War* (London: Oxford University Press, 1971).
50. Rogers, "Intuition," 50.

Select Bibliography

This does not include articles in periodicals or research reports. The sources listed here were particularly useful in preparing this essay. Sources with bibliographies are so indicated.

Ambrose, Stephen E. and James A. Barber, Jr., eds., *The Military and American Society: Essays and Readings.* New York: Free Press, 1972.
Andrzejewski, Stanislaw. *Military Organization and Society.* London: Routledge & Kegan Paul, 1954.
Bogart, Leo. *Social Research and the Desegregation of the U.S. Army: Two Original 1951*

Field Reports. Chicago: Markham, 1969.

Bourne, Peter G. *Men, Stress, and Vietnam*. Boston: Little, Brown, 1970.

Brodie, Bernard. *War and Politics*. New York: Macmillan, 1973.

Burns, James MacGregor. *Leadership*. New York: Harper Torchbooks, 1979.

Cherns, Albert. *Using the Social Sciences*, London: Routledge & Kegan Paul, 1970; bibliog.

Cochran, Thomas C. *The Inner Revolution: Essays in the Social Sciences in History*. Gloucester, Mass.: Peter Smith, 1970.

Deutsch, Karl W., Andrei S. Markovits, and John Platt, eds., *Advances in the Social Sciences: What, Who, Where, How?* Lanham, Md.: University Press of America, 1986.

Edmonds, Martin. *Armed Forces and Society*. Boulder, Colo.: Westview, 1990.

Enthoven, Allan C. and K. Wayne Smith. *How Much is Enough? Shaping the Defense Program 1961–1969*. New York: Harper Colophon, 1971.

Gordon, Scott. *The History and Philosophy of Social Science*. London: Routledge, 1991.

Henderson, Wm. Darryl. *Cohesion: The Human Element in Combat*. Washington, D.C.: National Defense University Press, 1985; bibliog.

Herr, Michael. *Dispatches*. New York: Avon, 1978.

Huntington, Samuel P. *The Soldier and the State: The Theory and Politics of Civil-Military Relations*. Cambridge, Mass.: Harvard University Press, 1957.

Janowitz, Morris. *The Professional Soldier: A Social and Political Portrait*. New York: Free Press, 1971.

Janowitz, Morris and Roger W. Little. *Sociology and the Military Establishment*. New York: Russell Sage Foundation, 1965; bibliog.

Johns, John H. *Cohesion in the US Military*. Washington, D.C.: National Defense University Press, 1984.

Just, Ward. *Military Men*. New York: Knopf, 1970.

Keeley, John B., ed., *The All-Volunteer Force and American Society*. Charlottesville: University Press of Virginia, 1978.

Leckie, Robert. *The Wars of America*, 2 vols. New York: Harper Perennial, 1993.

Littauer, Raphael and Norman Uphoff, eds., *The Air War in Indochina*, rev. ed. Boston: Beacon, 1972.

Little. Roger W., ed., *Handbook of Military Institutions*. Beverly Hills, Calif.: Sage, 1971; bibliog.

Marmion, Harry A. *Selective Service: Conflict and Compromise*. New York: Wiley & Sons, 1968.

Marshall, S. L. A. *Men Against Fire*. New York: Morrow, 1947.

Moskos, Charles C. *The American Enlisted Man*. New York: Russell Sage, 1970; bibliog.

Mylander, Maureen. *The Generals: Making It, Military Style*. New York: Dial, 1974.

Nachimas, David and Chava Nachimas. *Research Methods in the Social Sciences*. New York: St. Martin's, 1976.

Santoli, Al. *Everything We Had: An Oral History of the Vietnam War by Thirty-three American Soldiers Who Fought It*. New York: Ballantine, 1981.

Sarkesian, Sam C. *Beyond the Battlefield: The New Military Professionalism*. New York: Pergamon, 1981; bibliog.

———. *The Professional Army Officer in a Changing Society*. Chicago: Nelson-Hall, 1975.

Schell, Jonathan. *The Real War: The Classic Reporting on the Vietnam War*. New York: Pantheon, 1987.

Stouffer, Samuel A. et. al. *The American Soldier,* 4 vols. Princeton, N.J.: Princeton University Press, 1949.

The President's Commission on an All-Volunteer Armed Force. Washington, D.C.: Government Printing Office, February 1970.

Van Doorn, Jacques. *The Soldier and Social Change*. Beverly Hills, Calif.: Sage, 1975.

Weigley, Russell F. *The American Way of War: A History of United States Military Strategy and Policy*. Bloomington: Indiana University Press, 1977; bibliog.

Woodward, Bob. *The Commanders*. New York: Simon & Schuster, 1991.

Index